D. P. F. Moeller

Mathematical and Computational Modeling and Simulation

W0079213

Engineering ONLINE LIBRARY

Springer-Verlag Berlin Heidelberg GmbH   http://www.springer.de/engine/

Dietmar P. F. Moeller

Mathematical and Computational Modeling and Simulation
Fundamentals and Case Studies

With 187 Figures

 Springer

Prof. Dr.-Ing. Dietmar P.F. Moeller

University of Hamburg
Faculty of Computer Science
Chair Computer Engineering (AB TIS)
Vogt-Kölln-Str. 30
22527 Hamburg / Germany

and

California State University, Chico
College of Engineering, Computer Science and Technology
O`Connel Technology Center
Chico, California 95929-0410

ISBN978-3-540-40389-0   ISBN 978-3-642-18709-4 (eBook)
DOI 10.1007/978-3-642-18709-4

Cataloging-in-Publication Data applied for.
Bibliographic information published by Die Deutsche Bibliothek. Die Deutsche Bibliothek lists
this publication in the Deutsche Nationalbibliografie; detailed bibliographic data is available in
the Internet at <http://dnb.ddb.de>.

http://www.springer.de
© Springer-Verlag Berlin Heidelberg 2004
Originally published by Springer-Verlag Berlin Heidelberg New York in 2004

The use of general descriptive names, registered names, trademarks, etc. in this publication does
not imply, even in the absence of a specific statement, that such names are exempt from the
relevant protective laws and regulations and therefore free for general use.

Typesetting: Datenconversion by author
Cover-design: Medio, Berlin
Printed on acid-free paper   62 / 3020 hu – 5 4 3 2 1 0

Preface

The book addresses one of the most interesting topics in modern systems theory: *mathematical and computational modeling and simulation*. A recent White House report identified computational modeling and simulation as one of the key enabling technologies of the 21st century. Its application is universal.

Some choices have been made in selecting the material of this book. First we describe the fundamentals of modeling, as this represents the largest portion of system analysis. In addition, the mathematical background describing real-world systems is introduced on a basic level as well as on a more advanced one and its correspondence with the respective modeling methodologies is described. Secondly, we present the most interesting simulation systems at the language and logical level, and describe their use in several case study examples. However, a textbook can not describe all available simulation systems in detail, for this reason the reader is referred to the specific written material such as textbooks, reference guides, user manuals, etc., as well as the web-based information addressed to the several simulation languages. Thirdly, we present an algorithmic approach to ill-defined and distributed systems based on the respective mathematical frameworks.

The purpose of this book is to expose undergraduate and graduate students to the use of mathematical and computer modeling and simulation as a basis for developing an understanding of the response characteristics of a broad class of real-world systems. Mathematical and computational modeling is based on systems theory as a mathematical form of representation, while building models of real-world systems. The simulation methodology behind this is used for a better understanding of the time-dependent transient behavior of the complex models developed. The subject matter of the book can be considered to form an introduction to the methodology of mathematical and computational modeling of real-world systems, as well as into simulation languages, to gain experience, which results from the different application domains introduced as case study examples in this book.

The nature of the material in the book can be more or less difficult, if the reader is new to such an approach, which is also due to the fact that mathematical and computational modeling and simulation is a multi-disciplinary domain, founded in computer science, engineering, mathematics, physics, chemistry, biology, life science, etc. The material may not be read and comprehended either quickly or easily. This is why specific case study examples, from the various disciplines, have been embedded due to the related topics of system-theory representation of the material, to master the material, at least for most individuals of the several scientific disciplines. It is assumed that the reader has some previous background in mathematics through calculus, differential equations, Laplace transforms, and matrix fundamentals. The most common simulation software systems will be introduced and their performance will be discussed based on several case study examples. But real-world systems often are ill defined and the important parameters that should be known for modeling may not be known and/or not measurable, that calls for parameter-identification methods to estimate unknown parameters. Moreover, virtual-reality and soft-computing methods have arisen recently that are now added being as part of the methods used for modeling and simulating real-world systems, which is shown for the respective case studies and examples.

This book can be used in courses in various ways. It contains more material than could be covered in detail in a quarter-long (30-hour) or semester-long (45-hours) course, leaving instructors with the possibility of selecting their own topics and adding their own case study examples. Sections denoted with an asterisk in the contents report on advanced topics and can be skipped in a first reading or in undergraduate courses. The book can also be used for self-study or as a reference for graduate engineers, scientists and computer scientists for training on the job or in graduate schools.

It is noted, that this book was developed under a transatlantic grant from the European Commission Directorate General for Education and Culture, Brussels, with the University of Hamburg, Germany, as lead European University, and the US Government Federal Institute of Post Secondary Education, Washington, with the California State University, Chico, as lead US University for the USE-ME Project (United States Europe Multicultural Alliance in Computer Science and Engineering). The material on modeling and simulation in this book was designed for use in the respective classes and courses of the transatlantic USE-ME Project partners home institutions.

Chapter 1 on *Modeling Continuous-Time and Discrete-Time Systems* contains the introductory material on developing simulation models for real-world systems. The developed models usually take the form of a set of assumptions concerning the operation of real-world systems. These assumptions are expressed in mathematical, logical, and symbolic relationships between the entities or objects of interest of real-world systems. Once developed and verified, a model can be used to investigate a wide variety of problems and questions about the real-world system, which is shown in the respective case study examples for the several application domains such as biology, business, chemistry, electrical engineering, mechanical engineering, medicine, physics, etc.

Chapter 2 on *Mathematical Description of Continuous-Time Systems* focusses on the most important mathematical methods in the time domain and in the frequency domain that are used for the mathematical description of real-world systems. These methods are based on ordinary differential equations (ODEs) of n-th order, sets of n first-order ordinary differential equations, partial differential equations (PDEs), the superposition integral, the convolution integral, Laplace transforms, etc.

Chapter 3 on *Mathematical Description of Discrete-Time Systems* focusses on the general principles modeling queuing systems, discrete event concepts, Petri nets, and statistical models.

Chapter 4 on *Simulation Languages for Computational Modeling and Simulation* contains the introductory material on the most interesting simulation systems at the language and logic level, such as ACSL, AnyLogic, B2Spice A/D, CSMP, FEMLAB, GPSS, GPSS/H, MATLAB, Modelica, ModelMaker, SIMULINK, SIDAS, SIMAN V, SIMSCRIPT, SLX, PASION, and their application in the several case study examples.

Chapter 5 on *Parameter Identification of Dynamic Systems* contains a mathematical approach of ill-defined real-world systems for which the parameters of importance are not known or not measurable. Based on identification these unknown or unmeasurable parameter can be estimated, using the several methods such as gradient methods, direct search methods, least square methods, etc.

Chapter 6 on *Soft-Computing Methods* focusses on fuzzy sets and neural networks in modeling and simulation to generate the basic insight that categories are not absolutely clear cut, they belong to a lesser or greater degree to the respective category. Soft-computing breaks with the tradition that real-world systems can be precisely and unambiguously characterized, meaning divided into categories, for manipulation these formalizations according to precise and formal rules.

Chapter 7 on *Distributed Simulation* contains the introductory material of real-world systems that are distributed, and can be analyzed using the tie-breaking method, the critical time path method, and the High-Level-Architecture (HLA) concept. The methods are introduced and used for real-world traffic problems.

Chapter 8 on *Virtual Reality* contains the introductory material of computer-generated worlds that are based on real-time computer graphics, color displays, and advanced simulation software, etc. The topics of virtual reality are used for real-world applications in the medical and geological domains.

The textbook contains additional case study examples (see Appendix C).

I would like to thank my very good friend, Prof. Ralph Conrad Hilzer, California State University, Chico, for his constant help and generous support. Furthermore, I would like to thank Dr. Kenneth Derucher, Dean of the College of Engineering, Computer Science and Technology, California State University, Chico, for supporting my research work at California State University, Chico. I would also like to thank Dr. Roland E. Haas, Managing Director DaimlerChrysler Research Institute India, Bangalore, for his critical review of this book and his personal encouragement. Dr. Joachim Wittman, University of Hamburg, is also acknowledged for a critical review of parts of the book and several of the examples. I thank Dr. Dieter Merkle, Springer Publishing Inc., Heidelberg, for his help with the organizational procedures between the publishing house and the author.

Finally, I would like to deeply thank my wife Angelika for her encouragement, patience and understanding during writing of this book.

This book is dedicated to my parents Wilhelm Ch. and Hildegard Möller whose hard work and belief in me made my dreams a reality.

Dietmar P. F. Möller
Professor, University of Hamburg, Germany
Adjunct Professor, California State University Chico, USA
Spring 2003

Contents

List of Examples

List of Case Studies

1 Modeling Continuous-Time and Discrete-Time Systems

1.1 Introduction

Engineering is concerned with understanding and controlling the materials and forces of nature for the benefit of humankind. Therefore it is necessary for engineers and scientists to analyze and improve the performance of complex systems, when the components of which originate from different domains. Examples include either adapting existing systems to new demands and/or conditions, or designing new applications such as those in mechatronics, automotive, avionics, aerospace, robotics, traffic control, digital or microsystems, etc. These systems include components derived from many different engineering domains such as electrical, mechanical, hydraulic, and control. In many cases, solutions to problems have been found by applying appropriate mathematical models and computer simulation to them.

Computational modeling and simulation has grown to where it now represents one of the most powerful design tools available in industry, particularly when used to analyze and control dynamic systems. This has been primarily due to remarkable advances that have taken place in systems theory, computer science, and engineering, as well as other human activities in engineering and science. A recent White House report identifies computer modeling and simulation as one of the key enabling technologies of the 21st century. Its applications are virtually universal.

Modeling and simulation can be viewed as an iterative process consisting of successive mathematical model building and computer-assisted simulation steps. In this way, a dynamic system model can be manipulated in accordance with the scope of the simulation study, i.e. by changing the model structure, parameters, inputs, and outputs to accurately match the real-world system behavior. In fact, a derived model achieves its purpose when an optimal match is achieved between the simulation results, based on the mathematical model, and data sets gathered through real system measurements and experimentation.

The use of systems theory for solving problems that overlap several disciplines of science and engineering has improved cooperation among these disciplines and lowered previously rigid barriers between them. The most important step of systems theory, as applied to a particular dynamic system, is the translation of a real-world system into the mathematical systems theory language, which is universal while independent due to application and domain.

In order to develop suitable models of real-world systems and/or processes, a thorough understanding of the dynamic system and its operating range is necessary. Consequently, a dynamic system can be defined using the respective mathematical description, for which the mathematical model fulfills the following requirements:

$u_1(t) = u_2(t)$ *which show for all* $t \in (t_0, t_1)$, *hence it follows that*:

$v(t_1, t_0, x_0, u_2(\cdot))$;

$v(t_1, t_0, x_0, u_2(\cdot)) = x_0$;

$v(t_2, t_0, x_0, u_2(\cdot)) = v(t_2, t_1 \cdot v(t_1, t_0, x_0, u(\cdot)), u(\cdot))$;

$y(\cdot) = \gamma((\cdot), x(\cdot), u(\cdot))$,

which may be stated as transformation in t, x, *and* u,

with x_0 as the initial state of the system, $u(.)$ as the input function of the system for the time interval (t_0, t_1) if $t_1 > t_0$, and $y(.)$ as a p-dimensional output vector function.

The first equation characterizes the property of causality of a dynamic system, the second is the general equation of a consistent dynamic system, the third represents the so-called half-group property of dynamic systems, and the last equation is a transformation of a p-dimensional output vector of the dynamic system in t, x, and u.

Using this mathematical notation, a causal system is said to be a dynamic system if it holds for the expression $u_1(t) = u_2(t)$.

More generally, a system is assumed to conform to this definition if it contains the following objects:

- Elements
- Attributes
- Relations

Based on logical assumptions, these objects are part of the whole structure, called a system. Elements can be components, parts, and so forth, while relations are cooperatives, couplings, and so on, and attributes introduce properties, features, signatures, and so on. Moreover, attributes provide connections between the system and the system environment. An attribute that describes a system condition is called a system state. Attributes that interact with each other are called system-related internal descriptions.

Assuming that A is a nonempty set of attributes α and B are nonempty sets of relations, a system description β can be defined as

$$F := (\alpha \in A, \beta \in B).$$ (1.1)

The structural description of dynamic systems can be provided by matrix notation, which contains an:

- Input vector u
- Output vector y
- Operator matrix S.

In the following formula:

$$\underline{y} = \underline{S} \circ \underline{u} \tag{1.2}$$

the system operator matrix $\underline{S}$ can be expressed as the state matrix $\underline{X}$, shown in Fig. 1.1.

Fig. 1.1. Block diagram of a dynamic system

If a system represented by the state vector $x(t_1)$ is determined through the initial state $x(t_0)$ and the input function $u(\cdot)$ over the time interval (t_0, t_1) for all $t_1 > t_0$, it is said to be a dynamic system. A dynamic system is of finite order if the state vector $x(t_1)$ has a finite number of components.

The various methods used to build up a model of a dynamic system are usually combined in a more general view, containing the following information sources:

- Defining goals and purpose
- Determining boundaries
- Identifying relevant components
- Determining the necessary level of detail
- Establishing a priori knowledge
- Gathering data sets through experimentation
- Measuring system inputs and outputs
- Estimating nonmeasurable data, and/or state-space variables

It should be noted that the scopes using models in the various disciplines, such as science and engineering, can be different. In control-systems engineering they are concerned with understanding and controlling segments of systems in order to provide useful economic products for society. In engineering this way mostly deals with system synthesis and optimization, while engineers are primarily interested in a mathematical model of the system operated under normal operating conditions. Engineering scopes, while using models, will be to control systems optimally, or to keep systems at least in a relatively close vicinity of conditions that avoid danger of a possible drifting of the system out of the margins of safe operating conditions. In contrast, in life science, biomedical scientists are not solely interested in mathematical models of biomedical systems that are operated under normal operating conditions. Life science scientists prefer to develop mathematical models, that adequately describe the system behavior outside the normal operating range, which can be interpreted, in medical terms, as a disease case notation. This represents a dynamic system that is operated outside normal operating set points.

With respect to the spectrum of available models, the variety of levels of conceptual and mathematical representations is evident, which depend on the goals and purposes for which the models usage was intended, the extent of a priori knowledge available, data gathered through experimentation and measurements on the dynamic system, estimates of system parameters as well as system states. Hence a dynamic system can be seen as a system that is decomposed to a certain level of detail. From a more general point of view the mathematical representation of dynamic systems is based on the foundations of decomposed systems at any required level, which are the:

- Behavior level; at which one can describe the dynamic system as a black box, in which we record measurements in a chronological manner, based on a set of trajectories that characterize the behavior of the system. The behavior level is of importance because experimentation with dynamic systems addresses this level, due to the input-output relationship, which can be expressed for a black-box system as:

$$\underline{y}(t) = F(\underline{u}, t) , \tag{1.3}$$

with $u(t)$ as input set, and $y(t)$ as output set, and F as transfer function, for a state-structure level; at which one can describe the dynamic system, taking into account the system state structure that results, by iteration over time, in a set of trajectories, called behavior. The internal state sets represent the state-transition function that provides the rules for computing the future states, depending on the current states,

$$\underline{y}(t) = G(\underline{u}(t), \underline{x}(t), t) . \tag{1.4}$$

A state of a dynamic system represents the smallest collection of numbers, specified at time $t = t_0$, in order to be able to predict uniquely the behavior of the system for any time $t \geq t_0$ for any input belonging to the given input set, provided that every element of the input set is known for $t >\geq t_0$. Such numbers are called state variables.

- Composite structure level; at which one can describe the dynamic system by connecting elementary black boxes that can be introduced as a network description. The elementary black boxes are the components and each one must be given by a system description at the state-structure level. Moreover each component must have identified input and output variables as well as a specification determining the interconnection of the components, and interfacing the input and output variables.

Difficulties in developing mathematical models may arise because dynamic systems are, in general, extremely complex. In addition, sufficient amounts of operating data are often not available. Hence constructing a mathematical model for a given real-world system first requires the selection of a model structure and then some form of parameter estimation to determine acceptable model parameter values if they are not available. For this reason sometimes it may be better to develop simplified models that may eliminate intrinsic characteristics of the system, while an overcomplicated mathematical model will cause mathematical difficulties. From a more general point of view two major facts are important when developing mathematical models of real-world systems:

1. A model is always a simplification of reality, but should never be so simple that its answers are not true.
2. A model has to be simple enough to allow easy studying and working with it.

Hence a suitable model is a compromise between the mathematical difficulty attached to complicated equations and the accuracy in the final result. The corresponding relationships are shown in Fig. 1.2.

From Fig. 1.2 one can conclude that there is no reason to develop expensive models because the increment of quality is less than the increase in cost. This point is of importance because a mathematical model is a very compact way to describe dynamic systems. But a complex model not only describes the relations between the system inputs and outputs, it also allows detailed insights into the system structure and internal relationships. This is due to the fact that the main relations between the variables of the real-world system to be modeled are mapped into appropriate mathematical expressions. For instance, the relation between input and output variables of a dynamic system can be described, depending on the complexity of the dynamic system, and by a set of differential equations.

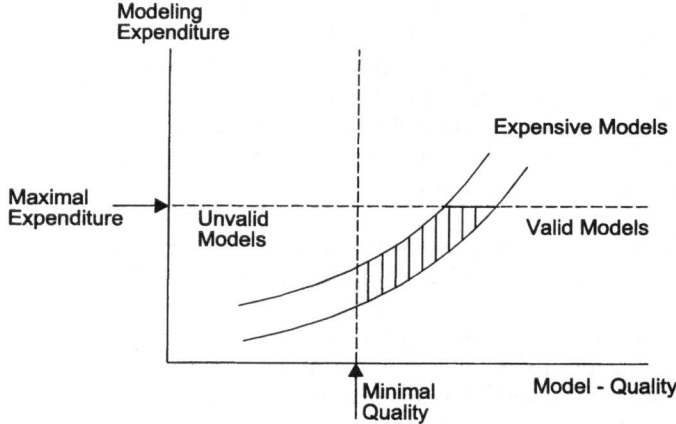

Fig. 1.2. Dependence of the modeling expenditure (costs) versus the degree of accuracy (model quality)

In principle, there are two different approaches to obtain a mathematical model of a real-world system: the deductive or theoretical approach, based on the derivation of the essential relations of the dynamic system, and the empirical one, based on experiments on the dynamic system itself. It should be noted that practical approaches often use a combination of both, which might be the most advantageous way. The two methods result in the

- Empirical method of experimental modeling; based on measures available on the inputs and outputs of a real-world system. Based on these measurements, the empirical model allows construction of a model for the given real-world system, as shown in Fig. 1.3. The characteristic signal-flow sequence of the experimental modeling process is used to determine the model structure for the mathematical description, based on a priori knowledge, which has to fit the used error criterion, which is chosen in the same way as the performance criterion for the deductive modeling method.
- Deductive method of theoretical or axiomatic modeling; represents a bottom-up approach starting at a high level of well-established a priori knowledge of system elements, representing the mathematical model. In real-world situations, problems occur in assessing the range of applicability of these models, the deductive-modeling methodology are supplemented by an empirical model validation proof step. Afterwards the model can be validated by comparing the simulation results with the data known from the dynamic system whether they match an error criterion or not. Let e be an error margin, which depends on the difference between measures on the real-world system y_{RWS} and data from the simulation of the mathematical model y_{MM}, as follows:

$$e := e(\underline{y}_{RWS}(t), \underline{y}_{MM}(t)),\qquad(1.5)$$

the error criterion can be determined by minimizing a performance criterion

$$J = \int_0^t e^2 \cdot dt \to \text{Min} \; \blacksquare.\qquad(1.6)$$

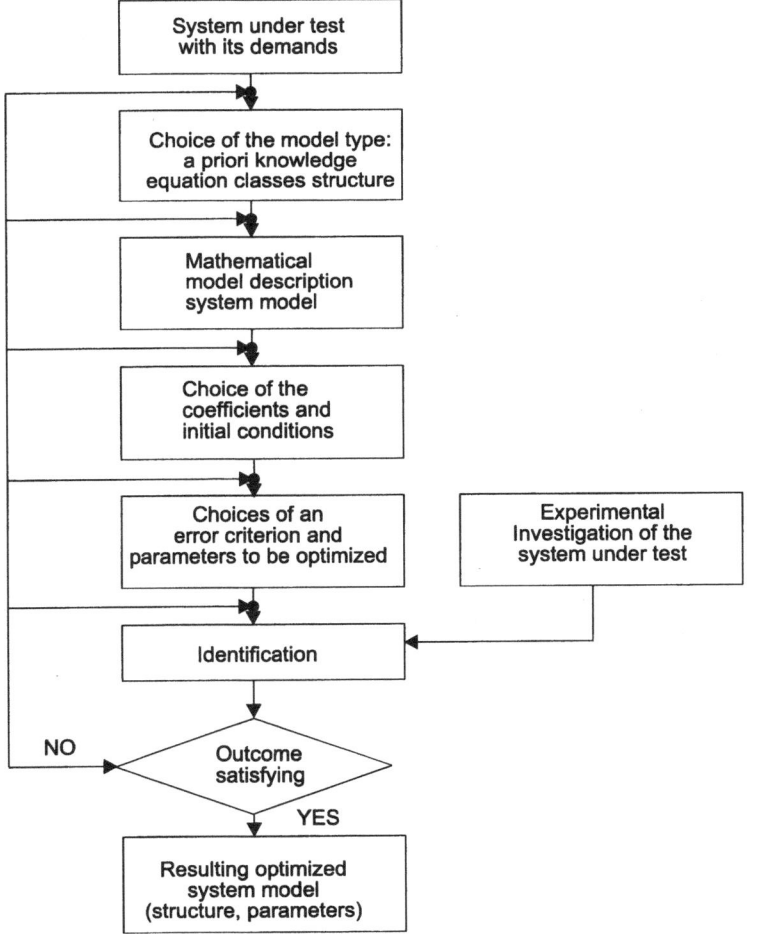

Fig. 1.3. Block diagram of the empirical-modeling methodology

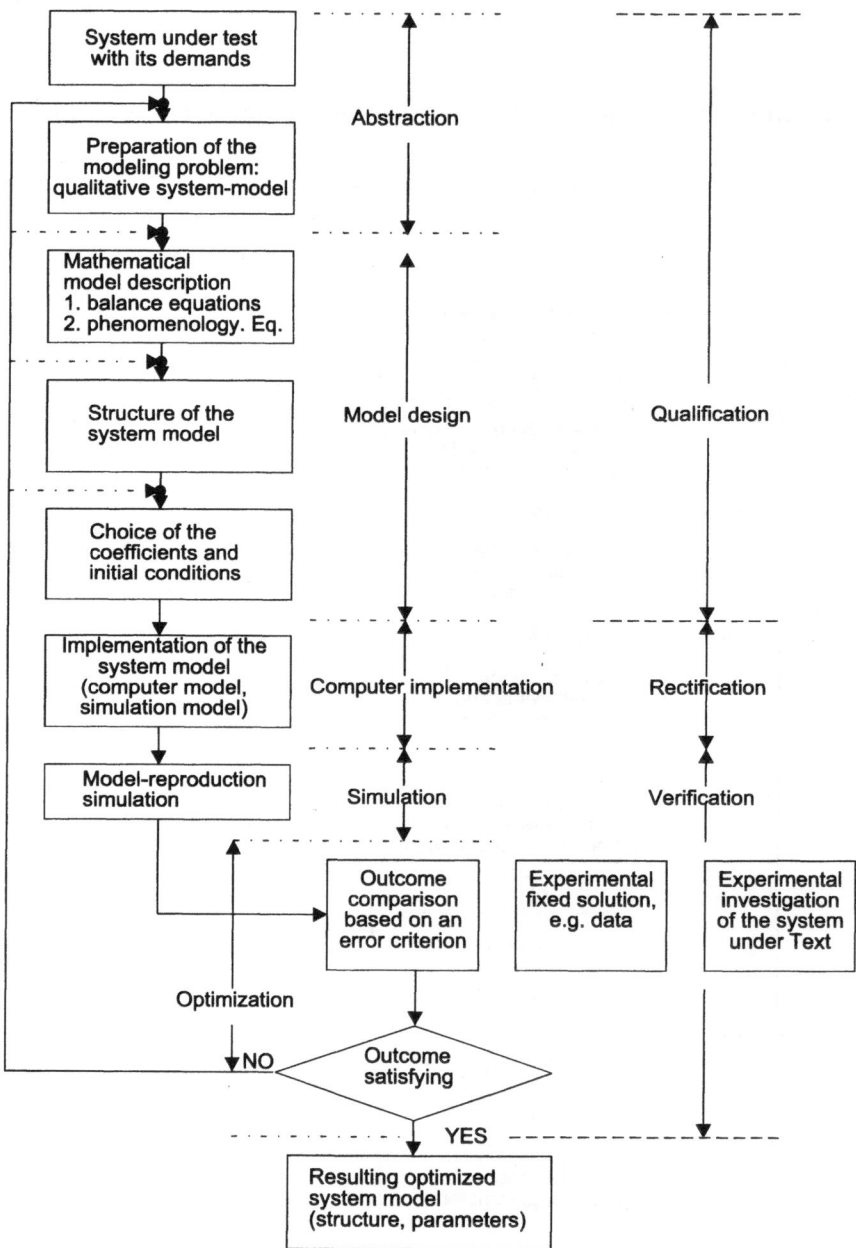

Fig. 1.4. Block diagram of the empirical modeling expanded by the deductive-modeling methodology

The model fits the chosen performance criterion which depends on the specific application, when the results obtained from the model by simulation are compared with the results from data on the real-world system are within the error margin of the error criterion. If the constructed model is unsatisfactory, a modification is necessary at the different levels of the deductive modeling scheme, as shown in Fig. 1.4. The result of the modification, a specific form of a model-validation procedure, is a model that fits within the boundaries of the intended application better than the previous model.

It should be noted that a model of a real-world system not only describes the relations between the system inputs and outputs – such as that for black box models – it also gives insight into the real-world systems structure and into some systems internal relationships for the respective levels of representation for non-black-box models. This is due to the fact that the main relationships between the physical variables of the dynamic system are mapped into appropriate mathematical expressions. For instance, the relation between input and output variables of a dynamic system can be described – depending on its complexity – by sets of ordinary differential equations, called ODEs, or by sets of partial differential equations, called PDEs, which represent the mathematical notation of the dynamic system.

To solve a systems-analysis problem it is necessary to understand the system to be analyzed, as well as its environment and the respective requirements. Typically this consists of the following steps:

- Abstraction
- Representation of the model, e.g. by mathematical notation
- Analysis, e.g. by simulation
- Design

which are shown in the block diagram representing the empirical-modeling methodology in Fig. 1.3, and for the deductive methodology of the theoretical or axiomatic-modeling methodology in Fig. 1.4.

The first step, abstraction, means searching for a model that resembles the dynamic system in its salient features but is easier to study. A dynamic system can be assumed as a real-world object, but its precise description is often unknown. We may, however, apply all kinds of test signals in order to get a deeper insight into the system, and from the measured data we are able to determine the description of the system. To study the system in an analytical manner, a model that resembles the behavior has to be determined, which can be based on the characteristics measured, obtained at least from the test signals, applied to the system inputs.

Applying test signals as system inputs, there are two types of deterministic test signals used, the unit step and the ramp function, which are related through the expressions:

$$
u_{us}(t) = \begin{cases} 0 & for\ t\ <\ t_0 \\ 1 & for\ t_0\ <\ t\ <\ t_1 \\ 1 & for\ t\ >\ t_1 \end{cases} \tag{1.7}
$$

for the unit step, and

$$
u_R(t) = \begin{cases} 0 & for\ t\ <\ t_0 \\ a\dfrac{t-t_0}{t_1-t_0} & for\ t_0\ <\ t\ <\ t_1 \\ a & for\ t\ >\ t_1 \end{cases} \tag{1.8}
$$

for the ramp function.

The unit step and the ramp function, as well as many other functions, prove to be particularly valuable in problems in which successive switching characteristics may occur. As described in the equations above, it is possible to clearly specify the order of the switching events. This will be the same in the case that the given functions are delayed in time.

From a more general point of view the unit step ($u_{us}(t)$) and the impulse function ($u_{if}(t)$) can be written using the general mathematical expressions:

$$
\frac{d}{dt}u_{us}(t) = u_{if}(t) \tag{1.9}
$$

and

$$
\int_{-\infty}^{+\infty} u_{if} \cdot dt = u_{us}(t). \tag{1.10}
$$

It is these forms that lack precision in the mathematical operations involved. Let a time function $f(t)$ be a regular continuous function with $t \geq 0$, but that is zero for $t \leq 0$. The notation $f(t) \cdot u_{us}(t)$ may define this function. When there is a time delay until $t = t_d$, before the function is switched, we get

$$
\Phi(t) = f(t-t_d) \cdot u_{us}(t-t_d) \tag{1.11}
$$

With this notation one obtains the mathematical formalisms

$$\Phi(t) = 0 \qquad \text{for } t \leq 0 \tag{1.12}$$

$$\Phi(t) = f(t) \qquad \text{for } t \geq 0 \tag{1.13}$$

Notation $u_{us}(t - t_d)$ characterizes that a unit step function is zero until $t = t_d$, after which the magnitude is unity. In other words, the unit step function is applied at time $t = t_d$. The form $f(t - t_d)$ is required to insure that the functionals are the same function $f(t)$ but applied at time $t = t_d$.

The impulse function as a test signal can be introduced in a similar way for the unit step test signal. Suppose that a function $f(t)$ is continuous at $t = 0$, we receive:

$$\int_{-\infty}^{+\infty} f(t) \cdot u_{IF}(t) \cdot dt = f(0). \tag{1.14}$$

This result is heuristically evident while the impulse function is zero everywhere except at $t = 0$, with a contribution only at this value of t, where the area under the unit impulse function is unity. Correspondingly, for the unit impulse, applied with a delay in time $t = t_d$, we obtain:

$$\int_{-\infty}^{+\infty} f(t) \cdot u_{if}(t - t_d) \cdot dt = f(t_d). \tag{1.15}$$

For the derivatives of this function we can write

$$\frac{d\Phi(t)}{dt} = \frac{df(t)}{dt} \cdot u_{us}(t) + f(t) \cdot u_{if}(t). \tag{1.16}$$

It can be noted, that the derivative function contains an impulse at time $t = 0$.

Example 1.1
The sinusoidal time function $\cos(t)$ is started at time $t = 0$. The function and its derivative are unknown. The solution is as follows

$$f(t) = \cos(t) \cdot u_{US}(t), \tag{1.17}$$

with the derivative

$$\frac{df(t)}{dt} = -\sin(t) \cdot u_{US}(t) + \cos(t) \cdot \frac{d}{dt} u_{US}(t) \tag{1.18}$$

which can be rewritten as

$$\frac{df(t)}{dt} = -\sin(t) \cdot u_{US}(t) + \cos(t) \cdot u_{if}(t). \tag{1.19}$$

The scope of the formalization introduced above is to show how functions used for system-excitation purposes can be separated by applying the superposition method for the singular test function.

1.2 Modeling Formalisms

Models are used for many different purposes, to explain behavior and data, and to provide a compact representation of data. Hence modeling is a simpler method of solving complex problems in science, technology, economics, and other domains, because the success in analyzing real-world systems depends upon whether or not the model is properly chosen. In order to develop suitable system models, a thorough understanding of the system and its operating conditions is of essential importance. Since a model is an abstraction it will only capture some properties of the real-world system. Therefore, it is often necessary to use many different models.

From a more general point of view we can introduce three types of modeling concepts for dynamic systems that can be stated as general systems analysis concepts. These concepts depend on the respective a priori knowledge based on

- Knowledge of inputs
- Knowledge of outputs
- Knowledge of system states

for the decision of the unknown, as shown in Fig. 1.5

After a system model structure is found for the dynamic system, the next step is to find the model formalisms, which means the mathematical equations. These describe the dynamic system on a much more abstract level. Depending on the different types of systems to be modeled one may model from data as well as from physics. The latter uses fundamental physical laws like Kirchhoffs circuit laws and Newtons laws of motion, for example, to determine model structures and parameter values. The equations obtained describing the systems behavior have many forms like:

- Linear equations
- Non-linear equations
- Integral equations
- Difference equations
- Differential equations
- Petri-net equations

- Bond graph equations
- Stochastic equations
- Tuning band equations
- etc.

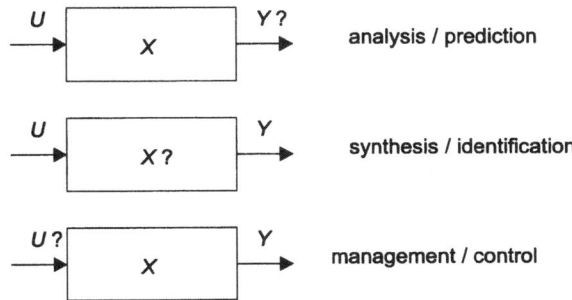

Fig. 1.5. Types of systems concepts that depend on a priori knowledge of the inputs, outputs, or system state

Mathematical models of physical systems, such as mechanical systems, can be derived from consideration of the physical laws and basic relationships governing the behavior of the system, such as the equation of motion, the dynamics of motion, linear springs, friction, levers and gears, as well as the existing boundary conditions.

Let us consider the dynamics of a body moving with a varying velocity, v. If the body moves an arbitrary distance x in a given time t, the average velocity of the body can be simply expressed by the equation

$$v = \frac{x}{t}.$$
(1.20)

While the body is moving linear, by the movement of the body could be examined over a small period of time Δt. While the velocity of the body has little possibility to change within this period the equation

$$v = \frac{\Delta x}{\Delta t}$$
(1.21)

is a very close approximation of the instantaneous velocity of the body when moving over small distances Δx. As we continue to reduce the size of the period we find that, as Δt approaches zero, the above equation becomes an exact expression of the instantaneous velocity of the body at any instant of time t', as follows:

$$v(t') = \lim \frac{\Delta x}{\Delta t}\Big|_{t=t'} ,$$

$$\Delta t \rightarrow 0$$

(1.22)

which can be rewritten as follows

$$v(t) = \frac{dx(t)}{dt} ,$$

(1.23)

which means that both, velocity v and arbitrary distance x are functions of time t, and that v is a measure of the instantaneous rate of change of the distance x with respect to time t.

Let us now consider a change of velocity with respect to time, which is the case when the body speeds up or slows down, and on these occasions the body is said to accelerate or decelerate. Hence the motion can be described as follows:

$$a(t) = \frac{dv(t)}{dt} ,$$

(1.24)

where $a(t)$ is the acceleration, a function of time t. Substituting the equation of $v(t)$ in this equation we obtain

$$a(t) = \frac{d\left(\dfrac{dx(t)}{dt}\right)}{dt} = \frac{d^2 x(t)}{dt^2} ,$$

(1.25)

which is the second derivative of distance with respect to time.

As an example, in this sense the relationships of systems, which could be the variables of biological, electrical, mechanical, physical systems, and so forth, can be described by a set of ordinary differential equations (ODEs) representing the mathematical system model. Hence a dynamic system can be described by the differential equation

$$a_n \cdot y^{(n)}(t) + a_{n-1} \cdot y^{(n-1)}(t) + \ldots + a_2 \cdot y''(t) + a_1 \cdot y'(t) + a_0 \cdot y(t)$$

$$= b_0 \cdot u(t); \qquad a_n \neq 0$$

(1.26)

where $y(t)$ is the system output variable and $u(t)$ is the system input variable. Dividing this equation and shifting by a_n results in:

$$y^{(n)} = -\frac{a_{n-1}}{a_n} \cdot y^{(n-1)} - \ldots - \frac{a_2}{a_n} \cdot y^{(2)} \cdot \frac{a_1}{a_n} \cdot y^{(1)} - \frac{a_0}{a_n} \cdot y + b_0 \cdot u(t). \qquad (1.27)$$

Rewriting this differential equation with, $x_0(t)$, $x_1(t)$, $\ldots$, $x_n(t)$ as substitution variables, and $y(t)$ as the output variable, we get n first order differential equations

$$x_0(t) := y(t) \qquad\qquad (1.28)$$

$$x_1(t) := \dot{x}_0(t) = y'(t)$$
$$x_2(t) := \dot{x}_1(t) = y''(t)$$

$$\cdot$$
$$\cdot$$
$$\cdot$$

$$x_n(t) := \dot{x}_{n-1}(t) = y^{(n)}(t)$$

which result in n differential equations of first order

$$x'_{n-1}(t) := x_n(t) = -\frac{a_{n-1}}{a_n} \cdot x_{n-1}(t) - \frac{a_{n-2}}{a_n} \cdot x_{n-2}(t) - \ldots - \frac{a_1}{a_n} \cdot x_1(\qquad (1.29)$$

$$-\frac{a_0}{a_n} \cdot x_0(t) + b_0 \cdot u(t)$$

$$\cdot$$
$$\cdot$$
$$\cdot$$

$$x'_1(t) := x_2(t)$$

$$x'_0(t) := x_1(t)$$

This type of differential equation can be solved by evaluating the integral

$$x(t) = x_0 + \int_0^t f(x(\tau), u(\tau), \tau)dt \qquad (1.30)$$

for successive values $t - t_1$ within the calculation interval with numerical integration methods, such as Euler, Adams Bashfort, Runge Kutta, etc. In general, the different numerical-integration methods approximate the integral by summing. Accuracy of the numerical-integration approximation can be achieved when the integration step length is in between the range of small and large step lengths, which means in between the region of decreased and increased step length, as shown in

Fig. 1.6. The step length chosen influences both the accuracy as well as the calculation time interval. For details see Appendix A.

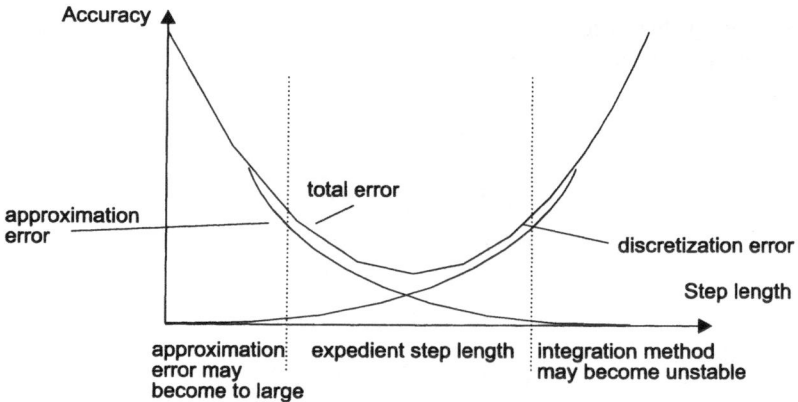

Fig. 1.6. Influence of the step length on accuracy

The most important step of modeling is the translation of our understanding of the dynamic system into the mathematical equations of systems theory that can be introduced as an abstract form of a model. This expresses the characteristics of the dynamic system in such a way that it selects a particular system from the set of all systems. Systems theory has its roots in the mathematical representation of dynamic systems that was strongly inspired by mechanics. In any case, it is usual to obtain a set of mathematical equations that describe the important physical variables of interest of the dynamic system. Describing a dynamic system, based on the translation of systems knowledge of the system into the mathematical model language, we are able to use the several possible descriptions, meaning the sets of mathematical equations. In the case of a time invariant, time continuous system, we can find a mathematical model MM_{IC} based on ordinary differential equations as a set of mathematical equations, of the form:

$$MM_{IC}: (U, X, Y, f, g, T) \tag{1.31}$$

with $u \in U$: set of inputs, $x \in X$: set of states, $y \in Y$: set of outputs, f : rate of the change function, g : output function, T: time domain, and

$$\begin{aligned} x' &= f(x,u) \\ y &= g(x,u) \end{aligned} \tag{1.32}$$

It may be convenient to place these mathematical equations in a standard or normal form; the important point here is to decide on variables, called state variables, which are essential in characterizing the dynamic system. Such a model formalism is a special case of a set structure. Let MM = Σ, then one can use the three notations: input, output, and state. Correspondingly we have a state X, a set U of input values, and a set Y of output values. A mathematical model of a system is called a dynamic one, if it can be defined as a set structure Σ:

$$\sum := (X, Y, U, v, t, a, b),$$ (1.33)

with state variable X, set of output values Y, set of input values U, set of admissible controls v, time domain T, state transition map a, read-out map b.

Assuming a dynamic system with state vector $x = [i, u_1, u_2]^T$ representing an electrical RCL network, the respective mathematical model equations of which show the state variable X, set of input variables U as well as the system-parameter matrix A. The electrical RCL network may be described as follows:

$$i = \frac{1}{L} u_1 - \frac{1}{L} u_2$$ (1.34)

$$\dot{u}_1 = \frac{1}{C_1} i + \frac{1}{C_1} i_{in} + \frac{1}{C_1} i_A$$ (1.35)

$$\dot{u}_2 = \frac{1}{C_2} i - \frac{1}{C_2} i_{out}.$$ (1.36)

The graphical representation is shown in Fig. 1.7. These equations can be simplified by supposing that $i_{in} = i_A = 0$, and the resistance R_A of the system is

$$R_A = \frac{u_2}{i_{out}}.$$ (1.37)

The state variable of this linear time-continuous system can be described as

$$x'(t) = A \cdot x(t),$$ (1.38)

whereby the (n, n)-system matrix A is given by

$$A = \begin{bmatrix} 0 & \dfrac{1}{L} & -\dfrac{1}{L} \\[2ex] -\dfrac{1}{C_1} & 0 & 0 \\[2ex] \dfrac{1}{C_2} & 0 & -\dfrac{1}{C_2 \cdot R_A} \end{bmatrix}$$

(1.39)

The variables i, u_1, u_2, represent the state variables of the state variable X. In a similar way, the values of L, C_1, C_2, and R_A are the parameters of the state variable model, which are delineated by the system matrix A, which is a (n,n)-matrix. The electrical representation of the network is given in Fig. 1.7.

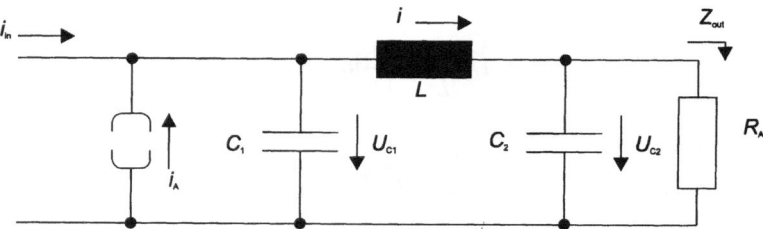

Fig. 1.7. Electrical representation of the network model

In some cases it can be necessary to specify unmeasurable and/or random inputs. These system disturbances can be described as impacts of uncontrollability and/or unobservablility of the dynamic system being modeled, which can mathematically be described by stochastic, continuous-time models, as follows:

$$\text{MM}_{SC} : (\, U,\ V,\ W,\ X, T, f,\ g),$$

(1.40)

with

$$x' = f(x,\ u,\ w,\ t)$$

(1.41)

$$y = g(\,x,\ v,\ t).$$

The vectors v and w are random model disturbances. In the case that v and w are random or stochastic vector processes, meaning that the stochastic properties of these vectors are not related to the model specification, then x and y will be the same process.

In many cases and especially in management and operational research, the dynamic system can be thought of as being built up of a collection of events. Even the state variables change at specific time instants. A mathematical description of which can be based on the notation of a mathematical discrete event model, yields

$$\text{MM}_{\text{DE}} : (V, S, Y, \delta, \lambda, \tau, T) \qquad (1.42)$$

with V: set of external events, S: sequence of states, Y: set of outputs, δ: transition function, λ: output function, τ: time function

Many dynamic systems have properties that vary continuously in space, which can be described based on distributed models. The mathematical expression for distributed models is based on partial differential equations, which result in the following mathematical model description:

$$\text{MM}_{\text{PDE}} : (U, \Theta, Y, F, r, g, z, T), \qquad (1.43)$$

with

$$0 = f(\Theta, \frac{\partial \Theta}{\partial t \partial z}, u, z, t)z \in Z; \ 0 = r(\Theta, z, t)z \ domZ; \ y = g(\Theta, z, t)z \in Z. \quad (1.44)$$

Apart from the independent variable t, the space coordinate z is introduced. Θ is the vector of the dependent variables that can vary in space and time. The equation(s) hold(s) in a spatial domain Z while conditions given by r are provided on the boundary of the domain $domZ$. There are input u and output y.

Dynamic systems may have different specific mathematical-modeling formalisms expressed in the several mathematical equations. Then finally applying system theory on dynamic systems, we obtain sets of state variables that are the mathematical model of the respective dynamic system. This can be interpreted as translating the equations of the dynamic system variables into state-variable equations in the standard form, if possible, the latter equations are called state equations for the particular dynamic system. In conclusion, the modeling procedure consists of defining a mathematical model of the dynamic system and translating it into a set of descriptive state equations that can be expressed in the following state-variable model description:

$$\begin{aligned} x'(t) &= f(x(t), \ u(t), \ t) \\ y(t) &= g\ (x(t), \ u(t), \ t), \end{aligned} \qquad (1.45)$$

with $x(t)$ as n-dimensional state vector, $x'(t)$ as its derivative, $y(t)$ as q-dimensional output vector, $u(t)$ as p-dimensional input vector, and f and g as nonlinear vector functions. The state variables in (1.45) are said to be related by a nonlinear transformation and hence they are called nonlinear state equations. The block diagram of the nonlinear state variable model is shown in Fig. 1.8.

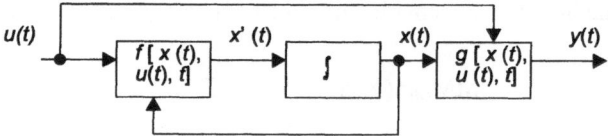

Fig. 1.8. Structural representation of the state-variable model of (1.45)

Hence we can rewrite (1.45) as a nonlinear system with small excitation Δx around the initial state x_0

$$\frac{d}{dt}(x_0 + \Delta x) = f(x_0, u_0, t) + \left(\frac{\partial f}{\partial x}\right)_0 \Delta x + \left(\frac{\partial f}{\partial u}\right)_0 \Delta u. \qquad (1.46)$$

Simplification of (1.46) results in a linear state-variable model

$$x'(t) = A \cdot x(t) + B \cdot u(t)$$
$$\qquad\qquad\qquad\qquad\qquad (1.47)$$
$$y(t) = C \cdot x(t) + D \cdot u(t),$$

with x as n-dimensional state vector, u as p-dimensional input vector, and y as q-dimensional output vector. A is a (n, n)-matrix, which is called the system matrix, B is a (n, p)-matrix, which is called the input matrix, C is a (q, n)-matrix, which is called the output matrix, and D is a (q, p)-matrix, which is called the transition matrix. The matrixes A, B, C, and D are based on constant elements.

The state equations for systems with only one input and one output variable are a special case of the state variable expression above, given in (1.47).

The state equations given in (1.47) are said to be related by a linear transformation. For given initial state values $x_1(0)$, ... , $x_n(0)$ and a given input function $u(t)$ defined for $t > 0$ there exists a unique solution of the state equations $x_1(t)$, ... , $x_n(t)$ defined for all $t > 0$, i.e the functions $x_1(t)$, ... , $x_n(t)$ satisfy the state equations exactly for all $t > 0$, and hence there exists a unique output function $y(t)$ defined for $t > 0$.

In the case that elements of the matrixes A, B, C, and D are time dependent, the state variables given in (1.47) are said to describe a linear time-variant system.

Assuming that the influence of $u(t)$ on $y(t)$ is normally indirect, we may neglect D, writing $D \equiv 0$, which results in a structural diagram of a linear state-variable model, as shown in Fig. 1.9.

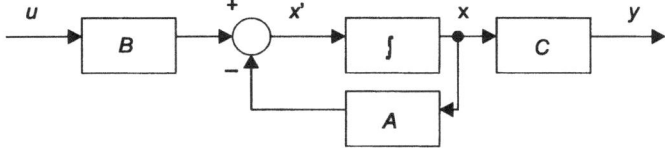

Fig. 1.9. Structural representation of the linear state-variable model with $D \equiv 0$

Once a mathematical description of a dynamic system is obtained, the next step in studying the dynamic system involves the analysis – quantitative as well as qualitative – as a way of understanding the transient responses. Our interest in the case of a quantitative analysis is the response calculation of the system to certain initial input conditions. This part of the analysis can easily be achieved by computer simulation using the various simulation software packages, as described in Chap. 4. For a qualitative analysis, our interests are the properties of the system under test such as stability, controllability, observability, and others, which were introduced by Kalman in the 1960s. These proofs of system characteristics are of importance, while it is possible to obtain a better understanding of the system by the respective proofs. For example, stability analysis illustrates that a system with an equilibrium state is roughly stable if for any initial state close to the equilibrium state the state response tends, in the limit as time increases, to go to the equilibrium state. In the same way, controllability and observability illustrate the dynamic properties of the system, which are examined in more detail in Chap. 2.

If the response of a modeled and simulated system are found to be unsatisfactory, the model of the system has to be improved or optimized. In some cases, the response of the model of the system can be improved adjusting certain parameters of the system model that can be successfully done by using parameter-estimation techniques, which are examined in more detail in Chap. 5, as well as adjusting the system model state itself. This can be done successfully by using state-estimation methods. The basic principle of parameter estimation are the adjustment of the parameter vector $\underline{p}$ of an identification model with the same structure as the true model of the dynamic system in such a way that its output $\underline{y} = g(\underline{u}, \underline{x}, \underline{p}, t)$ will coincide with the true model output $\underline{y} = g(\underline{u}, \underline{x}, \underline{p}_{TM}, t)$, as shown in the principle diagram of the parameter estimation scheme, in Fig. 1.10.

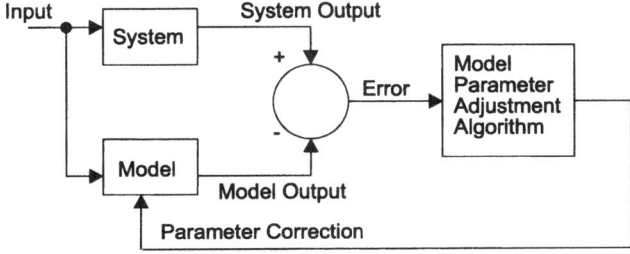

Fig. 1.10. Structural representation of the parameter-estimation method

1.3. System Elements and Models of Continuous-Time Systems

A number of systems are studied from a diverse range of scientific disciplines. As mentioned above, one of the advantages of the systems-theory approach in modeling and simulation is that systems theory is applicable to a wide range of complex dynamic systems. Due to this fact the basic principles of the most important classes of physical systems that are used for modeling and simulation of systems, can be introduced from a more general point of view. The most general are electrical elements that are used in most tutorials of simulation systems like Modelica, SIMULINK, and so forth. It should be noted that the complexity of model structures, as they are used for simulation studies, depends on the purpose for which the models are developed.

1.3.1 Electrical Elements

An important class of systems are those that can be described based on electrical elements like resistors, capacitors, inductors, and are expressed in terms of the physical variables, voltage, charge, and current. Resistors, capacitors, and inductors are the heart of any electrical circuit and/or network configuration, using the standardized symbols as shown in Fig. 1.11.

Fig. 1.11. Electrical representation of network components

In an ideal resistor the voltage drop V across a resistor R is related to the current I through the resistor R, expressed by Ohms law, named after the German physicist, born 1789 in Erlangen, Bavaria, as

$$V_R = R \cdot I ,$$
(1.48)

where R is a constant, called resistance, and depends on the physical material constants χ, ρ, length l, and cross-sectional area A, hence

$$R = \frac{l}{\chi \cdot A} = \rho \cdot \frac{l}{A}; \ \Omega,$$
(1.49)

with

$$\chi = -\mu_e \cdot \rho_e = -\mu_e(-n \cdot e), \tag{1.50}$$

where constant χ represents the conductivity of the material, constant μ is used for the mobility of the electrons, $n \cdot e$ give the elementary charge, ρ is a constant characterizing the density of electrons of the respective material, A is the cross-sectional area of the conductive material the current I passes through, and l is the length of the (copper) wire the resistor is made from. A resistor has a linear response characteristic over frequency, which means the resistor is independent of frequency.

In an ideal capacitor the charge Q on a capacitor is related to the voltage drop V across the capacitor C, given by the relationship

$$Q = C \cdot V_C, \tag{1.51}$$

where C is a proportional constant, called capacitance. The capacitance is a geometric factor given by the expression

$$C = \frac{\varepsilon \cdot A}{4 \cdot \pi \cdot d}; \ \text{F}, \tag{1.52}$$

where ε is the permittivity, called dielectric constant, of the material, A is the area of the plates of the capacitor, and d is the space between these plates. In general $\varepsilon = \varepsilon_r \varepsilon_0$ where $\varepsilon_0 = 1/36 \ \pi 10^9$ F/m, called free-space permittivity, and ε_r is called relative permittivity. For the current through the capacitor the mathematical formula is:

$$I = \frac{dQ}{dt} = \frac{dV}{dt} \cdot C; \ \text{A} \tag{1.53}$$

In cases such as the nonlinear one, C can be a complex relation between the energy stored in the electric field E of the capacitor, and the charge Q, given by a relationship like

$$E = \frac{\partial V_C}{\partial Q}. \tag{1.54}$$

For a linear relationship, the voltage is proportional to the charge, that is $Q \approx E$, thus

$$I = C \cdot \frac{dV_C}{dt}. \tag{1.55}$$

It should be noted that the response characteristic of a capacitor is nonlinear against frequency.

Example 1.2
If the voltage across a capacitor changes by 1000 V in 20 ms when a current of 50 mA passes into it, what is the value of the capacitor? With

$$I = \frac{C \cdot V}{t}$$
(1.56)

and after rearranging we obtain

$$C = \frac{I \cdot t}{V} = \frac{50}{10^3} \cdot \frac{20}{10^3} \cdot \frac{1}{10^3} = 1 \ \mu F$$
(1.57)

In an ideal inductor the voltage drop V across the inductor is related to the rate of change of current through the inductor $\frac{dI}{dt}$ by

$$V_L = L \cdot \frac{dI}{dt},$$
(1.58)

where L is a constant, called inductance. Assuming a linear inductor, the inductance is related to the flux linkage $\psi = N \cdot \phi$, where N is assumed as a single inductor circuit of N turns carrying a constant current I that is placed in a magnetic field from fixed external sources. Owing to the mechanical force of magnetic origin that acts on the system, we can imagine that the inductor is displaced from its zero- current position. Due the displacement of the inductor current there is a change of flux linkages. This flux linkage is accompanied by an electromagnetic force being induced into it, which according to Faradays law of induction, named after the English chemist and physicist Faraday, born 1791 in Surrey, England, is

$$E = -H \cdot \frac{d\Phi}{dt} = \frac{d\psi}{dt},$$
(1.59)

where $\frac{d\psi}{dt}$ denotes the time required to effect the displacement. For a linear (ψ, I) characteristic, $\psi \approx I$, the proportional constant is termed the inductance L

$$L = \frac{\psi}{I}; \ H.$$
(1.60)

In order to describe the inductor L as a network element, we can use a relation between current I and voltage V, which is given by

$$V_L = \frac{d\psi}{dt} = L \cdot \frac{dI}{dt}; \text{ V.} \tag{1.61}$$

This is the fundamental relation for the linear inductor when I is the dependent variable. The corresponding form, when V_L is the dependent variable, is:

$$I = \frac{\psi}{L} = \frac{1}{L} \cdot \int V_L \cdot dt . \tag{1.62}$$

Since resistors, capacitors, and inductors are typically connected in networks there are necessarily imposed relations between the physical variables; these electrical relationships are referred to as Kichhoffs voltage law and Kirchhoffs current law, discovered on 1854 by the German physicist Kirchhoff, born 1824 in Königsberg, Germany.

Simply stated, the voltage law says that the sum of all the voltage rises and voltage drops around a closed loop is equal to zero:

$$\sum V = \sum I \cdot R . \tag{1.63}$$

Kirchhoffs current law states that the sum of all the currents entering and leaving a node is equal to zero:

$$\sum I = 0 . \tag{1.64}$$

These two laws are important as they form the very heart of any electrical circuit and/or network. In most cases, electrical networks includes voltage sources and current sources.

Example 1.3
Application of Kirchhoffs circuital laws in an electrical network, see Fig. 1.12.

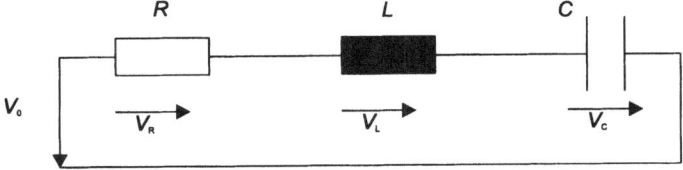

Fig. 1.12. Electrical network

The electrical network has a resistor R, an inductance L and a capacitor C in series with an electromotive force V_0. Applying Kirchhoffs circuital laws we find

$$V_0 = V_R + V_L + V_C \,, \tag{1.65}$$

and

$$I = I_R + I_L + I_C \,, \tag{1.66}$$

hence

$$V_0 = R \cdot I + j\omega L \cdot I + \frac{1}{j\omega C} I = (R + j\omega L + \frac{1}{j\omega C}) \cdot I \,, \tag{1.67}$$

where $\omega = 2 \cdot \pi f$. With $X_R = R$, $X_L = \omega L$, and $X_C = 1/\omega C$ we find, due to the frequency dependency in ω, that the inductance and the capacitance have a nonlinear characteristic, as shown in Fig. 1.13.

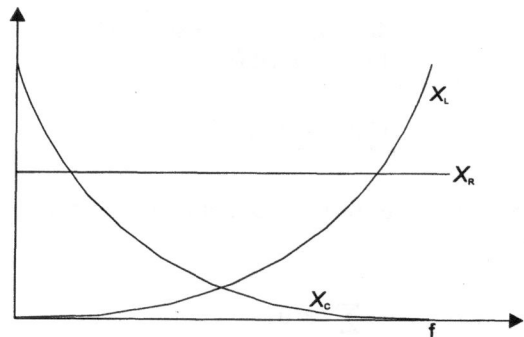

Fig. 1.13. Frequency dependency of X_R, X_L, and X_C

On the other hand, when applying Kirchhoffs circuit law, we may obtain

$$L \cdot \frac{dI}{dt} + R \cdot I + \frac{Q}{C} = V_0, \tag{1.68}$$

and since $\dfrac{dQ}{dt} = I$, we may differentiate the equation above, and obtain (when V_0 is assumed to be a constant):

$$L \cdot \frac{d^2 I}{dt^2} + R \cdot \frac{dI}{dt} + \frac{1}{C} = 0, \tag{1.69}$$

which is a second-order linear homogeneous differential equation with the auxilliary equation

$$x^2 + x \cdot \frac{R}{L} + \frac{1}{C \cdot L} = 0, \tag{1.70}$$

which has the roots

$$x = \frac{-R \pm (R^2 - 4\frac{L}{C})^{1/2}}{2 \cdot L}. \tag{1.71}$$

As Example 1.3 shows, real-world problems could easily be modeled based on the respective a priori knowledge (which can be given in physical laws), measurements of the system inputs and outputs, phenomenological knowledge, and so forth. The simplicity of Example 1.3 has no influence on the general impact of the methodology on mathematical and computer modeling and simulation. It only shows how to proceed when solving problems related to an electrical network. It is clear that most of the real-world problems that are solved using modeling and simulation are very complex ones, which means incorporate nonlinearities or various time dependencies, and so forth. In these cases the only possibility to obtain a good enough solution will be simulation, while an analytical solution is not possible.

1.3.2 Particle Dynamics

Another important class of physical processes in modeling and simulation are those that can be described in terms of motion of (ideal) particles. The important physical variables for each particle are its position as measured from a fixed reference, its velocity, and its acceleration. Also of importance is the force acting on the respective particle. The basic relations that characterizes the motion of a particle are based on the laws of mechanics, which are known as Newtons first, second and third law of motion, named after the English mathematician Newton, born 1643 in Woolsthorpe, Lincolnshire, England, which he was very secretive about;

- A body continues in the state of rest or of uniform motion in a straight line unless it is compelled to change that state by forces impressed upon it.
- The change of motion is proportional to the motive force impressed, and is in the direction of the straight line in which the force is impressed.
- To every action there is always a reaction equal in magnitude and opposite in direction.
- Two particles are attracted toward each other along a line connecting them with a force whose magnitude is directly proportional to the product of the masses and inversely proportional to the square of the distance between them.

The force F on a particle and the acceleration a of the particle are related by

$$F = m \cdot a, \tag{1.72}$$

where the mass m of the particle is assumed constant. The acceleration of the particle is related to its velocity v by

$$a \; = \; \frac{dv}{dt}, \tag{1.73}$$

and the velocity of the particle is related to its position H by

$$v \; = \; \frac{dH}{dt}. \tag{1.74}$$

Force, acceleration, velocity, and position can be considered as vector variables in the usual sense. In many situations it is desired to consider the motion of a collection of particles; it is often convenient to examine free-body diagrams, which are used to indicate the various forces on each particle in the collection. The motion of rigid bodies, consisting of an infinite number of such particles, is often of interest. The relevant equations for a rigid body can be obtained by using the relations given above, and integration over the body scope. The resulting motion is described by equations as above in terms of the center of gravity of the body; the rotational motion of the body is described in terms of the angular position of the body as measured from a fixed reference, its angular velocity, and its angular acceleration; the torque acting on the rigid body is also important. It can be shown that the torque T acting on a rigid body and the angular acceleration α of the rigid body are related by

$$T = I \cdot \alpha, \tag{1.75}$$

where the moment of inertia I of the rigid body is assumed constant. The angular acceleration of the rigid body is related to its angular velocity ω by

$$\alpha = \frac{d\omega}{dt}, \tag{1.76}$$

and the angular velocity of the rigid body is related to its angular position θ by

$$\omega = \frac{d\theta}{dt}. \tag{1.77}$$

Torque, angular acceleration, angular velocity, and angular position may also be considered as vector variables. The above relationships are essential ingredients in describing the motion of particles and rigid bodies.

The above is a very brief description of the basic elements in Newtonian mechanics. The principles of conservation of momentum and conservation of energy often lead to useful relationships for elastic collision between real objects. The simplest collision to model and simulate may involve two imaginary particles with mass and velocity, but no size. When such two particles collide, their new velocities are controlled, as mentioned, by the principle of the conservation of momentum and the principle of relative motion. Supposing that the momentum of a particle is the product of its mass and velocity, and given two particles with mass m_1 and m_2, and associated velocities v_1 and v_2,

$$m_1 \cdot v_1 + m_2 \cdot v_2 = \text{const.} \tag{1.78}$$

where const. represents a constant.

The principle of relative motion states that the relative velocity after an impact equals the relative velocity before the impact, but multiplied by a coefficient of restriction r. Hence the respective velocities of the two particles after the collision are v_{1ac} and v_{2ac}, which yields in

$$v_{1ac} - v_{2ac} = r(v_1 - v_2). \tag{1.79}$$

Example 1.4

As an example of particle dynamics we may suppose two particles are positioned at c_1 and c_2 at the time of impact. Their respective velocities are v_1 and v_2, and angles α_1 and α_2 relative to the line connecting their two centers. The components of velocity perpendicular to lines c_1 and c_2 are unaltered by the collision and are given by the relations $v_1 \cdot \sin\alpha_1$ and $v_2 \cdot \sin\alpha_2$ respectively. The components of the velocity parallel to line $c_1 c_2$ are subject to the conservation of momentum while the particular masses are accelerated. Hence:

$$m_1 \cdot v_1 + m_2 \cdot v_2 = m_1 \cdot v_{1ac} \cdot \cos\alpha_1 + m_2 \cdot v_{2ac} \cdot \cos\alpha_2 , \tag{1.80}$$

and by Newtons law of relative motion we obtain

$$v_{1ac} - v_{2ac} = -r(v_1 \cdot \cos\alpha_1 - v_2 \cdot \cos\alpha_2) , \tag{1.81}$$

and substituting the last two equations we find

$$v_{1ac} = \frac{(m_1 - r \cdot m_2) \cdot v_1 \cdot \cos\alpha_1 + m_2 \cdot v_2 \cdot \cos\alpha_2 (1+r)}{m_1 + m_2} \tag{1.82}$$

and

$$v_{2ac} = \frac{(m_2 - r \cdot m_1) \cdot v_2 \cdot \cos\alpha_2 + m_1 \cdot v_1 \cdot \cos\alpha_1 (1+r)}{m_1 + m_2}. \tag{1.83}$$

The final velocity of each particle, and its direction of motion, can be found from the perpendicular and parallel components.

Assuming that the second particle is at rest, the final velocities may be calculated as follows:

$$v_{1ac} = \frac{(m_1 - r \cdot m_2) \cdot v_1 \cdot \cos\alpha_1}{m_1 + m_2} \tag{1.84}$$

$$v_{2ac} = \frac{m_1 \cdot v_1 \cdot \cos\alpha_1 (1+r)}{m_1 + m_2}.$$

1.3.3 Mechanical Elements

Consider a source of kinetic energy E_k described by

$$E_k = \frac{1}{2} \cdot m \cdot v^2; \; \text{J}, \tag{1.85}$$

with v as the velocity. When considered in terms of Newtons law of motion, the definition results from

$$f = \frac{d}{dt}(m \cdot v); \; \text{N}, \tag{1.86}$$

which states that the time rate of change of momentum is equal to the applied force, f, and is in the direction of the force. Under ordinary circumstances the velocity of motion is sufficiently small so that the mass m remains substantially constant, and the above equation may be rewritten in the form

$$f = m \cdot \frac{dv}{dt} = m \cdot \frac{d^2x}{dt^2} = m \cdot a, \tag{1.87}$$

where $\dfrac{dv}{dt} = \dfrac{d^2x}{dt^2} = a$ is the acceleration of the mass element. In this form m may be introduced as the proportionality factor between force and acceleration.

Let us assume that a set of mechanical elements and rotational variables exist. In the rotational system, torque is the flow or through variable, and angular velocity (or angular displacement, or angular acceleration) is the motional or across variable. The corresponding fundamental quantities are:

J_x = polar moment of inertia, corresponds to m_x in translation,
D = rotational damping, corresponds to d in translation,
K_s = rotational spring constant, corresponds to k_s in translation,
$T(t)$ = torque, corresponds to f in translation,
ω = $\dfrac{d\Phi}{dt}$, angular velocity, corresponds to $v = \dfrac{dx}{dt}$ in translation.

When introducing inertia we are usually referring to the inertial mass of an object. For a body with a large amount of inertia it is difficult to start the body moving or to accelerate movement of the body and once the body is moving it is equally difficult to bring the body to rest again. Inertia in this sense means, the body could be moving in a straight line, could be rotating on a shaft, or could be a combination of both motions, as is realized in the wheel of cars, and so forth. However, for the moment of inertia of a rotating body, we refer to a specific property of the body and its manner of rotation. While the moment of inertia is a measure of the body's ability to resist any attempt to change its speed or rotation, and is dependent not only upon the body mass, but also upon how that mass is distributed about the axis around which the body rotates, which results in the moments of inertia of body mass m, about the axis x, y, and z, respectively. J_x is referred to as the polar moment of inertia, which is quite different in magnitude from J_y and J_z.

The rotary force required to start or stop a body spinning is called torque $T(t)$. Newtons second law of motion, when applied to rotating bodies, can be simply stated as a proportional relationship while the rate of change of angular momentum of a body is directly proportional to the torque acting on the body, where J, is the proportionality between torque and angular acceleration,

$$T(t) = J_x \cdot \frac{d\omega}{dt} .$$
(1.88)

The polar moment of inertia J_x of a rotational body depends upon the body mass m and the square of a characteristic distance of the body, which may be introduced as the radius of gyration, Ξ, which can be expressed as follows

$$J_x = m \cdot \Xi .$$
(1.89)

For a simple point mass rotating about an axis at a distance r from the center of mass, with $\Xi = r$, we obtain

$$J_x = m \cdot r^2.$$
(1.90)

Example 1.5
For a floppy disk of radius r, rotating about its center, with

$$\Xi = r \cdot \sqrt{2}, \tag{1.91}$$

we obtain the polar moment of the inertia as

$$J_x = m \cdot (r \cdot \sqrt{2})^2 = 2 \cdot m \cdot r^2. \tag{1.92}$$

Example 1.6
Suppose a planet of mass m_P is moving around the sun on a pathway with the position $r = r(t)$. Assuming Newtons inverse square gravitational law, the equation of motion for the planet may be described by a second-order differential equation

$$m_p \cdot \frac{d^2 r}{dt^2} = -\frac{g \cdot m_S \cdot m_P}{r^2} e_r, \tag{1.93}$$

where g is the gravitational constant, m_S is the body mass of the sun, and e_r is a unit vector in the direction of r. With $\Gamma = g \cdot m_S$, and assuming the motion is in a plane, with plane polar coordinates r and φ, we obtain

$$\frac{d^2 r}{dt^2} = -\frac{\Gamma}{r^2} e_r, \tag{1.94}$$

with $r = r \cdot e_r$, as illustrated in Fig. 1.14.

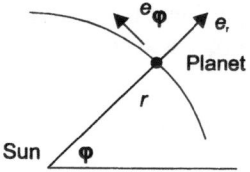

Fig. 1.14. Planetary motion around the Sun

With the first-order differential equation

$$\frac{dr}{dt} = r' \cdot e_r + r \cdot \varphi' \cdot e_\varphi \tag{1.95}$$

and the second-order differential equation

$$\frac{d^2 r}{dt^2} = (r'' - r \cdot \varphi'^2) e_r + (2 \cdot r' \cdot \varphi' + r \cdot \varphi'') e_\varphi \tag{1.96}$$

and substituting for $\dfrac{d^2}{dt^2}\, r$ in (1.94) results in

$$(r'' - r \cdot \varphi'^2)\, e_r + (2 \cdot r' \cdot \varphi' + r \cdot \varphi'')\, e_\varphi = -(\frac{\Gamma}{r^2}) \cdot e_{r.} \, . \qquad (1.97)$$

Equating coefficients e_r and e_φ we obtain the following equations:

$$r'' - r \cdot \varphi'^2 = -(\frac{\Gamma}{r^2}) \qquad (1.98)$$

and

$$2 \cdot r' \cdot \varphi' + r \cdot \varphi'' = 0 \, . \qquad (1.99)$$

Equation (1.99) can be written as:

$$\frac{1}{r} \cdot \frac{d}{dt}(r^2 \cdot \varphi') = 0 \cdot \qquad (1.100)$$

with $\dfrac{d}{dt}(r^2 \cdot \varphi') = 0$, and integrating

$$r^2 \cdot \varphi' = h \, , \qquad (1.101)$$

where h is the angular momentum per unit of the body mass of the sun, substituting for φ' in (1.98) results in

$$r'' - \frac{h^2}{r^3} = -(\frac{\Gamma}{r^2}) \, , \qquad (1.102)$$

which is a nonlinear second-order differential equation for the position vector r as a function of time t. Solving this equation by using the substitution $p = 1/r$, and solving the equation for p as a function of φ. Hence we receive, when using (1.101),

$$\frac{dp}{d\varphi} = \frac{dp}{dt} \cdot \frac{dt}{d\varphi} = \frac{d}{dt}\frac{\frac{1}{r}}{\varphi'} = -\frac{r'}{r^2 \cdot \varphi'} = -\frac{r'}{h} \, , \qquad (1.103)$$

and, using (1.101) again,

$$\frac{d^2 p}{d\varphi^2} = -\frac{1}{h} \cdot \frac{d}{d\varphi} r' = -\frac{1}{h} \cdot \frac{d}{dt} r' \cdot \frac{dt}{d\varphi} = -\frac{r''}{h \cdot \varphi'} = -r^2 \cdot \frac{r''}{h^2} \, . \qquad (1.104)$$

Thus we obtain for (1.102)

$$-h^2 \cdot p^2 \cdot \frac{d^2 p}{d\varphi^2} - p^3 \cdot h^2 = -p^2 \cdot \Gamma , \qquad (1.105)$$

and

$$\frac{d^2 p}{d\varphi^2} + p = \frac{\Gamma}{h^2} , \qquad (1.106)$$

which is a linear second-order differential equation, with the complementary solution $p = A \cdot \cos\varphi + B \cdot \sin\varphi$, and the one particular solution $p = \Gamma/h^2$. The axis can be chosen in such a way that $B = 0$, hence we can write the solution as

$$p = \frac{\Gamma}{h^2} + A \cdot \cos \varphi , \qquad (1.107)$$

and, with $p = r^{-1}$, we obtain

$$r = \frac{\dfrac{h^2}{\Gamma}}{1 + \dfrac{A \cdot h^2}{\Gamma} \cdot \cos \varphi} . \qquad (1.108)$$

For most planets eccentricity of the respective elliptic orbit can be assumed to be small, and as a good enough approximation we may assume a circular orbit with $r = \dfrac{h^2}{\Gamma}$. While $A = 0$ in (1.108) we obtain with (1.101)

$$\frac{d\varphi}{dt} = \frac{h}{r^2} = \frac{\Gamma^{1/2}}{r^{3/2}} . \qquad (1.109)$$

Integrating around a complete orbit results in

$$2 \cdot \pi = \frac{\Gamma^{1/2}}{r^{3/2}} \cdot T , \qquad (1.110)$$

with T the periodic time of the planet. Hence the planetary model shows the behavior

$$T = \frac{2 \cdot \pi}{\Gamma^{1/2}} \cdot r^{3/2} . \qquad (1.111)$$

Looking at the results by noting that the equation above can be written in a logarithmic form, we obtain

$$\ln T = K + 3.2 \ln r \qquad (1.112)$$

with K as a constant, hence a log T – ln r graph, which can be a straight line with slope $\dfrac{3}{2}$, the result of which is shown in Fig. 1.15

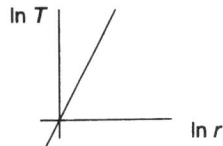

Fig. 1.15. Kepler's third law

Example 1.7
Applications of mechanical systems are diverse. In this example it may be used for modeling and simulation of soft-tissue deformation. Modeling and simulation of tissue deformation is of importance when designing the optimal implant shape and position of soft-tissue implants in order to achieve breast symmetry while using soft-tissue implants after a mastectomy in the case of breast cancer in women. Modeling may be done by using a mass damper spring model, which is a simple model type or using a more complex type of model, such as a finite element model. The mass damper spring model offers the possibility for computer simulation of tissue deformation under real-time constraints that can be described according to Newtons law by second-order differential equation (see Sect. 2.1)

$$M_i \cdot \frac{d^2 x}{dt^2} = -D_i \cdot \frac{dx}{dt} - K_i x + M_i \cdot g + f(t) \,. \qquad (1.113)$$

Rearranging (1.113) by placing terms involving x and its derivatives on the left gives

$$M_i \cdot \frac{d^2 x}{dt^2} + D_i \cdot \frac{dx}{dt} + K_i x + M_i \cdot g = f(t) \,, \qquad (1.114)$$

with M_i the mass of the soft-tissue implant under the musculus pectoralis major, D_i is the damping factor of the material of the soft-tissue implant, which can be silicone or saline filled, C_i is the strength constant of the separated upper part of the musculus pectoralis major while embedding the soft-tissue implant, x is the contractility of the separated upper part of the musculus pectoralis major.
    Neglecting $f(t)$ at the right side of (1.114), which represents an external force, and due to the downward forces as a result of gravity onto the mass of the tissue implant, we can solve the equation for the static operating point x_{SOP}, due to the gravitational force only, that is,

$$M_i \cdot \frac{d^2 x_{SOP}}{dt^2} + D_i \cdot \frac{dx_{SOP}}{dt} + K_i x_{SOP} = M_i \cdot g \,. \qquad (1.115)$$

Since x_{SOP} is a constant, which is proper to use while in a standing position, which show no movement of the breast, its derivatives are zero and (1.115) becomes

$$K_i x_{SOP} = M_i \cdot g \,. \tag{1.116}$$

Therefore, the static equilibrium position x_{SOP} of the soft-tissue implant is

$$x_{SOP} = \frac{M_i \cdot g}{K_i} \,. \tag{1.117}$$

If we define x as the sum of the constant x_{SOP} resulting from $M_i \cdot g$ and a variation δx resulting from the influence of $f(t)$, that is

$$x = x_{SOP} + \delta x \,. \tag{1.118}$$

Using (1.118) in (1.117) yields the linear differential equation for the variation δx as

$$M_i \cdot \frac{d^2(\delta x)}{dt^2} + D_i \cdot \frac{d(\delta x)}{dt} + K_i(\delta x) = f(t) \,, \tag{1.119}$$

which can be solved by using a simulation software package.

1.3.4 Fluid Mechanics

Some physical processes involve the motion of liquids and gases. Liquid systems are assemblies of liquid-filled tanks or vessels that are connected by pipes, tubes, orifices, and other flow-restricting devices. The analysis of such systems will proceed by using the fundamental laws that govern the flow of liquids. Assuming steady-flow conditions, the law states merely that accelerations, if they exist, will be small. To analyze flow phenomena, it is of importance to realize that two different types of flow exist, laminar flow and turbulent flow. A detailed study of fluid flow has yielded a criterion that allows an estimate of whether the flow is laminar or turbulent, called the Reynolds criterion: the flow is turbulent if the Reynolds number exceeds a certain figure, and is laminar if it is less than a somewhat smaller number. The Reynolds number, named after the physicist Reynolds, born 1842 in Manchester, England, which is deduced from dimensional analysis considerations, is a dimensionless quantity involving the important factors involved in flow problems, and is given by the quantity

$$\mathrm{Re} = \frac{\rho \cdot \upsilon \cdot D}{\eta} \,, \tag{1.120}$$

where ρ is the density, v is the stream velocity, D is a characteristic dimension (equals the diameter, if the duct is cylindrical), and η is the absolute viscosity.

The relationships that govern the macroscopic motions of fluids are conservation of mass and conservation of energy. It is usual to assume that liquids are incompressible, that is the volume of a liquid is a constant. Gases are usually considered to be compressible; the ideal gas law, called Boyle Mariottes law states that the pressure P of a gas multiplied by its volume V, divided by its absolute temperature T, is always constant

$$\frac{V \cdot P}{T} = const. \tag{1.121}$$

The Irish chemist and physicist Boyle was born 1627 in Lismore Castele, Munster, Ireland. The French physicist Mariotte was born 1629 in Chazeuil, France.

Example 1.8
Concept of liquid capacitance

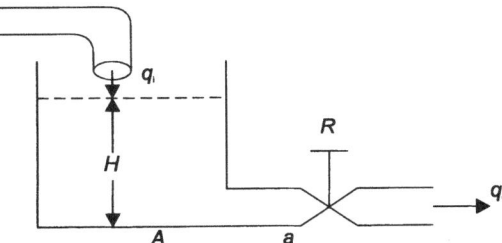

Fig. 1.16. Concept of liquid capacitance

The concept of liquid capacitance is a simple one to understand; it is simply a measure of the capacity of a tank to store the liquid. The situation is illustrated in Fig. 1.16. Given is a tank with the cross sectional area A [m²], and the water height in the tank H [m]. Assuming that the tank outlet q_o [m³/s] will be proportional to the tank inlet q_i [m³/s], we may calculate the liquid volume in the tank

$$\frac{dV}{dt} = q_i - q_o, \tag{1.122}$$

which simply relates the flow rate into the tank q_i with the flow rate out of the tank q_o. The water height in the tank will be influenced by the flow resistance R.

The state-variable equation as well as the output equation of the dynamic system has to be derived. The tank volume may be expressed as

$$V(t) = A \cdot H(t).$$
(1.123)

After differentiation we obtain:

$$\frac{dV}{dt} = A \cdot \frac{dH}{dt},$$
(1.124)

which results in the first-order differential equation:

$$\frac{dV}{dt} = A \cdot \frac{dH}{dt} = q_i - q_o.$$
(1.125)

The term q_o depends on the flow resistance, which can be expressed as follows

$$q_o = \frac{1}{R} \cdot H(t) \cdot$$
(1.126)

Hence we receive the first-order differential equation:

$$\frac{dV}{dt} = A \cdot \frac{dH}{dt} q_i - \frac{1}{R} \cdot H(t) \cdot$$
(1.127)

which results in the state equation:

$$\frac{dH}{dt} = \frac{1}{A} \cdot q_i - \frac{1}{R \cdot A} \cdot H(t) \cdot$$
(1.128)

and in the output equation

$$y(t) = \frac{1}{R} \cdot H(t).$$
(1.129)

The liquid capacitance is then defined by the relation

$$C = \frac{dV}{dH}; \quad \left[m^2 \right].$$
(1.130)

Several specific responses may now be determined, like the influence of the water level H in the tank due to the time needed to empty the tank, as shown in Fig. 1.17.

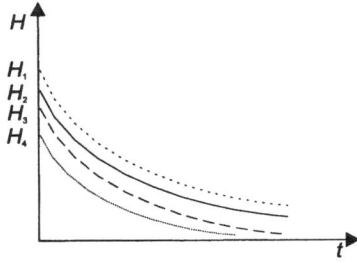

Fig. 1.17. Influence of the inflow-outflow relation on the water level H

1.3.5 Diffusion Dynamics

This is the inevitability of the transfer of mass from one region to another in homogeneous dynamic systems showing concentration gradients. The way in which the transfer of mass takes place can be assumed in the theoretical treatment of diffusion. The fundamental assumption of diffusion, given by the German physiologist Fick, in 1855, who was born 1829 in Würzburg, Germany, which stated that the rate of diffusion across any plane at right angles to the direction of diffusion bears a simple linear relation (quantitatively defined by a constant, called the diffusion constant) to the concentration gradient across the plane in the area investigated. Stated in mathematical terms the so-called Ficks law is as follows

$$dQ = -k \cdot A \cdot \frac{\partial u}{\partial x} \cdot dt , \qquad (1.131)$$

or

$$dQ = -D \cdot A \cdot \frac{\partial u}{\partial x} \cdot dt , \qquad (1.132)$$

where dQ represents the amount of mass (or material) diffusing in the time dt (during which all conditions may be considered to remain constant) across the plane of area A at right angles to the direction of diffusion, the concentration gradient at the plane being $\frac{\partial u}{\partial x}$. Instead of the symbol D in (1.132), Fick uses the symbol k assumed to be a constant for all values of u. Experimental work shows that this assumption is justified at best only as a somewhat rough approximation; for this reason the term diffusion coefficient is preferable. D evidently represents the amount of mass (or material) that in unit time and with unit concentration gradient would cross a plane of unit area at right angles to the direction of diffusion.

The unit of concentration, u, may be defined as one unit of quantity in unit volume. For the latter, the cubic centimeter is used rather than the liter, and if, as is generally the case, concentrations in a given problem are originally expressed in mol per liter they must first be divided by 1000 before being introduced into Ficks equation. Since the unit in which Q is measured also enters into the definition of the unit of concentration, the same numerical value of D must obviously apply to all cases of diffusion regardless of whether measurements are made in terms of mol, grams, number of individual molecules, etc.

Ficks equation contains four variables, Q, u, x and t. The number of these variables may be reduced to three, and a partial differential equation may be obtained whose solution for many popular problems may be much easier. The latter equation, sometimes known as Ficks second equation, may be introduced as the general diffusion equation.

1.3.6 Thermodynamics

Another important class of physical processes are those that involve the transfer of heat from one substance to another, which is a complex topic. The basic principle of thermodynamics is the principle of conservation of energy. Heat can be transferred between two bodies in such a way that the heat transfer rate Q to a body is related to the rate of change of the temperature of the body $\dfrac{dT}{dt}$ by the relation

$$m \cdot C_p \cdot \frac{dT}{dt} \;=\; Q, \tag{1.133}$$

where m is the body masss, and C_p is the specific body heat, T is the temperature, and Q as the heat flow is given by the Boltzmanns law for surface radiation, named after the Austrian mathematician and physicist Boltzmann, born 1844 in Vienna, which is

$$Q = \sigma \cdot A \cdot \varepsilon \cdot (\Theta^4_1 - \Theta^4_2), \tag{1.134}$$

where σ is the Boltzmann constant, A is the surface area, ε is the emissivity of the surface (which can range between 0 and 1) and Θ is the temperature in Kelvin. If the heat transfer depends on convection then the heat transfer rate is directly proportional to the temperature difference between the body and its surrounding. The temperature of a gas also depends on its volume and pressure. The above fourth power law, as compared with the square-root relation for turbulent fluid flow, and the linear relation for the electrical current, gives the radiation resistance as

$$R = \frac{d\Theta}{dQ} = \frac{1}{4 \cdot A \cdot \sigma \cdot \varepsilon \cdot \Theta^3{}_a}, \tag{1.135}$$

where Θ_a is the average of radiator and receiver temperatures.

Hence the thermal capacitance, expressed by C, can be formulated as

$$C = \frac{d\Theta}{dt} = Q .$$ (1.136)

Example 1.9

Let a heated hot water tank, measuring 1 m³, store water at a temperature level of 65°C. The tank is covered on all sides with an isolation iso_T of thickness 0.1 m. The thermal conductivity k is assumed to be $k = 0.5$ K/m, with K as temperature in Kelvin. If the surface temperature level of the isolation is assumed to be 25 °C, the heat transfer rate Q, which is in this case study the rate of heat loss through the isolation can be given by

$$Q = \frac{k \cdot A}{iso_T}\Delta\Theta = \frac{k \cdot A}{iso_T}(T_A - T_{iso}); \ W,$$ (1.137)

where A is the total surface of the tank, calculated as A = 4·1·1 + 2·1 = 6 m², which yields for (1.137)

$$Q = \frac{0.5 \cdot 6 \cdot (65 - 25)}{01} = 1200 \ W.$$ (1.138)

The concept of water heating is based, from a general point of view, on the assumptions that the heat is produced at a constant heat transfer rate Q for a time T so that the total heat supplied is given by

$$h = Q \cdot T .$$ (1.139)

Experiments show that the heat required to change the temperature of a mass m of liquid by the temperature difference $\Delta\Theta$ is proportional to both m and $\Delta\Theta$ i.e.

$$h \approx m \cdot \Delta\Theta .$$ (1.140)

Thus

$$Q \cdot T \approx m \cdot \Delta\Theta ,$$ (1.141)

and the proportionality constant for the liquid, called specific heat capacity or thermal capacity C, which yields

$$h = Q \cdot T = C \cdot m \cdot \Delta\Theta .$$ (1.142)

For a much more realistic model we have to take into account the casing of the cylinder and the surroundings. If the casing is assumed to be at the temperature of the water that it is in contact with and m is the mass of that part of the casing surrounding the heated water, then

$$h = C_w \cdot m_w \cdot \Delta\Theta + C_c \cdot m_c \cdot \Delta\Theta, \qquad (1.143)$$

where suffix w refers to the water and c to the casing, respectively, while C is the specific heat capacity of the casing. It should be noted that we have neglected heat lost to the surroundings in this model.

1.3.7 Chemical Dynamics

Another class of important processes in modeling and simulation are those that are characterized by chemical reactions. The chemical reaction can be described by the rate at which the reaction occurs, which can be quite complicated since the rate generally depends on the amount of reactants as well as the particular nature of the reactions. In general, chemical reactions can be described by using so called compartment models that are based on the rate exchange at which the reaction is based on. From a more general point of view the general principles of chemical reactions are governed by the law of mass action, which states that the rate of a reaction is proportional to the active concentration of the reactants. If a molecule each of A and B combine reversibly to form C we can write

$$A + B \underset{k_2}{\overset{k_1}{\rightleftharpoons}} C$$

where x_1, x_2, and x_3 are the concentrations of A, B, and C respectively, the law of mass action gives two nonlinear differential equations

$$\frac{dx_1}{dt} = \frac{dx_2}{dt} = k_2 \cdot x_3 - k_1 \cdot x_1 \cdot x_2, \qquad (1.144)$$

$$\frac{dx_3}{dt} = k_1 \cdot x_1 \cdot x_2 - k_2 \cdot x_3, \qquad (1.145)$$

with k_1, k_2 the reaction rates.

Example 1.10
A chemical reaction can be described by

$$A + A \underset{k_2}{\overset{k_1}{\rightleftharpoons}} A_2$$

Let us assume that x is the concentration of A, and y is the concentration of A_2 we get

$$\frac{dx}{dt} = 2 \cdot k_2 \cdot y - 2 \cdot k_1 \cdot x^2 ,$$

(1.146)

which results in the arbitrary equations:

$$y = \frac{1}{2 \cdot k_2} = (\frac{dx}{dt} + 2 \cdot k_1 \cdot x^2) ,$$

(1.147)

and

$$\frac{dy}{dt} = k_1 \cdot x^2 - k_2 \cdot y .$$

(1.148)

Substituting for y from (1.146) in (1.148) gives

$$\frac{d}{dt}(\frac{dx}{dt} + 2 \cdot k_1 \cdot x^2) = 2 \cdot k_2 \cdot k_1 \cdot x^2 - k_2(\frac{dx}{dt} + 2 \cdot k_1 \cdot x^2) ,$$

(1.149)

which can be written as

$$\frac{d^2 x}{dt^2} + \frac{dx}{dt} \cdot (4 \cdot k_1 \cdot x + k_2) = 0 .$$

(1.150)

Equation (1.150) is a nonlinear second-order differential equation, but of a specific type since there is no direct time dependency. Solve this equation based on the substitution

$$p = \frac{dx}{dt} ,$$

(1.151)

solving for p as a function of x, rather than t. We can now write

$$\frac{d^2 x}{dt^2} = \frac{dp}{dt} = \frac{dp}{dx} \cdot \frac{dx}{dt} = p \cdot \frac{dp}{dx} ,$$

(1.152)

hence (1.150) can be written as

$$p \cdot \frac{dp}{dx} + p \cdot (4 \cdot k_1 \cdot x + k_2) = 0 .$$

(1.153)

Assuming $p \neq 0$ we obtain for (1.153)

$$\frac{dp}{dx} = -4 \cdot k_1 \cdot x - k_2 , \qquad (1.154)$$

and, integrating

$$p = \frac{dx}{dt} = -2 \cdot k_1 \cdot x^2 - k_2 \cdot x + c_0 . \qquad (1.155)$$

Equation (1.155) is a first-order differential equation that can be solved by separation of the variables with the result

$$\int \frac{dx}{2 \cdot k_1 \cdot x^2 + k_2 \cdot x - c_0} = -\int dt , \qquad (1.156)$$

which can be written as

$$\frac{1}{2 \cdot k_1} \int \frac{dx}{(x - \alpha_1)(x - \alpha_2)} = -t + c_1 , \qquad (1.157)$$

where α_1, α_2 are the roots of

$$x^2 + \frac{k_2}{2 \cdot k_1} \cdot x - \frac{c_0}{2 \cdot k_1} = 0 , \qquad (1.158)$$

which gives

$$\alpha_1 = \frac{-k_2 + a}{4 \cdot k_1} \qquad (1.159)$$

$$\alpha_2 = -\frac{(k_2 + a)}{4 \cdot k_1} , \qquad (1.160)$$

with $a = (k_2^2 + 8 \cdot c_0 \cdot k_1)^{1/2}$.

Partial fractioning the integrand results in

$$\frac{1}{2 \cdot k_1 (\alpha_1 - \alpha_2)} \int \left[\frac{1}{(x - \alpha_1)} - \frac{1}{(x - \alpha_2)} \right] dx = -t + c_1 , \qquad (1.161)$$

with $\alpha_1 - \alpha_2 = \dfrac{a}{2 \cdot k_1}$. Integrating results in

$$\frac{1}{a} \log(\frac{x - \alpha_1}{x - \alpha_2}) = -t + c_1 \quad , \qquad (1.162)$$

and rearranging yields

$$x = \frac{(k_2 + a)c_2 \cdot e^{-a \cdot t} + a - k_2}{4 \cdot k_1 (1 - c_2 \cdot e^{-a \cdot t})} ,$$

(1.163)

where $c_2 = e^{a \cdot c}$. Substituting (1.153) into (1.164) gives

$$2 \cdot k_2 \cdot y = -k_2 \cdot x + c_0 ,$$

(1.164)

hence, when the initial concentrations x_0, and y_0 are given, we get

$$c_0 = 2 \cdot k_2 \cdot y_0 + k_2 \cdot x_0 ,$$

(1.165)

for c_0 and c_2.

From (1.163) we obtain

$$c_2 = \frac{4 \cdot k_1 \cdot x_0 - a + k_2}{4 \cdot k_1 \cdot x_0 + a + k_2} .$$

(1.166)

We may now obtain the complete solution for x and y. Assuming that $t \to 0$, we obtain from (1.163):

$$x \to \frac{a - k_2}{4 \cdot k_1} = \frac{(k_2^2 + 8 \cdot c_0 \cdot k_1)^{1/2} - k_2}{4 \cdot k_1} = x_e ,$$

(1.167)

with x_e as the x-equilibrium point of the chemical reaction, and from (1.164) we obtain for $t \to 0$:

$$y \to \frac{k_2 - (k_2^2 + 8 \cdot c_0 \cdot k_1)^{1/2}}{8 \cdot k_1} + \frac{c_0}{2 \cdot k_2} = \dot{y}_e ,$$

(1.168)

with y_e as the y-equilibrium point of the chemical reaction. Hence if time increases the reaction tends to reach an equilibrium state, which can be illustrated by an $x\,y$ graph as in Fig. 1.18.

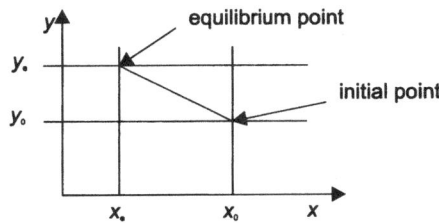

Fig. 1.18. Transient state-space response of a chemical reaction

1.4 Block Diagram-based Algebraic Representation of Systems

1.4.1 Introduction

There are numerous scientific areas where physical principles are well established and accepted in modeling and simulation. Such areas include topics like electro-mechanical energy conservation, nuclear reactions, operations research, physiology, population biology, economics, networking, and so forth. Certain principles have even been suggested in the fields of medicine, sociology, linguistics, anthropology, and so forth, which might serve as a basis for the use of the systems theory approach. However it should be noted that the correct understanding and application of physical principles and/or systems theory in the several domains usually requires a great amount of knowledge about the particular domain. For this reason a specified method is necessary to describe systems from a more general point of view. As we know any engineering system and for that matter any biological, social, or economic system can be represented by a combination of blocks. Each block has a single line into it and a single line out. Within the block is a statement of its operation, that is, the block indicates what happens to the input information before that information is passed on as the output. From engineering we also know a composite system can consist of two or more subsystems, each being represented by a block. There are many forms of composite systems, however, mostly, they are built from the following basic structures:

- Parallel
- Sequential
- Hybrid
- Feedback

The input output relationship of dynamic systems can be considered as a multivariable system (MVS), which can be described by:

$$Y_i(t) \int_{-\infty}^{+\infty} G_i(t,\tau), U_i(\tau) d\tau ,$$

(1.169)

where U_i and Y_i are the input and the output, and G_i is the impulse response matrix of the system MVS, which can be rewritten as follows:

$$G_i(s) = \frac{Y_{i_i}}{U_i} \quad i = 1,...,n ,$$

(1.170)

the algebraic notation of the input/output relationship of blocks, representing the subsystems of a dynamic system. The general structure of the equation above is shown in Fig. 1.19.

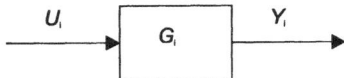

Fig. 1.19. Block diagram structure of subsystems of a dynamic system

Some elements that are often used as blocks for subsystems in dynamic systems are shown in Table 1.1, where K represents the proportional constant.

Table 1.1. Classification scheme

Block	Mathematical Description	Algebraic Operator
Proportional block	$Y(t) = KU(t)$	$G(s) = K$
Integral block	$Y(t) = K_{to}U_1(t)dt$ $t_0 : initial\ time$ $t : time$	$G(s) = \dfrac{K}{s}$
Differential block	$Y(t) = KU(t)$	$G(s) = Ks$
Exponential block	$Y(t) = e^x)$	$G(s) = \exp(U)$

The block diagram equations shown in Table 1.1 are often used to interconnect the individual terms of a proportional, integral, differential, and exponential blocks and thereby form a mathematical model. In Table 1.1 the algebraic operator represents the dynamic function such that variable $G(s)$, the transfer function, is a function of the input variable $U(t)$ and the output $Y(t)$. The notation of the algebraic operator in the tabulation is in the Laplace domain where s denotes the Laplace operator which is the first derivative – the operator $\dfrac{d}{dt}$ – and $\dfrac{1}{s}$ denotes the Laplace operator, which is the integral. For details see Appendix B.

The generalized structure of composite dynamic systems in Fig. 1.19 show that the block diagram structure illustrates that a mathematical operation is performed on the inputs U_i, multiplied by the constant, G_i. Therefore, for any input value the output will be given by $Y_i = G_i \cdot U_i$.

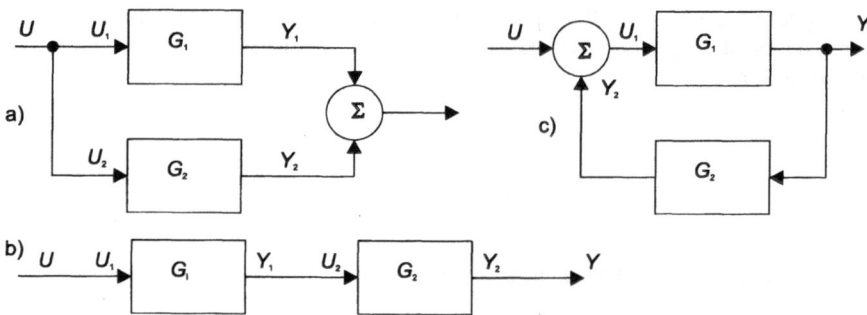

Fig. 1.20. Composite connections of two systems; (a) parallel, (b) feedback, (c) sequential (tandem)

From Fig. 1.20a we find for parallel connections $U = U_1 = U_2$, and with positive summary $Y = Y_1 + Y_2 = G_1 \cdot U_1 + G_2 \cdot U_2 = (G_1 + G_2)U$. By similar reasoning to the above we note that the feedback structure in Fig. 1.20c results in $U_1 = U \pm Y_2 = U \pm G_2 \cdot Y = U \pm G_2 \cdot G_1 \cdot U_1$. Now consider the effect of feeding the output of one block into the input of a second block, as shown in Fig. 1.20b, the resulting signal will be $Y = G_2 \cdot G_1 \cdot U$.

From Fig. 1.20a the impulse-response equation of the parallel connection can be derived as follows:

$$Y = G(t,\tau) \cdot u = (G_1(t,\tau) + G_2(t,\tau)) \cdot U \ . \tag{1.171}$$

For the feedback connection shown in Fig 1.20c, the impulse-response function is the solution of the integral

$$G(t,\tau) = G_i(t,\tau) - \int_\tau^t G_1(t_1,U) \int_t^v G_2(U,V)G(V,\tau)dudv \ . \tag{1.172}$$

For the sequential solution, shown in Fig. 1.20b, we obtain

$$G(t,\tau) = \int_\tau^t G_1(t_1,U)G_2(U,\tau)dU \ . \tag{1.173}$$

A block diagram illustrates the behavior of a system by depicting the action of the variables of the system. To extend the discussion, we focus on the fluid system, given in Example 1.7. Clearly, there is an inflow q_i and an outflow q_o: it is also assumed that fluid is being stored in the tank. The graphical representation also shows that the outflow has an associated resistance. A block diagram, which is appropriate to the simple fluid system, is shown in Fig. 1.21.

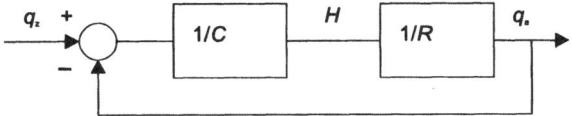

Fig. 1.21. Block diagram for the simple fluid system of Example 1.7

Three important block diagram symbols are introduced in this figure, the

- Summing point
- Splitting point
- Operator

Note that only one line enters and only one line leaves a block. Correspondingly, only two lines enter and only one line leaves a circle depicting a summing point. The block diagram figure can be used to show that the outflow is the important variable, since it is influenced by the inflow. The feedback line depicts the physical fact that a decreased output will increase the water level H in the tank. At the same time, an increased flow will increase the outflow, such as:

$$q_a = f(H,t).$$ (1.174)

The water level H can be expressed as follows:

$$H = \Phi[(q_i - q_o),t].$$ (1.175)

Thus, the overall outflow equation results in:

$$q_a = f\{\Phi[(q_i - q_o),t]\}.$$ (1.176)

This arrangement will be very useful for the explicit form of the functional relations, expressing the flow difference

$$q_i - q_o = C \cdot \frac{dH}{dt},$$ (1.177)

and

$$R \cdot q_o = H,$$ (1.178)

where C is the liquid capacitance, and R is the flow resistance.

Attention can be drawn to the fact that blocks and circles can be rearranged without destroying the validity of the representation. This offers the advantage that one may chose an arrangement that best satisfies the viewpoint of the analysis. For this reason the block diagram in Fig. 1.22 may be rearranged as follows:

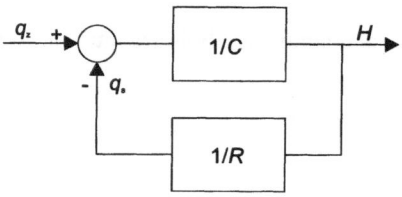

Fig. 1.22. Rearrangement of the block diagram of Fig. 1.21

This block diagram clearly shows that the inflow influence the water level H in the tank and that the outflow is a secondary variable. Now the overall relationship is

$$H = \Phi[(q_i - f(H,t),t].$$ (1.179)

It is important to note that a block diagram contains no more information than the differential equation. It provides a pictorial process of manipulating the differential equations. The advantage of the block diagram representation is that the operational relations in the system are emphasized rather than the physical system. Moreover, due to the possibilities of block arrangements, we are better able to interpret the function of the various elements than would be possible from the differential equations. For this reason a number of rules for the manipulation and reduction of block diagrams are being introduced.

1.4.2 Block Diagram Algebra

Based on the general description above the relevant relationships that are necessary when rearranging blocks of linear, as well as nonlinear, composite dynamic systems, based on a block diagram algebra, will be introduced. The relationships described by the respective algebraic equations show at first the original block diagram and thereafter the equivalent block diagram.

Combining parallel blocks:

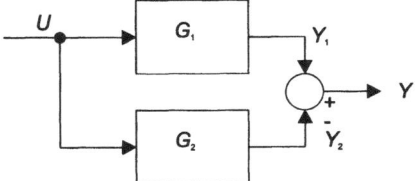

Fig. 1.23. Combining parallel blocks

The output variables Y_1 and Y_2 are multiplied by the respective transfer characteristic G_i, which results in $Y_1 = U{\cdot}G_1$, and $Y_2 = U{\cdot}G_2$. Adding the variables Y_1 and Y_2 results in the overall relation $Y = Y_1 \pm Y_2 = (G_1 \pm G_2){\cdot}U$. Thus the block diagram becomes

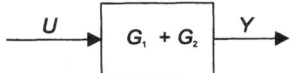

Fig. 1.24. Resulting block of Fig. 1.23

Feedback loop:

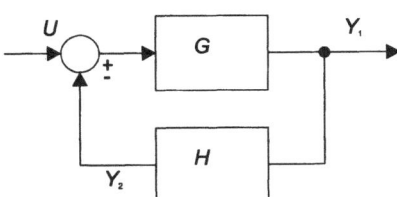

Fig. 1.25. Feedback loop

The block diagram shows the simplest form of a feedback control system, with G as the forward loop gain, and H as the feedback loop gain, and Y_1 as output variable, which can be described as $Y_1 = (U \pm Y_1){\cdot}G$, or $Y = U{\cdot}G \pm Y_1\,G{\cdot}H$. Rearranging result in $U{\cdot}G = Y_1 \pm Y_1{\cdot}G{\cdot}H$, hence $U{\cdot}G = Y_1(1 + G{\cdot}H)$, and the control ratio is as follows

$$Y_1 = \frac{G}{1 \pm G \cdot H}\,U\,. \tag{1.180}$$

The block diagram above can now be reduced to a single block diagram

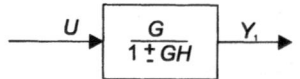

Fig. 1.26. Resulting block of Fig. 1.25

Cascaded blocks:

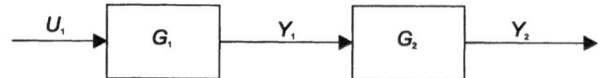

Fig. 1.27. Cascaded blocks

The output variables are $Y_1 = U_1 \cdot G_1$, and by similar reasoning $Y_2 = G_2 \cdot Y_1$. Therefore, if $U_1 \cdot G_1$ is substituted for Y_1 we find $Y_2 = G_2 \cdot U_1 \cdot G_1$. These equations, which are different forms of the same equation, demonstrate the very important idea, that simple blocks strung together in cascade fashion can be multiplied together, a fact that allows to be reduced such a string to a single block as follows:

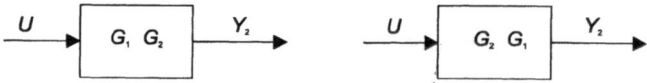

Fig. 1.28. Resulting blocks of Fig. 1.27

Permutation of blocks:

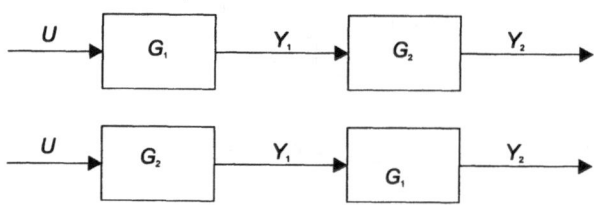

Fig. 1.29. Permutation of blocks

The output variable is $Y_2 = G_1 \cdot G_2 \cdot U = G_2 \cdot G_1 \cdot U$.

Moving a block before a summing junction:
It is sometimes necessary to feed more than one signal into a block at the same time, which is achieved by a summing network. In block diagram symbols it is achieved by means of a summing junction, as shown in Fig. 1.30.

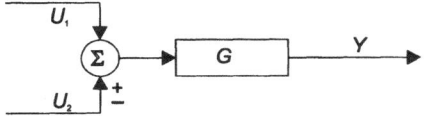

Fig. 1.30. Moving a block before a summing junction

The summing junction is shown as a circle with arrows into and out of the symbol. The arrows into the symbol are always identified with a plus or minus sign, indicating either a positive or a negative signal. Inside the circle sometime the Greek letter sigma (Σ) is used. As shown in Fig. 1.30, U_2 is added or subtracted from U_1 yielding U. U_1 and U_2 are simultaneously fed into the block G, the output Y is the weighted sum of the two inputs, $Y = (U_1 \pm U_2) \cdot G$, and for the rearranged case as shown below $Y = U_1 \cdot G \pm U_2 \cdot G$

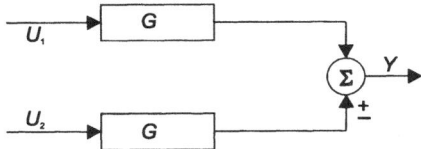

Fig. 1.31. Resulting blocks of Fig. 1.30

Moving a block behind a summing point:

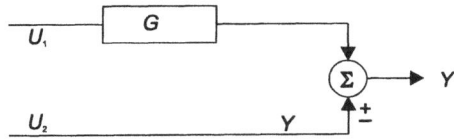

Fig. 1.32. Moving a block behind a summing point

with the equation $Y = U_1 \cdot G \pm U_2$. The output variable after moving is $Y = G(U_1 \pm G^{-1} \cdot U_2)$ which yields a multiplication with the inverse transfer function

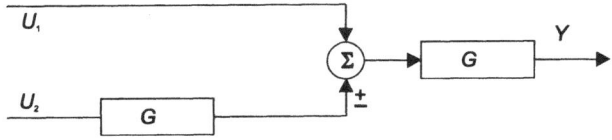

Fig. 1.33. Resulting blocks of Fig. 1.32

Moving a block before a branch point:

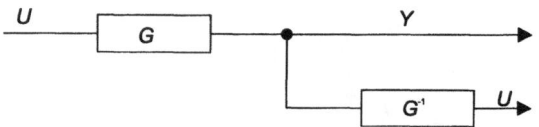

Fig. 1.34. Moving a block before a branch point

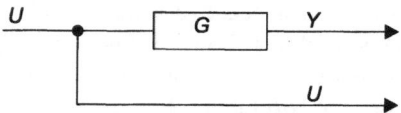

Fig. 1.35. Resulting blocks of Fig. 1.34

$$Y = G \cdot U.$$

Rearranging summing points:

$$Y = U_1 \pm U_2 \pm U_3.$$

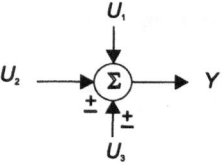

Fig. 1.36. Rearranging summing points

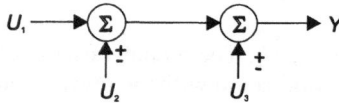

Fig. 1.37. Resulting blocks of Fig. 1.36

Inversion: $Y = G.$

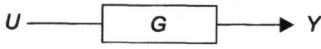

Fig. 1.38. Inversion

and the inverse function $U = G^{-1} \cdot Y$

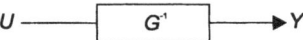

Fig. 1.39. Resulting block of Fig. 1.38

1.5 Basic Principles of Discrete-Time Systems

1.5.1 Introduction

As distinct from modeling and simulation of continuous-time systems the treatment of discrete-time systems follows a completely different modeling paradigm.
The difference depends on the appearance of trajectories of the respective system
variables. For comparison, Fig. 1.40 shows the typical graphs of continuous-time
and discrete-time model variables. In both cases the x-axis represents the time dependence, the y-axis marks the value of the model quantity. The characteristics for
continuous-time variables are continuous changes in value, which can be mathematically expressed by a differential equation. In contrast the value of a discrete-
time variable may be constant nearly all the time. But there are only a few points
on the time scale where the value changes. At these points, however, the value
changes abruptly and without any interim value.

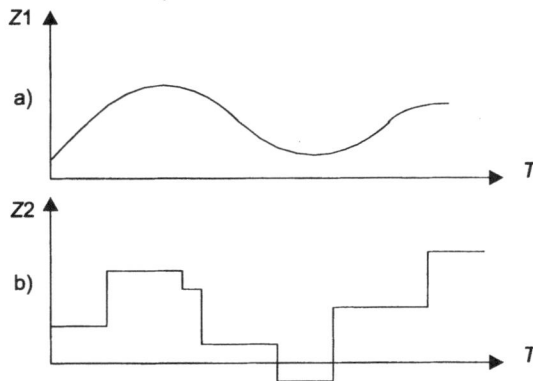

Fig. 1.40. Continuous-time and discrete-time representation of system variables

A characteristical example of a discrete-time transient of a model variable would
be the number of persons waiting for service in front of an information desk.
Changes in number are sudden: one person enters or leaves the queue. The process
of joining the others who are already waiting there is not differentiated in more detail: approaching, asking who is first and last.

The only intention of the model is to give a prognosis for the mean waiting time of the customers, the mean length of the queue and so on. Therefore, the abstraction during the process of model building reduces the dynamic behavior of the system to sudden changes in the number of people waiting. The number of people in a queue is a classical discrete-time model variable.

With this example in mind we may understand the two basic principles of building time-discrete models: First, the definition of an event in the course of a model variable. Second, the condition for its dynamic behavior between the events.

Remark 1.1
A discrete-time event is an instantaneous occurrence that changes the system state.

Remark 1.2
The value of a discrete-time model quantity is constant over the time interval defined by two consecutive events, which can be stated as a condition for the course of a variable between events.

Based on these two simple principles, systems from varying application areas can be sufficiently modeled. Characteristic examples of discrete event systems that follow these principles are:

- Queuing systems:
  These systems distinguish between stations that offer services, and mobile elements that request services, and are able to move from one service station to another. The main task is to organize the services, maximize their utilization, and minimize the waiting time for the mobile elements.
- Manufacturing systems:
  These systems are an important area of application of discrete-event modeling. The stations are the highly automated machines of the plant, the mobile elements are the raw materials, the semi finished products, and finally the assembled end product itself. In addition, highly automated transport systems or conveyors up to intelligent automotive units may complicate the systems behavior. However, the questions for simulation models are quite similar to those of the queuing systems: minimization of the production time, maximization of the utilization of the machines. But the stations and the strategies to move the mobile elements between them (through the transportation units) are much more complex and specialized in accordance with the technical realization of elements, stations, and transportation systems.
- High bay warehouses
  Highly automated management of warehouse systems is a very reasonable application field for simulation. Because the goods stored are discrete elements and the places in the warehouse are discrete as well, the model concentrates on discrete changes in state variables, as places are free or occupied and transitions by the autonomously guided vehicle system start or end. The

main results of such a model are: access time, optimal positioning of the goods, number of vehicles needed, etc.

- Computer systems:
  Historically the computer was the first application area for discrete simulation techniques. The main task is to optimize the architecture of a computer by simulation of its hardware components in relation to its operating system and observing the workload of the CPU, the bus system, the storage, and the peripheral devices. Typical parameters are the queuing strategies, strategies for sharing the processor and all other parameters of the operating system. The discrete modeling unit for the simulation is the task with its needs concerning CPU time, storage space, external devices, etc.

- Network systems
  The parameters of interest within a single computer and its dynamic behavior can easily be transferred to a network of computers: workload of its elements, dimensioning of buffers, strategies for routing, and so on. With the data package as the unit that moves between the nodes of the system, the simulation of network systems is a task for discrete-event simulation.

When modeling dynamic systems by means of discrete-time events, two main issues have to be mentioned:

- First, the term "event" implies a resolution of time that is related to infinite short time duration. An event happens without any consumption of time. In reality, however, every execution of an event will take (very little, but some) time. So we will have the resolution of the time axis as one problem to live with.

- Secondly, the problem of what to do when two (or more) events happen at the same time step simultaneously. These problems are caused by the definition of the event itself and have to be solved later when we discuss the simulation algorithm that executes a system description only consisting of events.

1.5.2 Modeling Concept of Discrete-Time Systems

With the definition of the event and the description of its semantics how to model discrete-time systems is obvious. The description of the system dynamics consists of a chronologically sorted list of events that occur between the start time and the end time of the observation. All knowledge about the system is represented in this list. As in continuous-time models an initial value for the model quantities influenced by the events must also be given.

When building a discrete-time system model, one has to specify these events and to put them into the correct order. If we look at the events used to model the very simple system shown in Fig. 1.41 we will find lots of very similar events. The example shows a single serving unit with a queue for the waiting customers. The customers are created randomly and receive a varying service time. After service the customers leave the system. This system is one of the simplest examples of discrete-event simulation and is called a "single-server system".

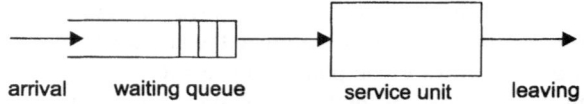

arrival   waiting queue   service unit   leaving

Fig. 1.41. Discrete-time representation of a system concept

Our objective now is to study the events for a single-server system, as shown in Fig. 1.42. Element $e1$ enters the queue at time $t1$, element $e2$ enters the queue at time $t3$, element $e3$ enters the queue at time $t4$, and so on. These are the events that describe the arrivals of customers. On the other hand, there are events describing the departures because the customers service time elapsed, which is the case for element $e1$, which finishes service at time $t2$, element $e2$, which finishes service at time $t6$, element $e3$ which finishes service at time $t9$, and so on.

For simplification of this task which specifies these events, a much more general specification scheme may be offered by the model description languages and the corresponding simulation system that is available for time-discrete event simulation.

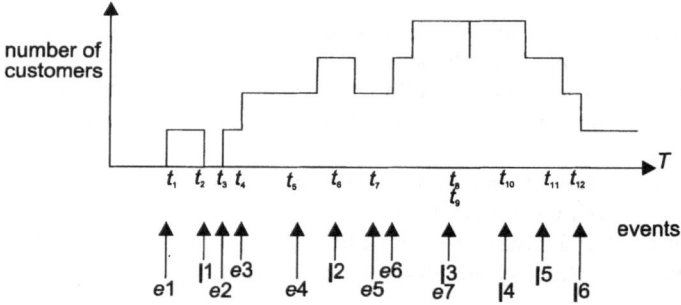

Fig. 1.42. Time events for a single-server system

The idea behind this is to build up classes of events that describe the dynamic system on a more abstract level as the particular events are introduced above. The main classes may be:

- Arrival of a customer
- Customer enters queue
- Start service
- End of service
- Customer leaves system

Using these more abstract event classes, all arrivals, all entering in queues, all services start up, etc., may be modeled by a single piece of model code. Therefore, the syntax of an event in a model description language consists of two defining parts:

1. The *condition* of the event which specifies *when* the event will be executed.

2. The *body* of the event which specifies *what changes* in the values of model quantities will happen.

It is possible to change the values of a set of model quantities in the body of one single event, e.g. if an element is taken from the queue to the service station the number of elements in the queue may decrease and the number of elements in service may increase for the same amount.

With respect to the event condition a further classification of events can be made:

- Time events, whose event condition exclusively uses the simulation time T and whose execution depends only on the course of T.
- State events, whose condition is a free Boolean expression that may include any model variable and whose execution depends on the state of the model variables they change, or even on the values of any other variables in the model.

Example 1.11
time event

```
WHENEVER    T >= T_enter
BEGIN_BODY
number_customers := number_customers + 1;
END_BODY
```

Example 1.12
state event

```
WHENEVER    number_customers / number_service_units >= 5
BEGIN_BODY
number_serverice_units := number_service_units + 1;
END_BODY
```

If the system dynamics follow some fixed rules such as iterations in time, depending on certain states of the model, the model builder has the possibility of formulating constructs like classes of events that represent more than one activity in the real world by a single event in the model description.

Example 1.13
state event

Whenever the value of the water level in a tank reaches its upper limit a quarter of its contents is taken away by the controller.

```
WHENEVER    tank_level >= level_max
BEGIN_BODY
tank_level := tank_level · 0.75;
END_BODY
```

Independently from the way the tank is filled (time-discrete by buckets or time-continuous by a water flow from a water tap) this event assures the level will not exceed the given limit.

Example: 1.14
time event

The following represents events that model the arrival of customers at a service station. The event is triggered by setting the next time the event will be active inside the body of the event itself. Thereby the variable *InterArrivalTime* may have a fixed value or may be represented by a random number to model a random arrival process.

```
WHENEVER    T >= T_NextArrival
BEGIN_BODY
customers_in_queue := customers_in_queue + 1;
T_NextArrival := T_NextArrival + InterArrivalTime;
END_BODY
```

For a single-server system the behavior may be modeled by a set of two event classes. The events have an implicit time condition for the next activation that is set by the procedures *schedule_arrival_event* and *schedule_departure_event*.

Example 1.15
complete set of events to simulate the single-server system

```
WHENEVER Arrival_event
    IF number_in_server == 1
            THEN (            number_in_queue := number_in_queue +1;
                    )
            ELSE (            number_in_server := number_in_server +1;
                              schedule_departure_event (T + T_service_time
                    ); )
        schedule_arrival_event (T + T_interarrival_time );
        protocol_state_changes ();
END Arrival_event
WHENEVER Departure_event
    IF number_in_queue == 0
            THEN (            number_in_server := 0;
                    )
```

> *ELSE (*   *number_in_queue := number_in_queue -1;*
> *schedule_departure_event (T + T_service_time*
> *);)*
> *protocol_state_changes ();*
> *END Departure_event*

1.5.3 Simulation Concept

Model specification of discrete-time systems has been briefly discussed. To run these models a simulation algorithm has to be chosen. The demands for such an algorithm has to fulfil some constraints that are known from the specification:

1. Execute the events that happen in the simulation period between T_start and T_end completely.
2. Execute them exactly at the point of time when their condition becomes true.
3. Execute them in the right order.
4. Execute them without consumption of simulation time.

Simulation is a very simple approach that demonstrates the advantages of the so-called next-event-simulation best. Assuming the simulation interval is given by the start time and the end time for the run. Furthermore, the resolution ΔT of the time axis is determined, for example by the representation of numbers on a computer. Hence the simplest simulation algorithm would be:

> *...*
> *Set ActTime := StartTime;*
>
> *WHILE ActTime <= EndTime*
> *DO*
> *WHILE NOT (all event conditions are false)*
> *DO*
> *<find an event_condition in model description that is true>*
> *<execute the corresponding event>*
> *END*
> *ActTime := ActTime + deltaT;*
> *END*
> *...*

This algorithm executes the simulation correctly but consumes a lot of CPU time while it checks all event conditions every time step ΔT. Due to the characteristics of discrete-event models in most cases nothing happens at the point of time under observation. It is typical for those systems to hold a given value constant for a certain period of time until the next event will change it. Checking the event conditions at every point of time mostly will be dispensable and causes an enormous consumption of calculation time. On the other hand, the algorithm is a very simple one and nothing is needed concerning the formulation of the events. Because of its run-time behavior the algorithm is refined and the result is the so- called *next event algorithm*. Its data structure consists of two elements:

- The current time
- The future-event-list: an ordered list of events that are to be executed in future

Each of these events has a time stamp that shows the point of time its condition becomes true. The list is ordered by a growing time stamp. Doing so, the event to be executed next makes the top of the list.

The advantage of this event list is that there will not be other events between two entries in that list. Hence there is no need for the algorithm to check all the conditions between two events and it knows exactly when the next change in value of a model quantity will happen.

The algorithmic version may be stated as follows:

> *# initialize*
>
>> *\<set start time\>*
>> *\<set end time\>*
>> *\<put an initial set of events into the event list\>*
>>
>> *T := T_start;*
>
> *# simulation loop*
>
>> *WHILE T \<= T_end*
>> *DO*
>>> *current_event := first entry of the event list;*
>>> *T := current_event.time_stamp;*
>>> *execute (current_event);*
>>> *delete_from_event_list (current_event);*
>>> *current_event := \<first entry of the event list\>;*
>>> *T := current_event.time_stamp;*
>> *END*

The disadvantage of this approach is that it needs the help of the model builder: somebody has to insert new entries in the next-event list. After initialization, this is done by expanding the body of the events. Within the event specification the model builder has to specify when the active event will be active again, or, if there is another event that is triggered by the active event and when it is set up for execution. These are the two types mentioned before: self-triggered events (e.g. by interarrival time) or condition triggered events at the same point of time (e.g. customer enters empty queue and is transmitted to the service unit at the same point in time).

More sophisticated solutions for discrete-event simulation algorithms are based on these two basic approaches. They modify the search for the next event in the list which means they allow parallelism by distributing the event list, and they integrate continuous model elements in the processing of the simulation algorithm.

1.6 Model Validation

Modeling is a complex procedure that contains several steps: the qualification, the rectification, and finally the verification.

- Qualification is the model-building process that is focused on the respective elements, relations, and attributes in order to describe the real dynamic system in an abstract manner as a so-called abstract model.
- Rectification is the model-building process behind the qualification, which means the abstract model will be transformed into a mathematical model, the so-called real model, of the dynamic system. Hence rectification decides the proper form for the realization, which may include implementation, iteration algorithms, programming, imitation/simulation based on mechanical, electrical, pneumatical elements, isomorphism, etc. From a more general point of view Fig. 1.43 shows the table of the respective correspondencies for rectification.
- Verification is the model-building process beyond rectification that is focused on fit or non fit of the model due to the respective dynamic behavior of the system. Verification includes the validation of the model, i.e. the quality of the model, and the falsification of the model, meaning less fitting.

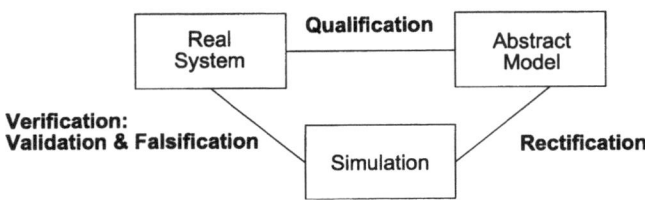

Fig. 1.43. Process of qualification, rectification, and verification of dynamic systems

VARIABLES	Physically System	General Description	Electrical	Hydraulical	Pneumatical	Thermal	Translational	Rotational
PRIMARY	Transversal Variable e(t)	Voltage, Pressure Velocity	U(t); Voltage	P(t); Pressure	P(t); Pressure	T(t); Temperature	V(t); Velocity	ω(t); Angular Velocity
PRIMARY	Transit Variable f(t)	Current, Flow Force, Momentum	I(t); Current	$\dot{V}(t)$; Volume Flow	$\dot{m}(t)$; Mass Flow	$\dot{q}(t)$; Heat Flow	f(t); Force	M(t); Torque
PRIMARY	e(t) Product	Power supplied to the element	$p(t)=u(t)\cdot I(t)$	$p(t)=P(t)\cdot \dot{V}(t)$	$p(t)=P(t)\cdot \dot{m}(t)$	$p(t)=\dot{q}(t)$	$p(t)=V(t)\cdot f(t)$	$p(t)=\omega(t)\times M(t)$
PRIMARY	e(t) Relation	Power Consumption $e(t)=R\cdot f(t)$	R; Electrical Resistance	$R=\frac{8 l \eta}{\pi r^4}$ Flow - resistance	identical to hydraulical	Thermal Resistance $R_F=\frac{1}{\lambda A}$ (Flow) $R_T=\frac{1}{\alpha A}$ (Transm.) $R_C=\frac{1}{mc}$ (Convect.)	d^{-1}; Damping - factor	d_φ^{-1}; Damping-factor
INTEGRAL	$\int e(t)\,dt$	$F(t)=1/L\cdot\int e(t)\,dt$	L; Inductor	$\frac{\rho l}{\pi r^2}$; Inertance	$\frac{\rho l}{\pi r^2}$; Inertance	———	c^{-1}; Spring - constant	c_φ^{-1}; Spring - constant
INTEGRAL	$\int f(t)\,dt$	$e(t)=1/C\cdot\int f(t)\,dt$	C; Capacitor	$\frac{A}{\rho g}$; Hydraulic Capacity	$\frac{m_e}{\delta_e}=\frac{V}{R\cdot T}$; Pneumatic Capacity	$m\cdot c$; Thermal Capacity	M; Mass	Θ; Moving Mass
INTEGRAL	$\int e(t)\,f(t)\,dt$	Energy done on system	E_M:Magnetic Energy of inductor E_E:Electric Energy of capacitor	E_K:Kinetic Energy of fluid flow E_P:Potential Energy of pressure head	E_K:Kinetic Energy of pneumatic flow E_P:Potential Energy of pressure	E_P:Thermal Potential Energy of stored heat	E_K:Kinetic Energy of moving mass E_P:Potential Energy of compressed Spring	E_K:Kinetic Energy of rotating mass E_P:Potential Energie of twisted spring
	Symbols		R L C			T_1 R T_2 T_1 C T_2	R L M C	L M C

Fig. 1.44. Correspondencies for modeling purposes

Model validation is a procedure that involves assessing the extent to which the model is focused, tractable, and fulfills the purpose for which the model has been formulated. From a more general point of view model validation is a multi dimensional procedure reflecting the model purpose, current theories, and experimental test data relating to the particular system of interest together with other important knowledge. Hence validation may be stated as a complex procedure that takes place at several levels

- Behavioral level, which means the model is able to reproduce the behavior of the dynamic system.
- State-structure level, which means the model is able to be synchronized with the dynamic system due to a state from which the prediction of future behavior may be possible.
- Composite-structure level, which means the model may be used to represent the internal interactions of the dynamic system.

A more straightforward validation method will be the deductive analysis, which shows the validity of the model, meaning its representation reflecting the model purpose which depends on the validity of the a priori knowledge. Validation due to deduction can be achieved in two ways:

- Investigation of the exactness of the premises validates the model.
- Checking other consequences of the premises validates that information and finally the model.

Moreover, the inductive analysis can be introduced as a straightforward validation, whether or not the induction procedure has been carried out in a mathematical and logically correct way. Assuming a model represents a source of data, a
valid model at a certain point in time has to have the equal signs specified.

From a practical point of view, a model is sufficiently valid if its goal can be
obtained. This means it fits the concept reflecting the model purpose. Hence a tru-
ly valid model would be a model that permits all possible objectives.

For Example 1.9, the respective hydraulic system representation may be transformed into an electrical system based on the table of correspondencies shown in
Fig. 1.44. The electrical network representation, shown in Fig. 1.45, gives essentially the same information as the hydraulic system representation above. In the
RC network system the real physical elements are replaced by a schematic representation of the actual hydraulic elements as they are shown in Fig. 1.16. The result is a model that is formulated by using the isomorphism concept reflecting the
real-model purpose. Hence we may state that the RC network model is well
founded due to its aims and scopes, and validity is given.

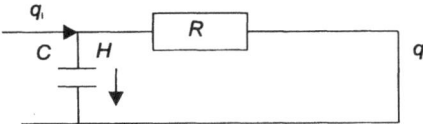

Fig. 1.45. Electrical network representation of the hydraulic system of Fig. 1.16

The criteria by which model validity is assessed can be divided into two criteria:

1. Internal criteria: enabling conditions within the model itself to be judged
    without external reference to the model purpose, theory and/or data, which
    can be:
    o   Consistency: requiring that the model formulated contains no logical,
        mathematical or conceptual contradictions.
    o   Algorithmic validity: requiring that the algorithm for analytical solu-
        tion or numerical simulation is appropriate and leads to accurate so-
        lutions.
2. External criteria: referring to the model itself, like the model purpose, the-
    ory an/or data, which can be:
    o   Empirical validity: requiring that the model formulated should corre-
        spond to the available data.
    o   Theoretical validity: requiring that the model should be consistent
        with accepted theories and/or models.
    o   Pragmatic validity: requiring that testing the extent to which the
        model satisfies the objectives for which it has been developed.

    o   Heuristic validity: requiring in connection with tests that are associ-
ated with the assessment of the heuristic potential of the model, e.g.
for scientific explanation, discovery, and/or hypothesis testing.

Considerations of validity are required from the very beginning of model building. Empirical and theoretical validity can be used by examining whether the respective validation criteria are met or not, which then can be used as a performance index. A performance index is a quantitative measure of the performance of a
model of a dynamic system and is chosen so that emphasis is given to the important real- world constraints.

A suitable performance index *PI* is the integral of the square of the error

$$PI = \int_0^T e^2(t)dt \,, \tag{1.181}$$

where the upper limit T is a finite time chosen somewhat arbitrarily so that the integral approaches a steady-state value of the transient behavior of the system model and e is a measure of the error between the real-world system and the system
model.

Another possible performance criterion is the integral of the absolute magnitude of the error, which can be written as:

$$PI = \int_0^T |e(t)|dt \,. \tag{1.182}$$

This performance index is particularly useful for computer-simulation studies.
In order to reduce the contribution of the large initial error to the value of the performance integral and to place an emphasis on errors occurring later in the response, another performance index has been proposed:

$$PI = \int_0^T t \cdot |e(t)|dt \,. \tag{1.183}$$

This performance index is designated the integral of the time multiplied by the
absolute error. Another similar performance index is the integral of time multiplied by the squared error, which is:

$$PI = \int_0^T t \cdot e^2(t)dt \,. \tag{1.184}$$

The general form of the performance index is:

$$PI = \int_0^T f[e(t),u(t),y(t),t]dt \,, \tag{1.185}$$

where f is a function of the error, input, output, and time (see Chap. 5).

1.7 References and Further Reading

Aström K, Albertos P, Blanke M, Isidori A, Schaufelberger W, Sanz E, (Eds.), (2001), Control of Complex Systems, Springer, London, Berlin, Heidelberg

Burghes DN, Borrie MS, (1981), Modelling with Differential Equations, John Wiley & Sons, New York

McClamroch HN, (1980), State Models of Dynamic Systems, Springer New York, Heidelberg, Berlin

McDonald AC, Loewe H, (1981), Feedback and Control Systems, Reston Publ. Comp. Inc. Reston

Dorf RC, (1986), Modern Control Systems, Addison-Wesley Publ. Reading

Möller DPF, Popovic' D, Thiele G, (1983), Modeling, Simulation and Parameter-Estimation of the Human Cardiovascular System, Vieweg Publ., Braunschweig, Wiesbaden

Ogata K, (1967), State Space Analysis of Control Systems, Prentice-Hall, Inc. Eaglewood

Seely S, (1964), Dynamic Systems Analysis, Reinhold Publishing Corporation, New York, Chapman & Hall Ltd., London

Thaler GJ, (1989), Automatic Control Systems, West Publ., St. Paul

van Wyk van Brievingh RP, Möller DPF, (Eds.), (1993), Biomedical Modeling and Simulation on a PC, Springer New York

1.8 Exercises

1.1 What is meant by the term modeling?
1.2 List and define the three main characteristics modeling a real-world problem.
1.3 What is meant by the term behavioral level of modeling?
1.4 What is meant by the term composite-structural level of modeling?
1.5 What is meant by the term empirical modeling?
1.6 What is meant by the term deductive modeling?
1.7 Test signals can be used for what?
1.8 Give the mathematical description for a unit step.
1.9 Give the mathematical description for a ramp function.
1.10 Give a graph for Example 1.1.
1.11 Define what is meant by the term simulation?
1.12 Differential equations are of importance for modeling real-world systems. Why?

1.13 Explain the structural representation of the state-variable model of Fig. 1.9.

1.14 Derive the mathematical equation for an electrical RCL Network model.

1.15 Derive the mathematical equation of the mechanical model being used for modeling tissue deformation.

1.16 Give a model for the concept of liquid capacitance.

1.17 Give the bock diagram for a feedback loop.

1.18 What is meant by the term queuing system?

1.19 What is meant by the term manufacturing system?

1.20 List and define the five main characteristics modeling a discrete-event system

1.21 Give an example for a time event.

1.22 Give an example for a state event.

1.23 What is meant by the term qualification while modeling?

1.24 What is meant by the term rectification while modeling?

1.25 What is meant by the term verification while modeling?

2 Mathematical Description of Continuous-Time Systems

2.1 Introduction

Based on the phenomenological and physical principles, relevant to describing a particular dynamic system, the equations that characterize the system are carried out in a number of ways, some of which are in the time domain, and others written in a transformed domain. In the time domain the methods for the analysis of the response of the dynamic system are ordinary differential equations (ODEs) of order n, sets of n first-order ordinary differential equations, partial differential equations (PDEs), the superposition integral, the convolution integral, and so on. Solving these equations can be done using numerical methods, based on suitable mathematical models, while more and more indispensable tools for advanced systems analysis and synthesis are in use, as well as for computer-aided engineering design. In conjunction with an experimental verification method, the numerical-simulation results of the suitable mathematical model can be proved. Moreover, the stability analysis of dynamic systems are quite useful when designing optimal control systems that are stable. For this purpose one has to know whether the roots of the system will be located near the equilibrium point or not. Stability analysis can be done, for example in the time domain, by means of the Routh Hurwitz criterion in conjunction with the differential equations relating the response to the excitation. It has to be said that this method is restricted to linear systems.

The time domain formulation can be transformed in the frequency domain by such transformations as the simple exponential function, the Laplace transform, named after the Frech astronomer Laplace, born 1749 in Beaumont-en-Auge. Normandy, France, as well as the Fourier transform, which is named after the French mathematician Baron Fourier, born 1768 in Auxerre, France. Stability analysis of the dynamic system in the frequency domain can be carried out by means of the Nyquist criteria, which is the imaginary or frequency axis for the particular system function. The Nyquist criteria is named after the American engineer Nyquist, born in Sweden 1889. The frequency-domain transformation, and their subsequent use, are also restricted to linear dynamic systems.

Definition 2.1
A dynamic system is said to be continuous in time if the time interval $I \subset \Re$ contains the definition range of the functions $u(\cdot)$, $x(\cdot)$, and $y(\cdot)$, discrete in time, if the time interval I contains the definition range of the functions $u(\cdot)$, $x(\cdot)$, and $y(\cdot)$. ∎

Remark 2.1
Continuous-time systems can be described by ordinary differential equations (ODEs), and/or partial differential equations (PDEs), respectively. Discrete-time systems are described by Petri-nets, named after the German mathematican Petri, born 1926 in Leipzig, Germany, queues, Markov-chains, named after the Russian mathematician Markov, born 1856 in Ryazan, Russia, and so forth.

Continuous-time systems, as they are considered in this book, are assumed to be described by ordinary differential equations of order n, or by a set of n first-order ordinary differential equations. Considering a differential equation of order n

$$x^{(n)}(t) = f(x, x', ..., x^{(n-1)}, u, t) , \tag{2.1}$$

where u is the control function. By defining new variables $x_1, x_2, ..., x_n$ such that

$$\begin{aligned}
x_1 &= x \\
x_2 &= x' \\
&\cdots \\
x_n &= x^{(n-1)}
\end{aligned} \tag{2.2}$$

(2.1) can be reduced to

$$\begin{aligned}
x'_1 &= x_2 \\
x'_2 &= x_3 \\
&\cdots \\
x'_{n-1} &= x_n \\
x'_n &= f(x_1, x_2, ..., x_n, u, t)
\end{aligned} \tag{2.3}$$

(2.3) represent n first-order differential equations, with f as a nonlinear function.

2.1.1 Representation of System Differential Equations in Terms of Vector-Matrix Notation

Consider that the system equations can be adequately described by a set of n first-order ordinary differential equations as follows

$$x'_i = f_i(x_1, x_2, \ldots, x_n; u_1, u_2, \ldots, u_m; t); \quad (i = 1, 2, \ldots, n), \tag{2.4}$$

where x_1, x_2, ..., x_n are state variables, and u_1, u_2, ,..., u_m are control variables. Assuming that the system outputs y_1, y_2, ..., y_k are related to the state variables x_1, x_2, ..., x_n and control variables u_1, u_2, ,..., u_m by the following equation

$$y_j = g_j(x_1, x_2, \ldots, x_n; u_1, u_2, \ldots, u_m; t); \quad (j = 1, 2, \ldots, k; k \leq n). \tag{2.5}$$

If the order of the dynamic system is greater, such a set of equations becomes notationally complicated. To simplify the notation it becomes necessary to use the vector and matrix notation. Rewritten with vector notation, (2.4) and (2.5) become, respectively

$$x' = f(x, u, t), \tag{2.6}$$

and

$$y = g(x, u, t), \tag{2.7}$$

where x, u, y, $f(x, u, t)$, and $g(x, u, t)$ are vectors defined by

$$x = \begin{bmatrix} x_1 \\ x_2 \\ \cdot \\ \cdot \\ \cdot \\ x_n \end{bmatrix}, \quad u = \begin{bmatrix} u_1 \\ u_2 \\ \cdot \\ \cdot \\ \cdot \\ u_m \end{bmatrix}, \quad y = \begin{bmatrix} y_1 \\ y_2 \\ \cdot \\ \cdot \\ \cdot \\ yk \end{bmatrix} \tag{2.8}$$

where vectors x, u, and y are, respectively, called the state vector, the control vector, and the output vector. The dynamic system is specified by the vector-valued functions f and g.

$$f(x, u, t) = \begin{bmatrix} f_1(x,u,t) \\ f_2(x,u,t) \\ \cdot \\ \cdot \\ \cdot \\ f_n(x,u,t) \end{bmatrix}, \quad g(x, u, t) = \begin{bmatrix} g_1(x,u,t) \\ g_2(x,u,t) \\ \cdot \\ \cdot \\ \cdot \\ g_k(x,u,t) \end{bmatrix}. \tag{2.9}$$

If the system is linear in x and u, a set of n first-order differential equations is

$$\begin{aligned} x'_1 &= a_{11}(t)x_1 + a_{12}(t)x_2 + \ldots + a_{1n}(t)x_n + b_{11}(t)u_1 + \ldots + b_{1m}(t)u_m \\ x'_2 &= a_{21}(t)x_1 + a_{22}(t)x_2 + \ldots + a_{2n}(t)x_n + b_{21}(t)u_1 + \ldots + b_{2m}(t)u_m \end{aligned} \tag{2.10}$$

...

$$x'_n = a_{n1}(t)x_1 + a_{n2}(t)x_2 + \ldots + a_{nn}(t)x_n + b_{n1}(t)u_1 + \ldots + b_{nm}(t)u_m$$

as well as m algebraic equations relating output variables, state variables, and control variables, which can be written as follows

$$\begin{aligned} y_1 &= c_{11}(t)x_1 + c_{12}(t)x_2 + \ldots + c_{1n}(t)x_n + d_{11}(t)u_1 + \ldots + d_{1m}(t)u_m \\ y_2 &= c_{21}(t)x_1 + c_{22}(t)x_2 + \ldots + c_{2n}(t)x_n + d_{21}(t)u_1 + \ldots + d_{2m}(t)u_m \end{aligned} \tag{2.11}$$

...

$$y_k = c_{k1}(t)x_1 + c_{k2}(t)x_2 + \ldots + c_{kn}(t)x_n + d_{k1}(t)u_1 + \ldots + d_{km}(t)u_m$$

The linear continuous-time system, can be rewritten in terms of a vector matrix

$$x' = A(t){\cdot}x + B(t){\cdot}u ; \quad x \in \mathfrak{R}^n ; u \in \mathfrak{R}^R ; \quad t > 0 \tag{2.12}$$

$$y = C(t){\cdot}x + D(t){\cdot}u ; \quad y \in \mathfrak{R}^R ; \quad t > 0.$$

The mathematical model, given in (2.12) is called linear, as shown in Definition 2.2. The matrixes of A(t), B(t), C(t), and D(t) are the transforms on the respective vector space, which is as follows:

$A(t)$: $\mathfrak{R}^n \to \mathfrak{R}^n$ as a (n, n)-matrix called the system matrix
$B(t)$: $\mathfrak{R}^m \to \mathfrak{R}^n$ as a (n, m)-matrix called the input matrix
$C(t)$: $\mathfrak{R}^n \to \mathfrak{R}^k$ as a (k, n)-matrix called the output matrix
$D(t)$: $\mathfrak{R}^r \to \mathfrak{R}^p$ as a (p, r)-matrix which called the transition matrix

Definition 2.2
Let a dynamic system have for time stamp t_0 with the initial state $x_{10} \in \mathfrak{R}^n$ and the input function $u_1(\cdot) \in U$ the solution $\{ x_1(t), y_1(t) \}$. Let for $x_{20} \in \mathfrak{R}^n$ and $u_2(\cdot) \in U$ the solution be $\{ x_2(t), y_2(t) \}$. Hence, for all $k_1, k_2 \in \mathfrak{R}^1$ we find with

$$x'(t_0) = k_1 \cdot x_{10} + k_2 \cdot x_{20}, \tag{2.13}$$

and

$$u(t) = k_1 \cdot u_1(t) + k_2 \cdot u_2(t), \tag{2.14}$$

the solution

$$x'(t) = k_1 \cdot x_1(t) + k_2 \cdot x_2(t)$$

$$y(t) = k_1 \cdot y_1(t) + k_2 \cdot y_2(t). \tag{2.15}$$

of the dynamic system, which is called a linear system. ■

For a differential equation of order n, describing a dynamic system, it is possible to reduce the order n to n first-order equations, while making the analysis of the dynamic system somewhat simpler, as shown in (2.2).

Example 2.1
Many systems in the physical world are oscillatory systems, with conversion of energy from one form to another. In a pendulum the potential energy of the bob in the gravitational force field is converted to kinetic energy as the bob swings from its highest position to the neutral position. Electrical RCL networks (see Chap. 1) allow energy to be changed between components, and oscillation may result. If no energy loss occurs, the oscillation continues at a constant amplitude; however, most real systems lose energy, i.e. through damping, and the oscillations eventually cease. A simple oscillatory system, such as the simple pendulum, can be described by nonlinear differential equations of second order

$$x'' = \frac{dx'}{dt} = k \cdot F(x) \cdot x' - x, \tag{2.16}$$

where x'' is the acceleration of the displacement, x' is the rate of change of displacement over time, x is the displacement, and k is the damping term. $F(x)$ is an algebraic function of x, which control the nature of the oscillation
The second term of the differential equation $(-x)$, when not dominant, causes acceleration of the oscillating object toward the neutral point.
A special form for $F(x)$ was suggested by the Dutch physicist van der Pol, born 1889 in Arnhem, Netherlands, when he investigated how to maintain the oscillations in a circuit that depends on continuous oscillations. The equation for maintaining the energy of the oscillating system becoming positive when $|x|$ is less than 1.0 and negative when $|x|$ is greater than 1.0. The second-order differential equation for the van der Pol oscillator can be written as

$$x'' + k \cdot (x^2 - 1) \cdot x' + x = 0. \tag{2.17}$$

Alternatively, one can describe the van der Pol oscillator by utilizing a set of two first-order differential equations as follows, the solution of which can be obtained by simulation.

$$
\begin{aligned}
x &= x_1 \\
x' &= x'_1 = x_2 \\
x'' &= x'_2
\end{aligned}
\tag{2.18}
$$

which results in

$$
x_2 = -k \cdot (x_1^2 - 1) \cdot x_2 - x_1, \tag{2.19}
$$

or

$$
\frac{d}{dt} = \begin{bmatrix} x_1 \\ x_2 \end{bmatrix} = \begin{bmatrix} 0 & 1 \\ -1 & k \end{bmatrix} \begin{bmatrix} x_1 \\ x_2 \end{bmatrix} - k \begin{bmatrix} x_2 & 0 \\ 0 & 0 \end{bmatrix} \begin{bmatrix} x_1^2 \\ x_2^2 \end{bmatrix}. \tag{2.20}
$$

The state vector $x(t)$ is defined as a minimal set of state variables which uniquely determines the future state of a dynamic system if their present values are given. Thus, if $x(t_0)$, the state at $t = t_0$ is known, then the state vector at any future time $x(t)$, for $t > t_0$, is uniquely determined by differential equations such

$$
x' = A(t) \cdot x + B(t) \cdot u \; ; \quad t > 0. \tag{2.21}
$$

This equation set may be rewritten more compactly in matrix form

$$
x' = A \cdot x + B \cdot u, \tag{2.22}
$$

which is in state-vector form, with x as state vector, u as the source or input vector, and A and B as the respective system and input matrix.

It is observed that we may write an output vector, which gives the output variables as linear combinations of the state variables and the inputs. The output vector has the general form

$$
y = C \cdot x + D \cdot u, \tag{2.23}
$$

where C is the output matrix, and D is the transition matrix.

The state-variable description of a linear multivariable system was given in (2.5) for the time-varying case. The corresponding time-invariant vector-matrix notation yields

$$x' = A \cdot x + B \cdot u$$
$$y = C \cdot x + D \cdot u. \tag{2.24}$$

By taking Laplace transforms (see Sect. 2.4) throughout in (2.24) and by setting all initial conditions to zero, we obtain

$$s\,X(s) = A \cdot X(s) + B \cdot U(s) \tag{2.25}$$

$$Y(s) = C \cdot X(s) + D \cdot U(s).$$

Solving the first equation in (2.25) for $X(s)$ and substituting into the second equation of (2.25) yields

$$Y(s) = [C(sI - A)^{-1} B + D]\,U(s) = G(s) \cdot U(s), \tag{2.26}$$

where $G(s)$ is the system transfer matrix of dimension n by m. For a single-input, single-output system, called SISO, the system matrix in (2.26) becomes a system transfer function given by

$$G(s) = \frac{Y(s)}{U(s)} = c^{T}(sI - A)^{-1} b + d, \tag{2.27}$$

which is shown in Fig. 2.1.

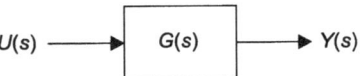

Fig. 2.1. Transfer characteristic of a SISO system

Example 2.2
Let us assume that the transfer matrix of a linear system can be described by

$$x'_1 = x_2 \tag{2.28}$$
$$x'_2 = x_3$$
$$x'_3 = -2x_1 - 4x_2 - 6x_3 + u(t)$$
$$y = x_1 + x_2 + u(t)$$

Solving for $(sI - A)^{-1}$ yields

$$(sI - A)^{-1} = \left[\begin{pmatrix} s & 0 & 0 \\ 0 & s & 0 \\ 0 & 0 & s \end{pmatrix} - \begin{pmatrix} 0 & 1 & 0 \\ 0 & 0 & 1 \\ -2 & -4 & -6 \end{pmatrix}\right]^{-T} = \begin{pmatrix} s & -1 & 0 \\ 0 & s & \\ 2 & 4 & s+6 \end{pmatrix}^{-1}. \qquad (2.29)$$

Forming $G(s)$ (2.27) gives

$$G(s) = \frac{Y(s)}{U(s)} = \begin{pmatrix} 1 \\ 1 \\ 0 \end{pmatrix}^{T} (sI - A)^{-1} \begin{pmatrix} 0 \\ 0 \\ 1 \end{pmatrix} + 1. \qquad (2.30)$$

2.1.2 Existence and Uniqueness of Solutions of Differential Equations

Considering a dynamic system defined by sets of first-order differential equations:

Definition 2.3
The set $Ax = b$ with m equations and n unknowns has solutions if and only if rank[A] = rank[Ab]. Let r = rank[A]. If condition rank[A] = rank[Ab] is satisfied and if $r = n$, then the existence of solutions is unique. ∎

Definition 2.4
The set $Ax = b$ with m equations and n unknowns has solutions if and only if rank[A] = rank[Ab]. Let r = rank[A]. If condition rank[A] = rank[Ab] is satisfied and if r < n, an infinite number of solutions exists and r unknown variables can be expressed as linear combinations of the other $n - r$ unknown variables, whose values are arbitrary. ∎

The check for existence and uniqueness of solutions requires that one form the augmented matrix [Ab]. The first n columns of the augmented matrix are the columns of A. The last column of the augmented matrix is the column vector b.

Example 2.3
Determine whether the following set has a unique solution

$$\begin{aligned} 6x + 3y + 2z &= 18 \\ -6x + 3y + 4z &= 12 \\ 6x + 3y + 4z &= 24 \end{aligned} \qquad (2.31)$$

The matrices A and b and x are

$$A = \begin{bmatrix} 6 & 3 & 2 \\ -6 & 3 & 4 \\ 6 & 3 & 4 \end{bmatrix} \qquad (2.32)$$

$$b = \begin{bmatrix} 18 \\ 12 \\ 24 \end{bmatrix} \qquad (2.34)$$

$$[Ab] = \begin{bmatrix} 6 & 3 & 2 & 18 \\ -6 & 3 & 4 & 12 \\ 6 & 3 & 4 & 24 \end{bmatrix} \qquad (2.35)$$

$$v = \begin{bmatrix} x \\ y \\ z \end{bmatrix}. \qquad (2.36)$$

Obviously, from Definitions 2.3 and 2.4, the rank of A and $[Ab]$ has to be proved. Rank $[A] = 3$ and rank $[Ab] = 3$. Because A and $[Ab]$ have the same rank, a solution exists. This rank equals the number of unknowns, the solution is unique, $x = 1$, $y = 2$, and z 3.

2.2 Controllability, Observability, and Identifiability

Controllability, observability, and identifiability are important properties of dynamic systems, written in the state-variables notation, which characterizes the systems dynamics. A linear system is said to be state controllable when the system input u can be used to transfer the system from any initial state to any arbitrary state in a finite time. A linear system is said to be observable if the initial state $x(t_0)$ can be determined uniquely when given the output $y(t)$ for $t_0 \leq t \leq t_1$ for any $t_1 > t_0$. The theory of controllability and observability was introduced in the 1960's by the Hungarian engineer and mathematician Kalman, born 1930 in Budapest, Hungary, the method of identifiability was later introduced by Astrom and Kalman. It is claimed that if a mathematical model of a dynamic system may be written in the state equations, or one may use this method to show whether the model can be used for predictive measures or not.

From a more general point of view the description of controllability can be given for the time-varying case as follows:

Definition 2.5

A linear dynamic system

$$x' = A \cdot x + B \cdot u$$

$$y = C \cdot x + D \cdot u$$

(2.37)

is said to be

- Controllable at time $t_0 \in T$, if for a finite $t_1 > t_0$, $t_1 \in T$ exist.
- Completely controllable, if for each $t_0 \in T$ a finite time $t_1 > t_0$, $t_1 \in T$ exist.
- Differential or particularly controllable, if for each $t_0 \in T$ and each finite $t_1 > t_0$, $t_1 \in T$, the matrix

$$w(t_0, t_1) = \int_{t_0}^{t_1} \Phi(t_0, \tau) B(\tau) B^T(\tau) \Phi^T(t_0, \tau) d\tau$$

(2.38)

is regular. ∎

The several descriptions of controllability represent the different characteristics of the dynamic system that are not the characteristics of the mathematical model. But the corresponding mathematical model has the same result, due to controllability, as the dynamic system.

An important approach in systems analysis involves systems being completely controllable, which allows the prediction of how the system may behave.

Definition 2.6

A linear dynamic system

$$x' = A \cdot x + B \cdot u$$

$$y = C \cdot x + D \cdot u$$

(2.39)

is said to be completely state controllable if there exists a control signal u, defined over the finite interval $t_0 < t < t_F$, which transfers the system from any initial state $x(t_0) = x_0$ to any desired final state $x(t_F) = x_F$ in the defined time interval. ∎

Definition (2.6) is said to be true if and only if the (n, np) controllability matrix

$$Q_{C:} = [B \ AB \ A^2B \ ... \ A^{n-1}B]$$

(2.40)

has full row rank n, which means that the vector elements $B, AB, ..., A^{n-1}B$ of Q_C are linear independent, which means that the controllability matrix Q_C has nonzero determinant.

Example 2.4
A dynamic system can be described by the state-equation model

$$x'(t) = A \cdot x(t) + B \cdot u(t) \tag{2.41}$$

$$= \begin{bmatrix} -3 & 1 \\ -2 & 1.5 \end{bmatrix} \begin{bmatrix} x_1 \\ x_2 \end{bmatrix} + \begin{bmatrix} 0 \\ 1 \end{bmatrix} [u],$$

with

$$A = \begin{bmatrix} -3 & 1 \\ -2 & 1.5 \end{bmatrix} \tag{2.42}$$

$$B = \begin{bmatrix} 0 \\ 1 \end{bmatrix}. \tag{2.43}$$

The dynamic system given in Example 2.4 is completely state controllable if B and AB are linear independent and the rank of the controllability matrix Q_C: $[B, AB] = 2$, with

$$B = \begin{bmatrix} 0 \\ 1 \end{bmatrix} \tag{2.44}$$

and

$$AB = \begin{bmatrix} 1.0 \\ 1.5 \end{bmatrix}, \tag{2.45}$$

hence

$$Q_C := [B, AB] = 2, \tag{2.46}$$

which means the dynamic system is completely state controllable.

Example 2.5
Suppose a dynamic system can be described by equations in the Laplace domain (see Sect. 2.4)

$$X_1 = \frac{1}{s+1} \cdot U \tag{2.47}$$

and

$$X_2 = \frac{1}{s+2} \cdot (X_1 + U),$$
(2.48)

which can be rewritten in the state-equation notation as

$$x'_1 = -x_1 + u$$
(2.49)

$$x'_2 = x_1 - 2 \cdot x_2 + u.$$
(2.50)

Assuming

$$b = \begin{bmatrix} 1 \\ 1 \end{bmatrix}$$
(2.51)

and

$$Ab = \begin{bmatrix} -1 & 0 \\ 1 & -2 \end{bmatrix} \begin{bmatrix} 1 \\ 1 \end{bmatrix} = \begin{bmatrix} -1 \\ -1 \end{bmatrix},$$
(2.52)

the dynamic system given above is not state controllable, while $b + Ab = 0$, which means that the vectors b, Ab are linear dependent.

Definition 2.7
A linear dynamic system

$$x' = A \cdot x + B \cdot u$$
(2.53)

$$y = C \cdot x + D \cdot u$$
(2.54)

is said to be
- Observable at time $t_0 \in T$, if for a finite $t_1 > t_0$, $t_1 \in T$ exist
- Completely observable, if for each $t_0 \in T$ and each finite $t_1 > t_0$, $t_1 \in T$ exist
- Differential or particularily observable, if for each $t_0 \in T$ and each finite $t_1 > t_0$, $t_1 \in T$, the matrix

$$m(t_0,t_1) = \int_{t_0}^{t_1} \Phi^T(t_1,t_0)C^T(t)C(t)\Phi(t_1,t_0)dt \qquad (2.55)$$

is a regular one.

The different ideas of observability are due to the properties of the dynamic system and not due to the properties of the mathematical model. But the mathematical model has the same results, due to observability, as the dynamic system.

Let us next consider that the system is completely observable, hence predictions with it are possible, and we may write:

Definition 2.8
A linear dynamic system

$$x' = A \cdot x + B \cdot u \qquad (2.56)$$

$$y = C \cdot x + D \cdot u \qquad (2.57)$$

is said to be completely observable within the finite interval $t_0 < t < t_F$, if any initial state $x(t_0) = x_0$ can be determined from the output y, observed over the same interval. ■

Definition 2.8 is said to be true if and only if the (n, nr) observability matrix

$$Q_O := [C^T \ C^T \cdot A^T \ ... \ C^T(A^T)^{n-1}] \qquad (2.58)$$

has full rank n, with C^T as the transpose of C.

Example 2.6
The state-variable description of a dynamic system is given for the time-varying case as follows

$$x'(t) = A \cdot x(t) + B \cdot u(t) \qquad (2.59)$$

$$y(t) = C \cdot x(t). \qquad (2.60)$$

The corresponding parameters are

$$\frac{d}{dt}x = \begin{bmatrix} 0 & 1 & 0 \\ 0 & 0 & 1 \\ -6 & -11 & 6 \end{bmatrix} \begin{bmatrix} x_1 \\ x_2 \\ x_3 \end{bmatrix} + \begin{bmatrix} 0 \\ 0 \\ 1 \end{bmatrix} [u] \qquad (2.61)$$

$$y(t) = \begin{bmatrix} 20 & 9 & 1 \end{bmatrix} \begin{bmatrix} x_1 \\ x_2 \\ x_3 \end{bmatrix}, \qquad (2.62)$$

with

$$A = \begin{bmatrix} 0 & 1 & 0 \\ 0 & 0 & 1 \\ -6 & -11 & -6 \end{bmatrix} \qquad (2.63)$$

and

$$C = \begin{bmatrix} 20 & 9 & 1 \end{bmatrix}, \qquad (2.64)$$

or

$$C^T = \begin{bmatrix} 20 \\ 9 \\ 1 \end{bmatrix}, \qquad (2.65)$$

which yields

$$C^T A^T = \begin{bmatrix} 20 \\ 9 \\ 1 \end{bmatrix} \begin{bmatrix} 0 & 0 & -6 \\ 1 & 0 & -11 \\ 0 & 1 & -6 \end{bmatrix} = \begin{bmatrix} -6 \\ 9 \\ 3 \end{bmatrix} \qquad (2.66)$$

and

$$C^T (A^T)^2 = \begin{bmatrix} 20 \\ 9 \\ 1 \end{bmatrix} \begin{bmatrix} 0 & -6 & 36 \\ 0 & -11 & 60 \\ 1 & -6 & 25 \end{bmatrix} = \begin{bmatrix} -18 \\ -39 \\ -9 \end{bmatrix} . \qquad (2.67)$$

The vectors C^T, $C^T A^T$, and $C^T (A^T)^2$ are linear independent and the rank of the observability matrix is

$$Q_O \colon = [C^T, C^T A^T, \text{ and } C^T (A^T)^2] = 3, \tag{2.68}$$

thus the dynamic system, given in Example 2.6 is completely observable.

Example 2.7
Suppose the output equation is

$$y(t) = \begin{bmatrix} 4 & 5 & 1 \end{bmatrix} \begin{bmatrix} x_1, x_2, x_3 \end{bmatrix}^T \tag{2.69}$$

instead of

$$y(t) = \begin{bmatrix} 20 & 9 & 1 \end{bmatrix} \begin{bmatrix} x_1, x_2, x_3 \end{bmatrix}^T \tag{2.70}$$

the vectors C^T, $C^T A^T$, and $C^T (A^T)^2$ are linearly dependent, and the rank of the observability matrix has the value

$$Q_O \colon = [C^T, C^T A^T, \text{ and } C^T (A^T)^2] = 2. \tag{2.71}$$

Thus the dynamic system is not observable.

Definition 2.9
A dynamic system is said to be identifyable in its parameters, within the time interval $t_0 < t < t_F$, if the parameter vector $\boldsymbol{\Theta}$ may be determined from the output y, observed over the same time interval $t_0 < t < t_F$. ■

Definition 2.10
A dynamic system is said, for the true model parameter vector $\boldsymbol{\Theta}_T$, to be

1. Parameter identifiable if there exists an input sequence $\{u\}$ such that $\boldsymbol{\Theta}$ and $\boldsymbol{\Theta}_T$ are distinguished for all $\boldsymbol{\Theta} \neq \boldsymbol{\Theta}_T$.
2. System identifiable if there exists an input sequence $\{u\}$ such that $\boldsymbol{\Theta}$ and $\boldsymbol{\Theta}_T$ are distinguishable for all $\boldsymbol{\Theta} \neq \boldsymbol{\Theta}_T$ but are a finite set.
3. Unidentifyable in all other cases. ■

The state-variable concept of dynamic systems completely characterizes the system's past, since the past input is not required to determine the future output of the dynamic system. This seemingly elementary notation of state equations is of importance in the systems state-variables approach. In fact, this mathematical notation, describing dynamic systems, is fundamentally based upon the state-variables concept.

On choosing state variables while describing a dynamic system mathematically, no prescription can be given for choosing the state variables in the sense of a general guideline, i.e., in electrical RCL networks (see Sect. 1.3.1), the charge on each capacitor and the current through each inductor in the network usually serve to define the state variable of the respective network; in a mechanical system (see Chap. 1) the force and mass of each body, usually serve to define the state variable of a mechanical system. In other domains choosing the state variables may be much more difficult. Once state variables are chosen, and the mathematical equations characterizing the state variables, the state equations can be derived.

Depending on the particular form of the equations used to describe the dynamic system the state equations can be in one of the many mathematical forms. It is possible to classify state equations on the basis of their mathematical structure.

Time is usually an independent variable in a state-variable model:

- Sometimes the time variable is considered as a discrete variable, in such cases the state-variable model is typically described by recursive equations.
- In other cases the independent variable time are considered to be real valued.
- Sometimes there can be additional independent variables in which case the state-variable model is said to be distributed; such state-variable models is given as partial differential equations.
- If time is the only independent variable then the state-variable model is said to be lumped, while the actual physical size will not really serve as a measure of lumpiness – moreover, we have to consider that we may have frequent occasions to distinguish between lumped and distributed elements –.
- Further, the state-equation model may include random effects in which case the state-equation model is said to be stochastic.
- If no such effects are included the state-equation model is said to be a deterministic one.

2.3 Time Domain Solution of the Linear State Equation System

We may note that the component state-variables equations, given in (2.27), express the uncoupled from, which in vector-matrix notation yield a diagonal matrix A in

$$x' = A \cdot x + b \cdot u \qquad (2.72)$$

and

$$y = c^T \cdot x, \qquad (2.73)$$

which are quite simple to solve for $x_i(t)$, where $i = 1,2,\ldots,n$. Consider a linear system with $u = 0$, described by

$$x' = A \cdot x. \qquad (2.74)$$

The time solution of (2.74) has the form

$$x(t) = \Phi(t, t_0) \, x(t_0) \qquad (2.75)$$

where $\Phi(t, t_0)$ is referred to as the state transition matrix. The Initial condition $x(t_0)$ is transferred to the state x at time by the matrix $\Phi(t, t_0)$. It is very obvious that $\Phi(t, t_0) = I$, the identity matrix, since the state $x(t)$ is equal to $x(t_0)$ at $t = t_0$. The transition characteristic of the state-transition matrix can be written as

$$\Phi(t_2, t_0) = \Phi(t_2, t_1) \, \Phi(t_1, t_0), \qquad (2.76)$$

which indicates that if an initial state vector $x(t_0)$ is transferred to $x(t_1)$ by $\Phi(t_1, t_0)$ and if $x(t_1)$ is then transferred to $x(t_2)$ by $\Phi(t_2, t_1)$, then $x(t_0)$ can be transferred in a direct way to $x(t_2)$ by $\Phi(t_2, t_0)$, which is the product of the two state transition matrices. If the matrix A in (2.74) is constant with time, then the state transition matrix $\Phi(t, t_0)$ is a function only of the distance t and t_0, that is,

$$\Phi(t, t_1) = \Phi(t - t_0). \qquad (2.77)$$

Consider the matrix-exponential solution of the linear dynamic system, as shown in (2.74), yields

$$x(t) = e^{At} \cdot x(0). \qquad (2.78)$$

This obviously means that the time response of the dynamic system is equal to the exponential function if the matrix differentiation rule is used to form

$$x' = \frac{d}{dt}[x(t)] = \frac{d}{dt}[e^{At} \cdot x(0)] = A \cdot e^{At} \cdot x(0).$$ (2.79)

Substituting (2.78) into (2.79) yields $x' = A \cdot x$. Comparing (2.78) and (2.74) yields

$$\Phi(t) = e^{At}.$$ (2.80)

We may now express the term e^{At} in a Taylor series about $t = 0$ to give

$$\Phi(t) = e^{At} = \sum_{k=0}^{\infty} \frac{1}{k!} A^k t^k = I + A \cdot t + \frac{1}{2!} A^2 \cdot t^2 + \dots\dots ,$$ (2.81)

which is the equation for solving the state-transition equation by series expansion.

Moreover, we may premultiply each term of the state-variables (2.74) by the exponential expression, which gives

$$x'(t) \cdot e^{-At} = A \cdot x(t) \cdot e^{-At} + B \cdot u(t) \cdot e^{-At},$$ (2.82)

that is, dropping the explicit notation for time dependence and rearranging the equation

$$x' \cdot e^{-At} - A \cdot x \cdot e^{-At} = B \cdot u \cdot e^{-At}$$ (2.83)

or

$$\frac{d}{dt}(e^{-At}x) = B \cdot u \cdot e^{-At}.$$ (2.84)

Multiplying by dt, and integrating over the time interval t_0 to t, as well as changing the variables, we may write

$$\int_{t_0}^{t} d/d\tau (e^{-A\tau}x) \, d\tau = \int_{t_0}^{t} B \cdot u(\tau) \cdot e^{-A\tau} d\tau,$$ (2.85)

or

$$x(t) \cdot e^{-At} - x(t_0) \cdot e^{-At_0} = \int_{t_0}^{t} B \cdot u(\tau) \cdot e^{-A\tau} d\tau. \tag{2.86}$$

Premultiplying all terms in this equation by $e^{A(t-t_0)}$ gives

$$x(t) = x(t_0) \cdot e^{A(t-t_0)} + \int_{t_0}^{t} B \cdot u(\tau) \cdot e^{-A(t-t_0)} d\tau, \tag{2.87}$$

that is, rewritten in terms of the state-transition matrix

$$x(t) = x(t_0) \cdot \Phi(t - t_0) = \int_{t_0}^{t} B \cdot u(\tau) \cdot \Phi(t - t_0) \cdot d\tau, \tag{2.88}$$

which is the matrix form of the convolution integral. The convolution integral in (2.88) involving the impulse-response function, and the superposition integral in terms of the state-transition matrix.

2.4 Solution of the State Equation using the Laplace Transform

The state-transition matrix, determined in Sect. 2.3, is used to yield the complete solution of the linear state-variable equations. Let the time-dependent behavior of a linear state-differential equation system be written in the notation:

$$x'(t) = A \cdot x(t) + B \cdot u(t) \tag{2.89}$$

and

$$y(t) = C \cdot x(t) + D \cdot u(t), \tag{2.90}$$

which means the dependence of the output vector $y(t)$ from the input vector $u(t)$, with system matrix A and input matrix B assumed to be constant matrices. To determine the solution of this state-differential equations the Laplace transform $\mathscr{L}[f(t)] = F(s)$ can be used as follows:

$$s \cdot X(s) = A \cdot X(s) + X(0) + B \cdot U(s), \tag{2.91}$$

where $X(0)$ denotes the initial-state vector, presumably a known quantity, and s denotes the Laplace operator, which is the first derivative. Solving (2.91) gives

$$X(s) = (sI - A)^{-1} X(0) + (sI - A)^{-1} B \cdot U(s), \tag{2.92}$$

with $X(s)$ as Laplace transform of $x(t)$, and I as (n, n)-unit matrix which is defined by $AI = IA = A$, that is

$$\begin{bmatrix} a_{11} & a_{12} \\ a_{21} & a_{22} \end{bmatrix} \begin{bmatrix} 1 & 0 \\ 0 & 1 \end{bmatrix} = \begin{bmatrix} 1 & 0 \\ 0 & 1 \end{bmatrix} \begin{bmatrix} a_{11} & a_{12} \\ a_{21} & a_{22} \end{bmatrix} = \begin{bmatrix} a_{11} & a_{12} \\ a_{21} & a_{22} \end{bmatrix}. \tag{2.93}$$

Using A^{-1} as the inverse matrix of A, that is $A A^{-1} = A^{-1}A = I$, and $(sI - A)$ as a matrix called the characteristic matrix, gives

$$(sI - A) X(s) = X(0) + B \cdot U(s), \tag{2.94}$$

with $L(s) = (sI - A)$ and $L^{-1}(s) = (sI - A)^{-1}$, where $L(s)$ is the adjoint of the characteristic matrix, and $\Delta(s)$ the determinant of the matrix, called the characteristic polynomial of matrix A, we can write

$$\mathscr{L}^{-1}(s) = (sI - A)^{-1} = \frac{1}{\Delta(s)} \cdot \mathscr{A}(s). \tag{2.95}$$

The roots of the polynomial above are called eigenvalues of the dynamic system. For a linear dynamic system with constant coefficients they will be simple relationships.

Obviously, the time-domain solution can be obtained by convolving the inverse Laplace transform of (2.95) for (2.92), that is,

$$X(s) = \frac{1}{\Delta(s)} \cdot \mathscr{A}(s) \cdot X(0) + \frac{1}{\Delta(s)} \cdot \mathscr{A}(s) \cdot U(s) \tag{2.96}$$

As for ordinary differential equations, we may expand $X(s)$ into a partial fraction expansion,

$$X(s) = X_1(s) \cdot \frac{1}{(s - s_1)} + X_2(s) \cdot \frac{1}{(s - s_1)} + \dots \qquad (2.97)$$

with the specialized form of the solution

$$X(s)_i = \mathscr{A}(s) \cdot X(0) + \mathscr{A}(s) \cdot U(s) = X_i + U_i, \qquad (2.98)$$

and the corresponding results

$$X_1 = \lim_{s \to s_1} \left\{ \mathscr{A}(s) \cdot \frac{1}{[(s - s_2)(s - s_3)\dots]} \right\} \cdot X(0) \qquad (2.99)$$

$$\Rightarrow X_1 = \left\{ \mathscr{A}(s) \cdot \frac{1}{[(s_1 - s_2)(s_1 - s_3)\dots]} \right\} \cdot X(0) \qquad (2.100)$$

$$X_2 = \lim_{s \to s_2} \left\{ \mathscr{A}(s_1) \cdot \frac{1}{[(s - s_1)(s - s_3)\dots]} \right\} \cdot X(0) \qquad (2.101)$$

$$\Rightarrow X_2 = \left\{ \mathscr{A}(s_2) \cdot \frac{1}{[(s_2 - s_1)(s_2 - s_3)\dots]} \right\} \cdot X(0) \qquad (2.102)$$

and then form x_i as

$$X_i = \lim_{S \to s_i} \left[\frac{(s - s_i)}{\Delta(s)} \right] \cdot \mathscr{A}(s) \cdot X(0) \qquad (2.103)$$

$$i = 1, 2, \dots , n, \qquad (2.104)$$

for each root of $L(s) \cdot U(s) \cdot \dfrac{1}{\Delta(s)}$. The resultant state-vector $x(t)$ can be written as

$$x(t) = X_1 \cdot e^{s_1 t} + X_2 \cdot e^{s_2 t} + , \dots , + X_n \cdot e^{s_n t} + U_1 \cdot e^{s_1 t} + , \dots , + U_m \cdot e^{s_m t}. \qquad (2.105)$$

Therefore, we can write the complete solution $X(t)$ by using the Laplace transform

$$X(t) = \mathscr{L}^{-1}\{(sI - A)^{-1}\}X(0) + \mathscr{L}^{-1}\{(sI - A)^{-1} B \cdot U(s)\}. \qquad (2.106)$$

Using the notation of

$$X(t) = \mathscr{L}^{-1}\{(sI - A)^{-1}\}X(0) = e^{At} \qquad (2.107)$$

and the respective correspondence of the Laplace transforms

$$\Phi(t) = \mathscr{L}^{-1}\{(sI - A)^{-1}\}X(0) = e^{At} \qquad (2.108)$$

we obtain the state-transition matrix for the linear system. The transition matrix determines the transient behavior of the dynamic system over all time, that is between time t_0 and time t_1, that is

$$\Phi(t) = e^{At} \qquad (2.109)$$

Assuming that the initial time is denoted $t = t_0$ instead of $t = 0$, we may write

$$x(t) = x(t_0) \cdot e^{A(t-t_0)} \qquad (2.110)$$

as well as

$$x(t) = x(0) \cdot \Phi(t - t_0). \qquad (2.111)$$

2.5 Eigenvalues of the Linear Vector-Equation Systems*

Consider a linear system with $u = 0$, described in (2.74) as

$$x' = A \cdot x, \qquad (2.112)$$

where A hold a $n \times n$ matrix. Taking Laplace transforms and solving for $X(s)$, we have

$$sX(s) = A \cdot X(s) \tag{2.113}$$

with

$$0 = (A - sI) X(s) \tag{2.114}$$

where s is a scalar. The scalars s are called eigenvalues of A, and the vectors $X(s)$ are called eigenvectors of A. The complete set of all eigenvalues are called the spectrum of A.

(2.114) gives, for nontrivial cases of $X(s)$

$$\det(A - sI) = 0. \tag{2.115}$$

the characteristic equation, with $\det(A - sI)$ as tha characteristic polynomial of A. It is a n-th-degree polynomial in s. The characteristic equation is given by

$$det(A - sI) = \begin{bmatrix} a_{11} - s & a_{12} & a_{13} \\ a_{21} & a_{22} - s & a_{23} \\ a_{31} & a_{32} & a_{33} - s \end{bmatrix} \tag{2.116}$$

$$= (-1)^n (s^n + \alpha_1 s^{n-1} + \ldots + \alpha_{n-2} s^2 + \alpha_{n-1} s + \alpha_n) = 0. \tag{2.117}$$

The n roots of the characteristics (2.116) are the eigenvalues of A. They are called the characteristic roots. Note that a real $n \times n$ matrix A does not necessarily possess real eigenvalues. But since $\det(A - sI) = 0$ is a polynomial with real coefficients, any complex eigenvalues must occur in conjugate pairs, namely if $\alpha+j\beta$ is an eigenvalue, then $\alpha-j\beta$ is also an eigenvalue of A.

Example 2.8
Assuming a dynamic system described by the system matrix A as below

$$A = \begin{bmatrix} 1 & 2 & -1 \\ 1 & 3 & 2 \\ 2 & -1 & 0 \end{bmatrix}. \tag{2.118}$$

Due to the Laplace transform we can write

$$\mathscr{L}(sI\text{-}A) = \begin{bmatrix} s-1 & -2 & 1 \\ -1 & s-3 & -2 \\ -2 & 1 & s \end{bmatrix}. \qquad (2.119)$$

Hence we find the eigenvalues of A as follows

$$\Delta(s) = (s-1)(s-3)s - 8 - 1 + 2(s-3) - 2s + 2(s-1) = s^3 - 4s^2 + 5s - 17 \qquad (2.120)$$

which can be rewritten as

$$L^{-1}(s) = \frac{1}{\Delta(s)} \cdot L(s). \qquad (2.121)$$

Example 2.9
Consider a linear system described by the differential-equation system

$$x_1' = -k_1 \cdot x_1 + u \qquad (2.122)$$

and

$$x_2' = k_1 \cdot x_1 - k_2 \cdot x_2. \qquad (2.123)$$

If the system matrix A is given by

$$A = \begin{bmatrix} -k_1 & 0 \\ k_1 & -k_2 \end{bmatrix}, \qquad (2.124)$$

then the eigenvalues of A are found from

$$\det(A - \lambda I) = 0 \qquad (2.125)$$

$$\det \begin{bmatrix} -k_1 - \lambda & 0 \\ k_1 & -k_2 - \lambda \end{bmatrix} = (-k_1 - \lambda)(-k_2 - \lambda) = 0 \qquad (2.126)$$

The eigenvalues can be found as $\lambda_1 = -k_1$ and $\lambda_2 = -k_2$.

2.6 Stability Analysis*

The transient response of a dynamic system is of primary interest and must be investigated. A very important characteristic of the transient performance of a dynamic system for this reason is the stability of the system. A dynamic system is said to be stable if the system remains near the equilibrium state, i.e. if the variables x and y remain bounded as $t \rightarrow \infty$. If the dynamic system tends to return to the equilibrium state, it is said that the dynamic system is asymptotically stable. An equilibrium state x_e is said to be asymptotically stable at large if it is asymptotically stable for any initial state vector $x(0)$, such that every motion converges to x_e as $t \rightarrow \infty$. The stability analysis is inherently related to the design problem for linear time invariant systems.

For the linear time invariant systems, described by

$$\frac{dx}{dt} = f(x, y), \tag{2.127}$$

$$\frac{dy}{dt} = g(x, y), \tag{2.128}$$

where f and g are continuous functions of x and y and have continuous partial derivatives, will have a unique solution $x = \Phi(t)$, $y = \Psi(t)$ for $t \geq 0$. The concept of stability can be illustrated as a curve in the $x\,y$ plane, called the phase plane, with t as parameter. The solutions are indicated as curves, referred to as trajectories.

Consider a mass m moving on a horizontal level, attached by a spring to a fixed point at a wall. Neglecting the resistive element and external force, the resulting equation will be

$$\frac{d^2 x}{dt^2} = -w^2 \cdot x. \tag{2.129}$$

Defining

$$y = \frac{dx}{dt}, \tag{2.130}$$

we have a system of differential equations

$$\frac{dx}{dt} = y, \tag{2.131}$$

$$\frac{dy}{dt} = -w^2 \cdot x, \tag{2.132}$$

which gives

$$x = \alpha \cdot \cos(\omega \cdot t + \beta) \tag{2.133}$$

$$y = -\alpha \cdot \omega \cdot \sin(\omega \cdot t + \beta). \tag{2.134}$$

These equations define the trajectories

$$x^2 + (\frac{y}{\omega})^2 = \alpha^2. \tag{2.135}$$

in the $x\,y$ plane, as shown in Fig. 2.2 as a simulation result using the ModelMaker simulation software (see Chap. 4), with the x-axis as the displacement, and the y-axis as the speed, starting at the initial conditions $x(0) = 1$, and $y(0) = 1$.

Obviously, the trajectories shown in Fig. 2.2 are directly defined in terms of (2.30), given as

$$\frac{dy}{dx} = \frac{\dfrac{dy}{dt}}{\dfrac{dx}{dt}} = -\frac{\omega^2 \cdot x}{y}. \tag{2.136}$$

In order to investigate the behavior of a dynamic system near equilibrium we may first assume that the equilibrium points are $x_0 = 0$, and $y_0 = 0$. Hence we may define the equilibrium points as

(i)   Stable if x and y remain bounded as $t \rightarrow \infty$
(ii)   Asymptotically stable if $x, y \rightarrow 0$ as $t \rightarrow \infty$
(iii)   Unstable in any other case

As mentioned earlier, stability analysis of a dynamic system is very important. For this reason we present different criteria to determine the stability of dynamic systems. A procedure for determining the stability of a linear time-invariant system by examining its characteristic polynomial, called Routh Hurwitz criterion, will be introduced. If the time-domain formulation is transformed into the frequency domain by such transformations as the simple exponential function, the stability of the system in the frequency-domain can be studied by means of the Nyquist criteria, which is the imaginary or frequency axis for the particular system function. The frequency-domain transformations, and their subsequent use, are also restricted to the linear systems.

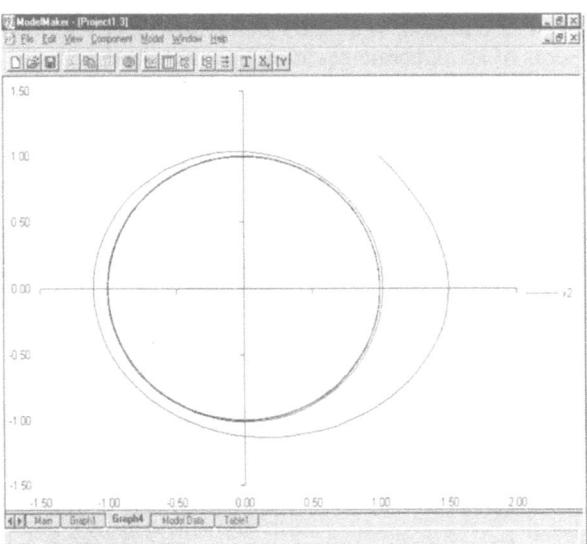

Fig.: 2.2. Simple trajectories in the *x y* plane

2.6.1 Routh Hurwitz Criterion*

Let the characteristic polynomial of a systems transfer function be written as

$$d(s) = s^n + \alpha_0 \qquad\qquad (2.137)$$

$$d(s) = s^n + \alpha_{n-1} \cdot s^{n-1} + ..., + \alpha_2 \cdot s_2 + \alpha_1 + s + \alpha_0 , \qquad (2.138)$$

with α_{n-1}, ... , α_0 as real constants. A necessary condition for asymptotic stability is that all coefficients must have the same sign and none may be zero. This test by inspection is referred to as the Hurwitz test.

The German mathematician Hurwitz, born 1859 in Hildesheim, Germany, published in 1895 a necessary and sufficient condition for asymptotic stability based on the evaluation of certain determinants involving the characteristic polynomial coefficients. These numerical computations are quite cumbersome, especially for systems of order n. Earlier, in 1877, the Canadian mathematician Routh, born 1831 in Quebec, Canada, had developed a tabular form that involves simple sequential calculation for inspection of the dynamics of systems. The interpretation of the Hurwitz determinants by means of calculations in a Routh table is referred to as Routh Hurwitz criterion. With the Routh Hurwitz criterion one can determine whether any roots of an algebraic equation lie in the right half s-plane.

The Routh Hurwitz criterion may consider the following algebraic equation

$$a_n \cdot s^n + a_{n-1} \cdot s^{n-1} +,...,+ a_2 \cdot s_2 + a_1 + s + a_0 = 0. \tag{2.139}$$

The algebraic schedule of the Routh Hurwitz criterion is provided in Table 2.1. The coefficients of the characteristic polynomial are arranged in the first two rows

Table 2.1.   General form of the Routh table

s^n	a_n	a_{n-2}	a_{n-4}	a_{n-6}	a_{n-8}
s^{n-1}	a_{n-1}	a_{n-3}	a_{n-5}	a_{n-7}	
	b_{n-1}	b_{n-3}	b_{n-5}	b_{n-7}	
	c_{n-1}	c_{n-3}	c_{n-5}		
	d_{n-1}	d_{n-3}	d_{n-5}		
	e_{n-1}	e_{n-3}			
	f_{n-1}	f_{n-3}			
	g_{n-1}				
	h_{n-1}				

The elements of this schedule are given as

Row 1: Alternate coefficients of the original equation

$$a_n \qquad a_{n-2} \qquad a_{n-4} \qquad a_{n-6} \qquad .... \tag{2.140}$$

Row 2: Remaining coefficients of the original equation

$$a_{n-1} \qquad a_{n-3} \qquad a_{n-5} \qquad a_{n-7} \qquad .... \tag{2.141}$$

Row 3: Take the appropriate cross product terms, as shown

$$b_{n-1} = \frac{a_{n-1}a_{n-2} - a_n a_{n-3}}{a_{n-1}} \tag{2.142}$$

$$b_{n-3} = \frac{a_{n-1}a4 - a_n a_{n-5}}{a_{n-1}} \tag{2.143}$$

$$b_{n-5} = ... \tag{2.144}$$

Row 4: Take cross products of the elements of the second and third rows exactly in the manner to obtain the elements of row 3. Thus

$$c_{n-1} = \frac{b_{n-1}a_{n-3} - a_{n-1}a_{n-4}}{b_{n-1}} \tag{2.145}$$

$$c_{n-3} = \frac{b_{n-1}a_{n-5} - a_{n-1}a_{n-6}}{b_{n-1}} \tag{2.146}$$

$$c_{n-5} = ... \tag{2.147}$$

Row 5, and so forth.

Once any row has been completed, the elements of the following row may be determined from the previous two rows, e.g. for rows 3 and 4, until all elements of a row are zero. Hence the interpretation of the array in Table 2.1 is as follows: The number of roots of the original equation that lie in the right half s-plane is equal to the number of sign changes in the first column of the final array. This condition is known as the Routh Hurwitz criterion.

Finally we can determine: All of the characteristic zeros have a negative real part if

(a) $a_{n-1} > 0,\ a_{n-2} > 0,\ a_{n-3} > 0,\ \ldots,\ a_{n-8} > 0$
(b) all of the n-1 numbers in the first column are positive

Obviously, the Routh Hurwitz criterion is satisfied if both conditions (a) and (b) are satisfied, in which case the characteristic zeros all have negative real parts. If either condition (a) or condition (b) is not satisfied then there must be a characteristic zero that has a real part that is zero or positive.

Example 2.10
For the given characteristic polynomial the stability proof should be done using the Routh Hurwitz criterion

$$d(s) = s^3 + s^2 + 4 \cdot s + 30 \cdot s . \tag{2.148}$$

The schedule for the Routh Hurwitz criterion is easily determined to be

$$\begin{array}{ccc} 1 & 4 & 0 \\ 1 & 30 & 0 \\ -26 & 0 & 0 \\ 30 & 0 & 0 \end{array} \tag{2.149}$$

Condition (a) is satisfied but condition (b) is not while the first entry in the third row is negative. Finally from the Routh Hurwitz criterion it follows that there is at least one characteristic zero with positive real part. In this case, the characteristic zeros can be determined to be $1 + j\,3$, $1 - j\,3$ and $-3 + j\,0$, which verifies the result.

Example 2.11
The stability and any limitations of the parameter κ for stability of the following equation has to be examined

$$(2s^2 - s + \kappa)(s+1)^2 = 2s^4 + 3s^3 + \kappa s^2 + (2\kappa - 1)s + (\kappa + 1) = 0 . \tag{2.150}$$

Observe that for all orders of s present in the polynomial, and for non-negative coefficients, κ must be $> \frac{1}{2}$. The schedule for the Routh Hurwitz criterion can be determined as

$$
\begin{array}{ccc}
2 & \kappa & \kappa + 1 \\[2mm]
3 & 2\kappa - 1 & \\[2mm]
\dfrac{2 - \kappa}{3} & \kappa + 1 & \\[2mm]
\dfrac{2\kappa^2 + 4\kappa + 11}{\kappa - 2} & &
\end{array}
\tag{2.151}
$$

Avoiding sign changes from the second to the third row κ must be < 2, and sign changes from the third row to the fourth row κ must be > 2. The conditions imposed on parameter κ cannot be fulfilled simultaneously, and there is no value of κ that allows stability.

2.6.2 Nyquist Criterion*

In 1932, the American engineer Nyquist, born in Sweden 1889, published his stability criterion test which is deduced by a graphical procedure. The Nyquist criterion allows determination of whether any of the roots of the equation

$$1 + G(s) = 0 \tag{2.152}$$

lie in the right half s-plane, where $G(s)$ is the forward transfer function. For the feedback system, shown in Fig. 2.3, the overall system transfer function $T(s)$ is

$$T(s) = \frac{Y}{U} = \frac{G(s)}{1 + G(s) \cdot H(s)} , \tag{2.153}$$

with $G(s)$ as forward transfer function, and $H(s)$ as feedback transfer function, U as the input, and Y as the output.

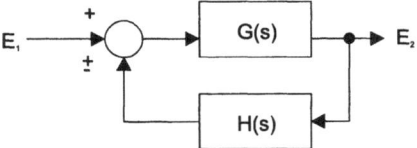

Fig. 2.3. Simple feedback system

When the feedback loop shown in Fig. 2.3 is opened, the open loop gain then is $G(s) \cdot H(s)$, which is directly related to (2.32), and also it will be the function in the denominator of (2.33).

Obviously, the zeros of the denominator of (2.33) are poles of $T(s)$. Since the right half-plane poles specify the instability margin of the feedback system, while the right-half plane zeros of the denominator are important in determination of the stability of the dynamic system.

Applying the Nyquist criterion for stability analysis requires:

1) The magnitude and phase angle of G in the $G(j\omega)$ plane; in the more general case $G(j\omega) \cdot H(j\omega)$.
2) The behavior of G in the s-plane and the poles of $G(s)$ that lie on the imaginary axis or at the origin of the s-plane.
3) The number of poles of $G(s)$ in the right half s-plane.

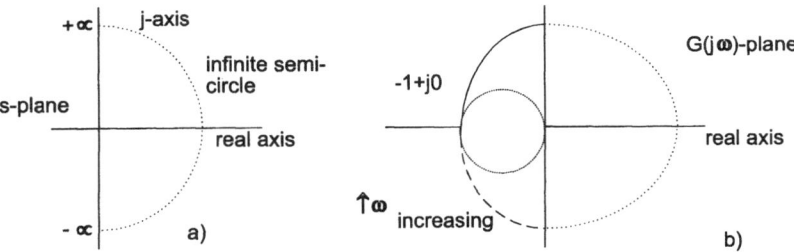

Fig. 2.4. Locus of s for the Nyquist criterion (a); assumed $1 + G(j\omega)$ for the system (b)

The Nyquist criterion may be expressed mathematically as $Z = N + P$ with

- Z: number of zeros of $1 + G(s)$ that lie in the right half s-plane.
- N: number of clockwise encirclements of the point $-1 + j\,0$ by the locus of $G(s)$ as s describes the path shown in Fig. 2.4 a,
- P: number of poles of $G(s)$ that lie in the right half s-plane.

For stability Z must be zero, that is $P = -N$. If $P \neq -N$, the system is unstable

Example 2.12
Determine for what range of K the dynamic system $1 + G(s)$ will have stable roots, with

$$G(s) = K \frac{1+s}{s(1-s)} .$$ (2.154)

A table of the pertinent information of solution of (2.34) shows

Table 2.2. Numbers of the Nyquist criterion

Z	N	P	Nature of response
1	0	1	Unstable
0	– 1	1	Stable
2	1	1	Unstable

It should be noted that stable roots exist for K in the range $-\infty < K < -1$.

2.6.3 Ljapunov Stability Theorem*

The stability of a dynamic system can be analyzed using the Ljapunov stability theorem, named after the Russian mathematician, born 1857 in Yaroslavl, Russia. Consider the unforced ($u = 0$) linear dynamic system described by

$$x' = A \cdot x .$$ (2.155)

Suppose a Ljapunov function of the from

$$V(x,t) = x' \cdot P \cdot x,$$ (2.156)

where P is positive-definite, and x^t is the transpose of x. Then

$$V'(x,t) = x' \cdot P \cdot x' + x' \cdot P \cdot x.$$ (2.157)

Combine (2.74) and (2.157) gives

$$V'(x,t) = x' \cdot P \cdot A \cdot x + A \cdot x \cdot P \cdot x.$$ (2.158)

which is, since $(Ax)' = x'A'$

$$V'(x,t) = x'(P \cdot A + A \cdot P) x. \tag{2.159}$$

If the vector $-O$ is defined by

$$A'P + PA = -O. \tag{2.160}$$

Consider O is negative definite, then $V(x,t)$ is negative definite, and $V(x,t)$ is defined by the Ljapunov function

$$V(x,t) = x' \cdot P \cdot x. \tag{2.161}$$

Example 2.13
Consider a second-order system

$$\frac{d}{dt}\begin{bmatrix} x_1 \\ x_2 \end{bmatrix} = \begin{bmatrix} a_{11} & a_{12} \\ a_{21} & a_{22} \end{bmatrix}\begin{bmatrix} x_1 \\ x_2 \end{bmatrix}. \tag{2.162}$$

Assuming O is an arbitrary symmetric positive-definite matrix, we choose $O = I$, the unit matrix. Equation $A'P + PA = -O$ thus becomes

$$\begin{bmatrix} a_{11} & a_{12} \\ a_{21} & a_{22} \end{bmatrix}\begin{bmatrix} p_{11} & p_{12} \\ p_{21} & p_{22} \end{bmatrix} + \begin{bmatrix} p_{11} & p_{12} \\ p_{21} & p_{22} \end{bmatrix}\begin{bmatrix} a_{11} & a_{12} \\ a_{21} & a_{22} \end{bmatrix} = \begin{bmatrix} -1 & 0 \\ 0 & -1 \end{bmatrix}. \tag{2.163}$$

Multiplying the matrix equation and rewriting gives

$$\begin{bmatrix} 2a_{11} & 2a_{21} & 0 \\ a_{12} & a_{11}+a_{22} & a_{21} \\ 0 & 2a_{12} & 2a_{22} \end{bmatrix}\begin{bmatrix} p_{11} \\ p_{12} \\ p_{22} \end{bmatrix} = \begin{bmatrix} -1 \\ 0 \\ 1 \end{bmatrix}, \tag{2.164}$$

which can be solved for p s having the general form

$$\begin{bmatrix} p_{11} & p_{12} \\ p_{21} & p_{22} \end{bmatrix} = \tag{2.165}$$

$$-\frac{1}{2}[(\mathrm{tr}A)\Delta A]^{-1}\begin{bmatrix} \Delta A + (a_{21})^2 + (a_{22})^2 & -(a_{12}a_{22} + a_{21}a_{11}) \\ -(a_{12}a_{22} + a_{21}a_{11}) & \Delta A + (a_{11})^2 + (a_{12})^2 \end{bmatrix},$$

where $trA = a_{11} + a_{22}$ (the sum of the diagonal terms of the matrix A). The vector P is positive-definite such as

$$p_{11} = \frac{-\Delta A + (a_{21})^2 + (a_{22})^2}{2(trA)\Delta A} > 0 \qquad (2.166)$$

and

$$\Delta A = \frac{(a_{11} - a_{22})^2 + (a_{12} - a_{21})^2}{2(trA)^2\Delta A} > 0 . \qquad (2.167)$$

These inequalities hold if

$$\Delta A = a_{11} \cdot a_{22} - a_{12} \cdot a_{21} > 0 \qquad (2.168)$$

and

$$trA = a_{11} + a_{22} > 0 , \qquad (2.169)$$

which are the required stability conditions of the second-order dynamic system.

2.7 First-Order Linear State-Equation Models

Linear dynamic systems are generally in the form of first-order systems that require only one state variable that describes the system response. First-order linear state systems are dependent on one of the state-space variables x, y, z, or dependent of the state time stamp variable t. Assuming the state equations to be linear, Laplace transforms can be applied.

Let a linear first order state model given as

$$\frac{dx}{dt} = ax + bu, \qquad (2.170)$$

$$y = cx$$

where u is the input variable, y the output variable, x the state variable, and t is the time-dependent variable. The state variable model is defined by the constants a, b, and c, where a is the system parameter, b is the input parameter, and c the output parameter. The block diagram is shown in Fig. 2.5, where the Laplace transform $\frac{1}{s}$ is used to denote the integration operator, while s is used to denote the differential operator $\dfrac{d}{dt}$. Notice that the feedback structure of the state equations is obvious from the block diagram.

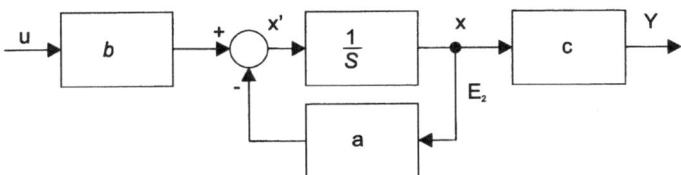

Fig. 2. 5. Block diagram of a first-order linear state systems model

As the state equation is simple the explicit representation of the solution response is determined as a function of the initial state and the input function. Suppose that $U(s)$ denotes the Laplace transform of $u(t)$, $Y(s)$ denotes the Laplace transform of $y(t)$, and $X(s)$ denotes the Laplace transform of $x(t)$. Taking Laplace transforms and solving (2.171) gives

$$sX - X(0) = a \cdot X + b \cdot U \qquad (2.171)$$

$$y = c \cdot X, \qquad (2.172)$$

with $X(0)$ as the initial state. Thus

$$X = \frac{X(0)}{s-a} + \frac{b \cdot U}{s-a}, \qquad (2.173)$$

and

$$Y = \frac{c}{s-a} \cdot X(0) + \frac{c \cdot b}{s-a} \cdot U. \qquad (2.174)$$

Using the notation of the convolution integral in (2.88) gives for $t > 0$

$$x(t) = e^{at} \cdot X(0) + \int_o^t e^{a(t-\tau)} \cdot b \cdot u(\tau) \cdot d\tau \qquad (2.175)$$

and

$$y(t) = ce^{at} \cdot X(0) + \int_o^t ce^{a(t-\tau)} \cdot b \cdot u(\tau) \cdot d\tau, \qquad (2.176)$$

which are functions of the state and the output transient response corresponding to the initial state and the input function. The first term in the expression above is the zero-input response while the second term is the zero-state response. In this case, the overall response is the sum of the zero-input response and the zero-state response. Using the expression

$$G(s) = \frac{c \cdot b}{s-a},$$

(2.177)

called the transfer function of the dynamic system, which leads to the transform of the zero-state output response, which is the product of the transfer function and the transform of the input function. Once $G(s)$ has been found, we have the expression

$$g(t) = c \cdot e^{at} \cdot b,$$

(2.178)

called the weighting or impulse-response function of the dynamic system, whereby the transfer function is the transform of the weighting function. The denominator polynomial of the transfer function

$$d(s) = s-a$$

(2.179)

tends to zero as t $\rightarrow$ 0 for any initial state, then the state equations are said to be stable. If $a > 0$ the state equations are said to be unstable.

Obviously, the state equation indicated with one state variable is a minimal realization for a dynamic system, assuming that $c \cdot b \neq 0$. Any state model for the same dynamic system with more than one state variable would certainly not be minimal.

Example 2.14
Electrical network models usually are based on the assumptions that the components R, L, and C are constant (see also Sect. 1.3.1). The input signal of the electrical network, shown in Fig. 2.6, is the voltage V_0. It is assumed that the applied voltage source V_0 has zero impedance.

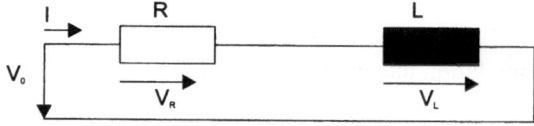

Fig. 2.6. Simple electrical RL network

If Kirchhoffs voltage law is applied, we can say that

$$V_R + V_L - V_0 = 0,$$

(2.180)

with

$$V_R = R \cdot I,$$ (2.181)

and

$$V_L = L \cdot \frac{dI}{dt},$$ (2.182)

that is,

$$R \cdot I + L \cdot \frac{dI}{dt} = V_0,$$ (2.183)

which gives

$$\frac{dI}{dt} + \frac{R}{L} \cdot I = \frac{V_0}{L},$$ (2.184)

a first-order linear differential equation with the integrating factor $e^{\frac{R}{L} \cdot dt} = e^{\frac{R \cdot t}{L}}$, which gives a specialized form,

$$\left(\frac{dI}{dt} + \frac{R \cdot I}{L} \right) \cdot e^{\frac{R^* t}{L}} = \frac{V_0}{L} \cdot e^{\frac{R^* t}{L}},$$ (2.185)

and combining terms, we have

$$(e^{\frac{R^* t}{L}} \cdot I) \cdot \frac{d}{dt} = \frac{V_0}{L} \cdot e^{\frac{R^* t}{L}}.$$ (2.186)

Integrating this equation gives

$$(e^{\frac{R^* t}{L}} \cdot I) = \int \frac{V_0}{L} \cdot e^{\frac{R^* t}{L}} \cdot dt + C,$$ (2.187)

which may also be expressed as

$$I(t) = e^{\frac{-R \cdot t}{L}} \int \frac{V_0(t)}{L} \cdot e^{\frac{R \cdot t}{L}} \cdot dt + C \cdot e^{\frac{R \cdot t}{L}},$$ (2.188)

where C is the constant of integration. Assuming $V_0(t) = V_0$, gives

$$I(t) = \frac{V_0}{R} + C \cdot e^{-\frac{R \cdot t}{L}} \ , \tag{2.189}$$

and if $I(0) = I_0$, gives

$$I(t) = \frac{V_0}{R} + \left(I_0 - \frac{V_0}{R}\right) \cdot e^{-\frac{R \cdot t}{L}} \ . \tag{2.190}$$

The time response of $I(t)$ is shown in Fig. 2.7.

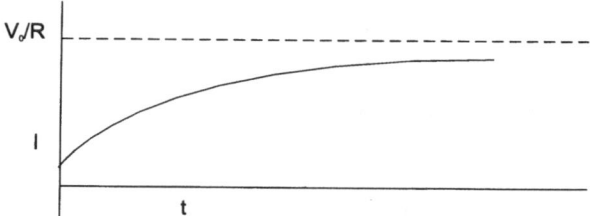

Fig. 2.7. Changes in current of the network shown in Fig. 2.6

From Fig. 2.7 it can be seen that the current tends to a steady state.

Example 2.15

The kidneys perform two major functions: firstly they excrete most of the final products of the body's metabolism, called waste products (toxins), and secondly, they control the concentration of the body fluids. The kidneys each contain about 2400000 nephrons, and each nephron is capable of producing urine itself. Therefore, as in many cases, it is not necessary to discuss the entire kidney but merely the activities in the single nephron to explain the function of the kidney. The basic function of the nephron is to clean the blood plasma of unwanted substances, as it passes through the kidney. The substances that must be cleared include particularly the final products of metabolism, the unwanted substances, such as urea, creatinine, uric acid, and urates. In addition many other substances, such as sodium ions, potassium ions, chloride ions, and hydrogen ions tend to accumulate in the body in excess quantities; hence it is the function of the nephron to clear the plasma of these excesses. The principal mechanism by which the nephron clears the plasma of unwanted substances is:

- It filters a large proportion of the plasma through the glomerular membrane into the tubules of the nephron.
- Then, as the filtered fluid flows through the tubules, the unwanted substances fail to be reabsorbed while the wanted substances, especially the water and many of the electrolytes, are reabsorbed back into the plasma of the peritubular capillaries.

The major kidney function is that the useful portions of the tubular fluid are returned to the blood, while the unwanted portions pass into the urine. For this purpose the tremendous permeability of the glomerular membrane is due to its specific structure by which the fluid is filtered between the fenestra of the capillary endothelial cells, that are so small in diameter that they prevent the filtration of all particles which an average size greater that 160 Å. Then, outside the endothelial cells is a basement membrane composed mainly of a meshwork of mucopoly-saccharide fibrillae. A final layer of the glomerular membrane is a layer of epithelial cells with slit pores, which prevent filtration of all particles with diameter greater than 70 Å.

In the case of a dysfunction of the kidneys, the clearance fails to function properly, and unwanted substances increase to toxic levels. To avoid this dangerous situation for the human body, the unwanted substances removal process can be performed through an artificial kidney machine, called dialyser. In this case blood is taken from the body and passed into the dialyser. A cleaning fluid, called the dialyse, flows in the opposite direction in an adjacent compartment to the blood, being separated by a membrane with small pores, which are large enough to allow the passage of the relatively small molecules of the unwanted substances. The flow rate of the unwanted substances through the menbrane of the dialyser is determined by the differences in concentration on either side, the flow being from high to low concentration. A schematic diagram of the dialyser is shown in Fig. 2.8.

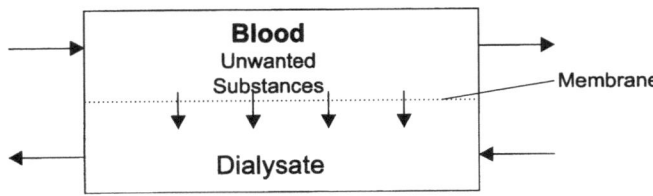

Fig. 2.8. Block diagram of a dialyser (modified after Burghes and Borrie)

The important quantity of the dialyser is the removal rate, which depends on the flow rates of the blood and dialysate through the dialyser, the size of the dialyser and the permeability of the membrane. Assuming that the last two factors are fixed, we may than concentrate on finding the dependence of the removal rate on the flow rate. Let x denote the distance along the dialyser. Hence we may consider what happens in a small section of the dialyser from x to $x + \partial x$, as shown in Fig. 2.9

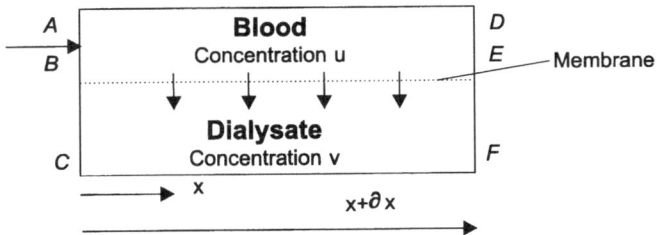

Fig. 2.9. Detailed block diagram of a dialyser (modified after Burghes and Borrie)

The state variables of the dialyser are the concentration of unwanted substances in the blood as well as in the dialysate. Denoting these by u and v respectively, we may assume that these quantities are functions of x, the distance along the dialyser, that is

$$u = u(x)$$
$$v = v(x). \tag{2.191}$$

The law govering the amount of unwanted substances passing through the membrane of the dialyser corresponds to Ficks law (see also Sect, 1.3.5) which states that the amount of substance passing through the membrane is proportional to the difference in concentration.

The difference in concentration across AC is $u(x) - v(x)$, hence the mass transfer through a section of the membrane with unit width and length ∂x from blood to dialyse in unit time is given approximately by

$$k \cdot [u(x) - v(x)] \cdot \partial x. \tag{2.192}$$

The proportional constant is assumed to be independent of x. To derive the equations of the model we consider that mass changes in the element $ADFC$ are in unit time. Hence we can notify in unit time,

mass flow across AB   *mass passing through*   *mass flow across DE*

$=$ $+$

into element   *membrane BE*   *out of element*

Converting this into mathematical terms gives

$$Q_B \cdot u(x) = k \cdot [u(x) - v(x)] \cdot \partial x + Q_B \cdot u(x + \partial x), \tag{2.193}$$

where Q_B is the flow rate of blood through the dialyser machine, which can be expressed as follows

$$Q_B \cdot [\frac{u(x + \partial x) - u(x)}{\partial x}] = -k \cdot [u(x) - v(x)]. \tag{2.194}$$

Considering $\partial x \to 0$, gives

$$Q_B \cdot \frac{du}{dx} = -k \cdot [u - v]. \tag{2.195}$$

Mathematically, we obtain (2.37) by considering a small blood flow through the dialyser machine. In a similar way, if we consider a small dialyse, we obtain

$$-Q_D \cdot \frac{dv}{dx} = -k \cdot [u - v]. \tag{2.196}$$

Equations (2.195) and (2.196) are called coupled differential equations, describing the dynamic model of the dialyser. Adding these equations and rearranging the terms yields

$$\frac{du}{dx} - \frac{dv}{dx} = -\frac{k}{Q_B} \cdot (u - v) + \frac{k}{Q_D} \cdot (u - v) \tag{2.197}$$

hence, if $z = u - v$,

$$\frac{dz}{dx} = -\alpha \cdot z , \tag{2.198}$$

where $\alpha = \dfrac{k}{Q_B} - \dfrac{k}{Q_D}$. For (2.198) we may use the e^λ solution as follows:

$$z = A \cdot e^{-\alpha \cdot x} \tag{2.199}$$

where A is an arbitrary constant. From (2.195) we know

$$\frac{du}{dx} = -\frac{k}{Q_B} \cdot z = -\frac{k}{Q_B} \cdot A \cdot e^{-\alpha \cdot x} \tag{2.200}$$

while integrating gives

$$u = B + \frac{k \cdot A}{\alpha \cdot Q_B} \cdot e^{-\alpha \cdot x} \tag{2.201}$$

where B is an arbitrary constant. We may obtain v from (2.199) using (2.201), since

$$u - v = A \cdot e^{-\alpha \cdot x} , \tag{2.202}$$

the solution of v is given by

$$v = B + \frac{k \cdot A}{\alpha \cdot Q_D} \cdot e^{-\alpha \cdot x} . \tag{2.203}$$

The overall solution depends on the boundary conditions chosen. Consider that the blood has initial concentration u_0 on entry and the dialysate has almost zero concentration on entry, giving

$$u = u_0 \quad at \;\; x = 0 \tag{2.204}$$

$$v = 0 \quad at \;\; x = L$$

where L is the length of the dialyser. The most important factor of the dialyser is the amount of unwanted substances removed (in unit time). This quantity may be represented as

$$\int_0^L k[u(x) - v(x)]dx = -Q_B \int_0^L \frac{du}{dx}dx = -Q_B \int_{u_0}^{u(L)} du = -Q_B[u_0 - u(L)]. \qquad (2.205)$$

These conditions determine the clearance function Cl of the dialyser, given by

$$Cl = \frac{Q_B}{u_0}[u_0 - u(L)]. \qquad (2.206)$$

Applying these condition and after some algebraic transforms we get

$$Cl = Q_B[\frac{1 - e^{-\alpha \cdot L}}{1 - (\frac{Q_B}{Q_D}) \cdot e^{-\alpha \cdot L}}], \qquad (2.207)$$

where

$$\alpha \cdot L = \frac{k \cdot L}{Q_B}[1 - (\frac{Q_B}{Q_D})]. \qquad (2.208)$$

The key parameters for this dialyser model are

$$(i) \frac{Q_B}{Q_D}, \qquad (2.209)$$

the flow-rate ratio

$$(ii) \frac{k \cdot L}{Q_B}. \qquad (2.210)$$

Typical operating conditions of dialysers showing a variation of the respective values. Q_B varies from 100 to 300 ml/min, and Q_D from 200 to 600 ml/min, whilst the ratio $\frac{k \cdot L}{Q_B}$ varies between 1 and 3.

2.8 Second-Order Linear State-Equation Models

Real-world systems are complex and in most cases, they cannot be described by first-order differential equations. This requires the formulation of higher-order differential equations for more accurately description of real-world systems. Consider the mathematical model of a dynamic system given by

$$\frac{d^2x}{dt^2} + a \cdot \frac{dx}{dt} + b \cdot x = f(t), \qquad (2.211)$$

which is a second order linear differential equation with constant parameters a and b, and function $f(t)$ is a specified function of t. Solving differential equations of this type involves finding one particular solution $x_p(t)$ of (2.211). We will consider the difference between the so-called general solution $x(t)$ of (2.211) and the so-called particular solution $x_p(t)$. Let

$$y(t) = x(t) - x_p(t), \qquad (2.212)$$

then the second-order model, which is a sufficiently accurate representation of the real-world system, yields

$$\frac{d^2y}{dt^2} + a \cdot \frac{dy}{dt} + b \cdot y = \frac{d^2x}{dt^2} + a \cdot \frac{dx}{dt} + b \cdot x - (\frac{d^2x_p}{dt^2} + a \cdot \frac{dx_p}{dt} + b \cdot x_p), \qquad (2.213)$$

while both, $x(t)$, and $x_p(t)$, satisfy (2.211). Hence the function $y(t)$ satisfies the associated homogenous equation

$$\frac{d^2x}{dt^2} + a \cdot \frac{dx}{dt} + b \cdot x = 0. \qquad (2.214)$$

We may write $y(t)$ as $x_c(t)$ – this function is called the complementary function – thus the general solution of (2.211) yields

$$x(t) = x_c(t) + x_p(t). \qquad (2.215)$$

Example 2.16
Determine the general solution of

$$\frac{d^2x}{dt^2} + a \cdot \frac{dx}{dt} + b \cdot x = 0, \qquad (2.216)$$

by using the Lagrange criterion $x(t) = e^{\lambda t}$, where λ is a constant to be determined. To satisfy the differential equation

$$e^{\lambda t}\left(\lambda^2 + a \cdot \lambda + b\right) = 0, \tag{2.217}$$

should be true for all appropriate t, we obtain

$$\lambda^2 + a \cdot \lambda + b = 0, \tag{2.218}$$

which is a quadratic equation in λ, called the auxilliary equation, with two solutions, λ_1 and λ_2, gives $e^{\lambda_1 \cdot t}$ and $e^{\lambda_2 \cdot t}$ which are linearly independent. Hence the general solution of (2.214) has the form

$$x(t) = A \cdot e^{\lambda_1 \cdot t} + B \cdot e^{\lambda_2 \cdot t}. \tag{2.219}$$

Depending on whether the square of (2.218) is positive, zero or negative, we obtain three cases:

Case 1: $a^2 - 4b > 0$:
For this case exist two real solutions, given by

$$\frac{d^2 x}{dt^2} + 3 \cdot \frac{dx}{dt} + 2 \cdot x = 0, \tag{2.220}$$

with the auxilliary equation

$$\lambda^2 + 3 \cdot \lambda + 2 = 0, \tag{2.221}$$

which gives

$$(\lambda + 2)(\lambda + 1) = 0, \tag{2.222}$$

with $\lambda_1 = -2$, and $\lambda_2 = -1$, and the general solution is

$$x(t) = A \cdot e^{-2t} + B \cdot e^{-t}. \tag{2.223}$$

Case 2: $a^2 - 4b = 0$:

For this case we have a repeated root $\lambda = -\dfrac{a}{2}$, and a second linear independent

solution may be given by $x \cdot e^{\lambda t}$, yields

$$\frac{d^2x}{dt^2} - 4 \cdot \frac{dx}{dt} + 4 \cdot x = 0, \tag{2.224}$$

with the auxilliary equation

$$\lambda^2 - 4 \cdot \lambda + 4 = 0, \tag{2.225}$$

which gives

$$(\lambda - 2)^2 = 0, \tag{2.226}$$

with $\lambda_1 = e^{2t}$ as a solution, and it may be verified that $\lambda_2 = x \cdot e^{2t}$ is a second solution. The general solution is given as

$$x(t) = A \cdot e^{2t} + B \cdot e^{2t}. \tag{2.227}$$

Case 3: $a^2 - 4b < 0$:
For this case we have complex roots, given as

$$\lambda_1 = \alpha + j\beta \tag{2.228}$$

and

$$\lambda_2 = \alpha - j\beta, \tag{2.229}$$

for which we may write the general solution as

$$\begin{aligned} x(t) &= A \cdot e^{(\alpha + j\beta)t} + B \cdot e^{(\alpha - j\beta)t} = e^{\alpha t} \left(A \cdot e^{+j\beta t} + B \cdot e^{-j\beta t} \right) \\ &= e^{\alpha t} \left[(A + B)\cos(\beta \cdot t) + j(A - B)\sin(\beta \cdot t) \right], \end{aligned} \tag{2.230}$$

since $e^{j\Phi} = \cos\Phi + j\sin\Phi$. Introducing the new arbitrary constants

$$\begin{aligned} C &= A + B \\ D &= j(A - B), \end{aligned} \tag{2.231}$$

yields

$$x(t) = e^{\alpha t} \left(C \cdot \cos(\beta \cdot t) + D \cdot \sin(\beta \cdot t) \right). \tag{2.232}$$

The second-order differential equation

$$\frac{d^2x}{dt^2} + 2 \cdot \frac{dx}{dt} + 2 \cdot x = 0, \tag{2.233}$$

has the auxilliary equation

$$\lambda^2 + 2 \cdot \lambda + 2 = 0, \tag{2.234}$$

which gives

$$\lambda = -1 \pm j, \tag{2.235}$$

where $\alpha = -1$, $\beta = 1$, gives the solution as

$$x(t) = e^{-t}\left(C \cdot \cos(t) + D \cdot \sin(t)\right). \tag{2.236}$$

We have seen how to solve (2.214) for several possible cases. In order to solve (2.211) we need to know only one particular solution, x_p of (2.211). Unfortunately, it is not straightforward to find x_c, and we have to use trial and error methods.

We may also write a second-order differential equation system as a set of first-order state differential equations like

$$\frac{dx_1}{dt} = a_{11}x_1 + a_{12}x_2 + b_1u, \tag{2.237}$$

$$\frac{dx_2}{dt} = a_{21}x_1 + a_{12}x_2 + b_2u, \tag{2.238}$$

$$y = c_1x_1 + c_2x_2, \tag{2.239}$$

with u as the input variable, y as the output variable and x_1 and x_2 as the state variables. The state model, given by the state equations and the output equation, is defined by the constants $a_{11}, a_{12}, a_{21}, a_{22}$, b_1, b_2, c_1, c_2.

Considering the state equations are linear, the Laplace transform can be applied. Supposing that $U(s)$ denotes the Laplace transform of $u(t)$, $Y(s)$ denotes the Laplace transform of $y(t)$ and $X(s)$ denotes the Laplace transform of $x(t)$, giving

$$sX_1 = a_{11}X_1 + a_{12}X_2 + b_1U_1 \tag{2.240}$$

$$sX_2 = a_{21}X_1 + a_{22}X_2 + b_2U_2 \tag{2.241}$$

$$Y = c_1X_1 + c_2X_2. \tag{2.242}$$

where the capital letters denote the Laplace transforms of the corresponding variables, we can write

$$X_1 = \frac{[(s - a_{22})b_1 + a_{12}b_{12}]U}{(s - a_{11})(s - a_{22}) - a_{12}a_{21}}, \tag{2.243}$$

and

$$X_2 = \frac{[a_{21}b_1 + (s - a_{11})b_2]U}{(s - a_{11})(s - a_{22}) - a_{12}a_{21}} \tag{2.244}$$

$$Y = \left\{ \frac{c_1[(s - a_{22})b_1 + a_{12}b_2] + c_2[a_{21}b_1 + (s - a_{11})b_2]}{(s - a_{11})(s - a_{22}) - a_{12}a_{21}} \right\} U. \tag{2.245}$$

Therefore, the transfer function for the linear second-order case is given by

$$G(s) = \frac{c_1[(s - a_{22})b_1 + a_{12}b_2] + c_2[a_{21}b_1 + (s - a_{11})b_2]}{(s - a_{11})(s - a_{22}) - a_{12}a_{21}}, \tag{2.246}$$

where the transform of the zero state output response is the product of the transfer function, and the transform of the input function. The denominator polynomial of the transfer function is the characteristic polynomial of the linear second-order system

$$d(s) = (s - a_{11}){\cdot}(s - a_{22}) - a_{12}{\cdot}a_{21}. \tag{2.247}$$

The zeros of this polynomial are called characteristic zeros of the system; they are also the so-called poles of the transfer function, and are part of the complex domain. Using this polynomial one can show that if the real part of the characteristic zeros are negative, then the input response always tends to zero, and the system is said to be stable. If this does not hold for the characteristic zeros, then the system is said to be unstable.

Example 2.17
Consider the second-order linear mechanical system, as shown in Fig. 2.10. The viscous damping force may be expressed by the equation

$$F_d = c \cdot x ,$$

(2.248)

where c is a constant of proportionality. From the free-body diagram, we obtain the equation of motion

$$m \cdot x'' + c \cdot x' + k \cdot x = F(t) .$$

(2.249)

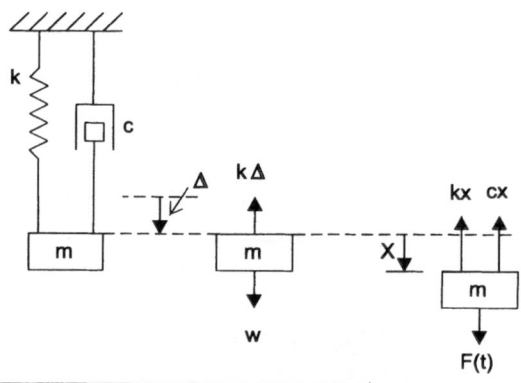

Fig. 2. 10. Second-order linear mechanical system model

The solution of (2.249), determined by several methods in this section, is used here to yield the complete solution of the second-order state-variable equation. If $F(t) = 0$, we obtain the solution of the homogeneous differential equation whose solution corresponds physically to that of free damped vibration. With $F(t) \neq 0$, we obtain the particular solution that is due to the excitation irrespective of the homogeneous solution. In this case we are considering the homogeneous equation in which the forces are equated to zero, gives

$$m \cdot x'' + c \cdot x' + k \cdot x = 0 .$$

(2.250)

The force acting on the mass is

$$F = m \cdot a ,$$

(2.251)

where a is the acceleration and is defined as the second derivative of the state variable x''

$$F = m \cdot x'' .$$

(2.252)

The force acting on the damper is

$$F = c \cdot x' . \tag{2.253}$$

The force due to the spring is

$$F = k \cdot x . \tag{2.254}$$

Thus the net force is equated to zero in (2.250). For the homogeneous equation we may assume a solution, given by

$$x = e^{\lambda t} , \tag{2.255}$$

where λ is a constant, as introduced in Example 2.16. Substituting (2.255) into (2.250) gives

$$\left(m \cdot \lambda^2 + c \cdot \lambda + k \right) \cdot e^{\lambda t} = 0 , \tag{2.256}$$

which is satisfied for all values of t when

$$\lambda^2 + \frac{c}{m} \cdot \lambda + \frac{k}{m} = 0 , \tag{2.257}$$

known as the characteristic equation and has two roots :

$$\lambda_{1,2} = -\frac{c}{2m} \pm \sqrt{(\frac{c}{2m})^2 - \frac{k}{m}} . \tag{2.258}$$

Hence, the overall solution in its general form can be expressed as

$$x(t) = A \cdot e^{\lambda_1 t} + B \cdot e^{\lambda_2 t} , \tag{2.259}$$

where A and B are constants to be evaluated from the initial conditions $x(0)$ and $x'(0)$. Substituting (2.258) into (2.219) gives

$$x(t) = e^{-\left(\frac{c}{2m}\right)t} \left[A \cdot e^{\left(\sqrt{\left(\frac{c}{2m}\right)^2 - \frac{k}{m}} \right)t} + B \cdot e^{\left(\sqrt{\left(\frac{c}{2m}\right)^2 - \frac{k}{m}} \right)t} \right] . \tag{2.260}$$

The first term,

$$e^{-\left(\sqrt{\frac{c}{2m}}\right)\cdot t}$$

(2.261)

is simply an exponentially decaying function over time. The behavior of the terms in parentheses, however, depends on whether the value of the radical is positive, zero or negative, giving three cases:

Case 1: Overdamped oscillation

If the damping term $\left(\dfrac{c}{2m}\right)\cdot t$ is larger than $\dfrac{k}{m}$, the exponents in the previous equation are real numbers, that is no oscillation is possible. We refer to this case as being an overdamped system.

Case 2: Underdamped oscillation

When the damping term $\left(\dfrac{c}{2m}\right)\cdot t$ is less than $\dfrac{k}{m}$, the exponent becomes an imaginary number,

$$\pm j\sqrt{\frac{k}{m}-\left(\frac{c}{2m}\right)^2}\cdot t\,,$$

(2.262)

and here the terms in parentheses are oscillatory. We refer to this case as being an underdamped system, what yields

$$e^{\pm\left(\sqrt{\frac{k}{m}-\left(\frac{c}{2m}\right)^2}\right)\cdot t} = \cos\sqrt{\frac{k}{m}-\left(\frac{c}{2m}\right)^2}\cdot t \pm j\cdot\sin\sqrt{\frac{k}{m}-\left(\frac{c}{2m}\right)^2}\cdot t\,.$$

(2.263)

Case 3: Critical damping
The case between the oscillatory and nonoscillatory motion yields

$$\left(\frac{c}{2m}\right)^2 = \frac{k}{m}\,,$$

(2.264)

and the radical is zero. The damping corresponding to this case is called critical damping, and thus

$$c_c = 2m\sqrt{\frac{k}{m}} = 2m\omega_n = 2\sqrt{km}\,.$$

(2.265)

Any damping can be expressed in terms of the critical damping by a nondimensional number z called the damping ratio, which is shown as the simulation result for several cases of oscillatory motion in Fig. 2.11, 2.12 and 2.13 (see Chap. 4).

Figure 2.11 shows the general nature of oscillatory motion with the respective relations

$$\frac{c}{2m} = \xi\left\{\frac{c_c}{2m}\right\} = \xi\omega_n \tag{2.266}$$

$$\omega_d = \frac{2\pi}{\tau_d} = \omega_n\sqrt{1-\xi^2} . \tag{2.267}$$

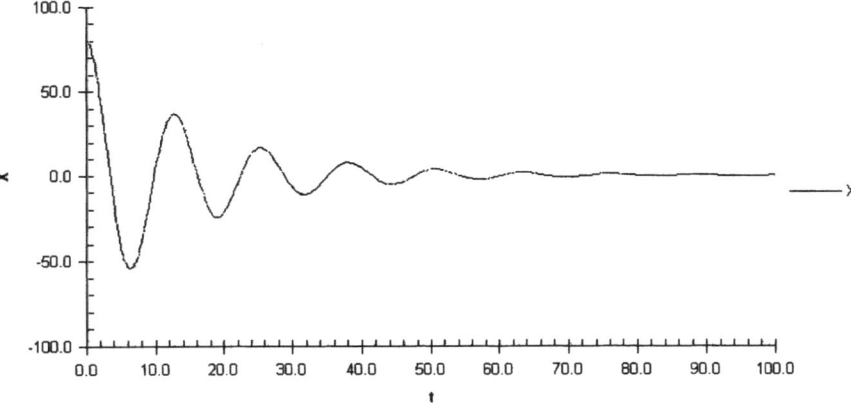

Fig. 2.11. Damped oscillation $z < 1$

Figure 2.12 shows the general nature of the nonoscillatory motion that holds for $z > 1$ the overdamped case with the relation

$$x = A \cdot e^{\left(-\xi+\sqrt{\xi^2-1}\right)\cdot\omega_n \cdot t} + B \cdot e^{\left(-\xi-\sqrt{\xi^2-1}\right)\cdot\omega_n \cdot t} . \tag{2.268}$$

The motion is an exponentially decreasing function over time as shown in Fig. 2.12, for a periodic input with $z > 1$

$$x = (A+B)\cdot e^{\omega_n \cdot t} . \tag{2.269}$$

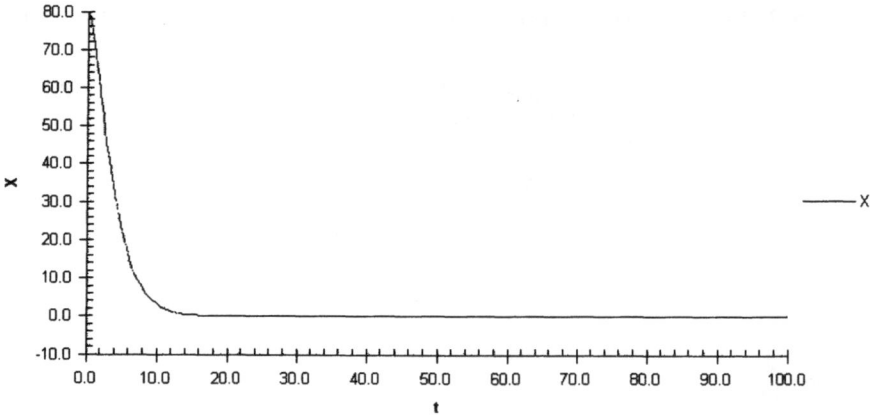

Fig. 2.12. Exponentially decreasing motion for $z > 1$

Figure 2.13 shows critically damped motion for $z = 1.0$ for the three types of response with initial displacement $x(0)$.

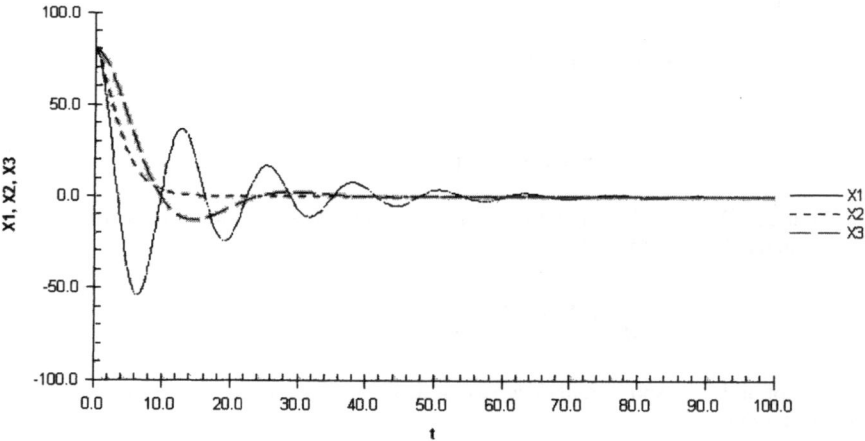

Fig. 2.13. Critical damped motion for $z \geq 0 \leq 1$

The mass damper spring system shown in Fig. 2.10, given by (2.260) in the notation of differential equations can also be represented in terms of isomorphisms, introduced in Sect. 1.6, by an electrical RCL network representation, as shown in Fig. 2.14.

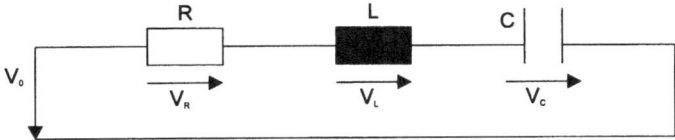

Fig. 2.14. RCL network

The behavior of the RCL network can be described by the differential equation

$$L \cdot q'' + R \cdot q' + \frac{q}{C} = \frac{V(t)}{C},$$
(2.270)

where L is the inductance, R is the resistance, C is the capacitance, q is the charge, and $V(t)$ is the time dependent voltage source.

Comparing the differential equations, (2.219) and (2.270), shows that they have common state variables and parameters which ensures that the dynamic systems are based on the same general mathematical equations, given by

$$A \cdot x'' + B \cdot x' + C \cdot x = D(t),$$
(2.271)

which results in the correspondences given in Table 2.3.

Table 2.3. Correspondences of linear systems

Mathematical model	Mechanical system	RCL network
x: state-space	x: oscillation	q: charge
x': derivative of x	x': velocity	i: current
A: system parameter	m: mass	L: inductance
B: system parameter	c: damper	R: resistance
C: system parameter	k : spring	C: capacitance
$D(t)$: input function	$F(t)$: force	$E(t)$: voltage

As Table 2.3 shows, the mechanical and the electrical RCL network description are based on the general mathematical models of the physical systems. Hence the one system can be used in place of the other one describing the one system. Due to this we have a transform formalism for a dynamic system description, which can be expanded, if necessary.

(2.271) can be rewritten as

$$x'' + \frac{B}{A} \cdot x' + \frac{C}{A} \cdot x = \frac{D(t)}{A}$$
(2.272)

Rewriting this differential equation, which is of second-order, using n first-order differential equations gives

$$x = x_1$$
$$x' = x_1' = x_2$$

(2.273)

$$x'' = x_2' = -\frac{B}{A} \cdot x' - \frac{C}{A} \cdot x + \frac{D(t)}{A},$$

(2.274)

which can be solved easily using simulation, as shown in Chap. 4, since each equation represents a single mode of dynamic behavior, we often refer to the form of state variables as the model-coordinate representation, given by

$$\frac{dx_1}{dt} = x_2,$$

(2.275)

$$\frac{dx_2}{dt} = -2 \cdot c \cdot \omega \cdot x_2 - \omega^2 \cdot x_1 + b \cdot u,$$

$$y = x_1.$$

where c, ω, and b are constant parameters. The parameter c is usually referred to as the damping ratio of the system and ω is referred to as the natural frequency of the system. The transfer function of the dynamic system yields

$$G(s) = \frac{b}{s^2 + 2 \cdot c \cdot \omega \cdot s + \omega^2},$$

(2.276)

and the characteristic polynomial is given by

$$d(s) = s^2 + 2 \cdot c \cdot \omega \cdot s + \omega^2.$$

(2.277)

Assuming $c > 0$ the system is said to be stable. If $0 < c < 1$ the characteristic zeros of the polynomial are real.

The zero input response is determined by using the Laplace transform, which gives

$$sX_1 - X_1(0) = X_2$$

(2.278)

$$sX_2 - X_2(0) = -2 \cdot c \cdot \omega \cdot X_2 - \omega^2 \cdot X_1,$$

(2.279)

hence

$$X_1 = \frac{(s + 2 \cdot c \cdot \omega) \cdot X_1(0) + X_2(0)}{s^2 + 2 \cdot c \cdot \omega + s + \omega^2} \tag{2.280}$$

$$X_2 = \frac{s \cdot X_2(0) - \omega^2 \cdot X_1(0)}{s^2 + 2 \cdot c \cdot \omega \cdot s + \omega^2}.$$

The time-dependent state responses can be determined using the inverse Laplace transform. The state response $x_1(t)$, if $0 < c > 1$, can be expressed by

$$x_1(t) = X_1(0) \cdot e^{-c \cdot \omega t} \cdot \cos\left(\omega \cdot \sqrt{c^2 - 1} \cdot t\right) \tag{2.281}$$

$$+ \frac{c \cdot \omega \cdot X_1(0) + X_2(0)}{\omega \cdot \sqrt{c^2 - 1}} \cdot e^{-c \cdot \omega t} \cdot \sin\left(\omega \cdot \sqrt{c^2 - 1} \cdot t\right),$$

in case that $c > 1$, the time dependent state response can be expressed by

$$x_1(t) = -\frac{\omega \cdot \left(c - \sqrt{c^2 - 1}\right) X_1(0) - X_2(0)}{2 \cdot \omega \cdot \sqrt{c^2 - 1}} \cdot e^{-\omega\left(c + \sqrt{c^2 - 1}\right) \cdot t} \tag{2.282}$$

$$+ \frac{\omega \cdot \left(c + \sqrt{c^2 - 1}\right) \cdot X_1(0) + X_2(0)}{2 \cdot \omega \cdot \sqrt{c^2 - 1}} \cdot e^{-\omega\left(c - \sqrt{c^2 - 1}\right) \cdot t}.$$

$x_2(t)$ can be derived similar to $x_1(t)$. Even in this case the zero-input response is a rather complicated function of the initial state $x_1(0)$ and $x_2(0)$. Instead of calculating, plotting $x_2(t)$ versus $x_1(t)$ gives a better intuitive understanding of the complex system behavior. The x_1 versus x_2 plane is called the phase plane or the state plane. The curves in this plane corresponding to solutions of the state equations are called trajectories.

2.9 Higher-Order Linear State-Space Models*

We now consider a specific class of dynamic systems that are of higher order, meaning they will have state variables higher than $n = 2$, with n as the order of the dynamic system. Furthermore, it will be assumed that the state equations are linear so that Laplace transforms can be used with benefit.

Linear higher-order state-space models can be expressed in the form

$$\frac{dx_1}{dt} = a_{11}x_1 + a_{1n}x_n + b_1 u, \tag{2.283}$$

....

$$\frac{dx_n}{dt} = a_{n1}x_1 + a_{nn}x_n + b_n u,$$

$$y = c_1 x_1 + c_n x_n$$

where u is the input variable, y is the output variable and x_1 and x_2 are the state-space variables. The state-space model is defined by the constants $a_{11}, \dots, a_{1n}, a_{n1}, \dots, a_{nn}, b_1, \dots, b_n$, and $c_1, \dots, c_n$. The function in the $\mathscr{L}$-domain is necessarily the ratio of two polynomial functions

$$G(s) \;=\; \frac{n(s)}{d(s)}. \tag{2.284}$$

Example 2.18
We now consider as an example of a higher-order dynamic system an automobile rear-end suspension system. Mathematical modelbuilding can be used to help identify the effects of changing the struts on the automobile. For simplification, the mass of the tires as well as the axle are neglected.

The automobile rear-end suspension system is a very complex real-world system that behaves like the translation system shown in Fig. 2.15. A bump in the road causes a displacement of the tire, thereby causing a force to be transmitted through the axle to the strut of the vehicle. Since the spring and dampening coefficients of the tire are already set, the strut is the only factor that can be modified. The system can be analyzed using modeling techniques that include free body diagrams, as well as the D'Alembert theorem. Hence it is possible to substitute the various values for the struts and determine the best possible configuration.

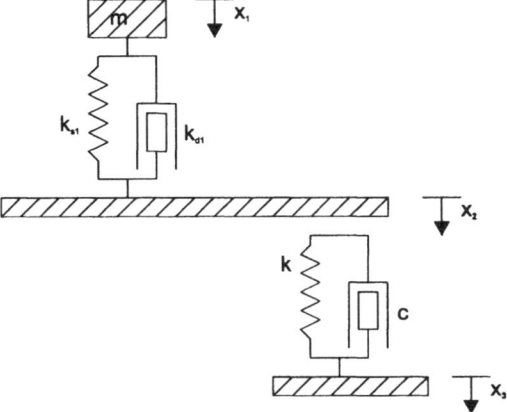

Fig. 2.15. System model of the automobile rear-end suspension system

The schematic diagram, shown in Fig. 2.15, represents the model of an automobile rear-end suspension system, where the mass is the weight of the car and the first spring damper pair is a shock absorber, while the second pair is the tire itself. The force F, represents the reaction force of the road. The suspension system can be easily modified by letting a mass represent the vehicle and connecting it to the axle by a spring damper combination, representing the strut. A mass-less lever represents the axle, which links the strut to the road via the tire. A spring damper pair also represents the tire. The damping coefficient is a fixed property of the rubber while the spring coefficient is a property of the tire pressure.

As shown in Fig. 2.15, a mass is attached to the lever through a spring damper pair. As the lever deflects, forces are exerted through another spring-damper pair to produce the reaction force F. In order to make this model manageable, we neglect the mass of the lever. We also assume that the lever moves through small angles. System analysis can be done based on a free-body diagram showing the forces acting on the mass M_1 as given in Fig. 2.16.

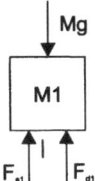

Fig. 2.16. Free-body diagram of mass M_1

If we assume the mass to be moving downward, viscous damping produces an opposing upward force proportional to the velocity $\dfrac{dx_2}{dt}$ as $K_{d1}\left(\dfrac{dx_2}{dt}\right)$, and the spring yields an upward force proportional to the position difference $(x_2 - l)$ as , $K_{s1}(x_2 - x_1)$, where x_1 is the unstretched length of the spring. Downward forces are the result of gravity ($M \cdot g$) and the external force $f_{x_3}(t)$. Using D'Alemberts law, a French mathematician, born 1717 in Paris,

$$\sum F = M \cdot x" , \tag{2.285}$$

we can sum these forces algebraically with the downward direction considered positive because x_2 is increasing in that direction and set that sum equal to the product of the mass and its acceleration $\dfrac{d^2 x_2}{dt^2}$ to yield

$$\sum F = M \cdot \frac{d^2 x_2}{dt^2} = -K_{d1} \cdot \frac{dx_2}{dt} - K_{s1} \cdot (x_2 - x_1) + M \cdot g + f_{x_3}(t) , \tag{2.286}$$

where K_{d1} is the damping coefficient of the first damper, K_{s1} is the spring stiffness of the first spring and $f_{x_3}(t)$ is an external force.

Rearranging (2.286) by placing terms involving x_2 and its derivatives on the left gives

$$M \cdot \frac{d^2 x_2}{dt^2} + K_{d1} \cdot \frac{dx_2}{dt} + K_{s1} \cdot (x_2 - x_1) = M \cdot g + f_{x_3}(t) . \tag{2.287}$$

Neglecting the external force $f_{x_3}(t)$ on the right side of (2.288), we solve this equation for the static operating point x_{2sop} due to the gravitational force only, yields

$$M \cdot \frac{d^2 x_{2sop}}{dt^2} + K_{d1} \cdot \frac{dx_{2sop}}{dt} + K_{s1} \cdot (x_{2sop} - x_1) = M \cdot g . \tag{2.288}$$

Since x_{2sop} is a constant, its derivatives are zero and (2.288) becomes

$$K_{s1}(x_{2sop} - x_1) = M \cdot g . \tag{2.289}$$

Therefore, the static equilibrium position x_{2sop} is

$$x_{2sop} = x_1 + \frac{M \cdot g}{K_{s1}} . \tag{2.290}$$

If we form x_2 as the sum of the constant x_{2sop} resulting from $M{\cdot}g$ and a variation of $\varDelta x$ resulting from $f_{x_3}(t)$ we get

$$x_2 = x_{2sop} + \Delta x_2 . \tag{2.291}$$

Using (2.291) in (2.287) with x_{2sop} from (2.289) yields the linear differential equation for the variation of Δx_2 as

$$M \cdot \frac{d^2(\Delta x_2)}{dt^2} + K_{d1} \cdot \frac{d(\Delta x_2)}{dt} + K_{s1} \cdot \Delta x_2 = f_{x_3}(t) . \tag{2.292}$$

Solving (2.293) for $\dfrac{d^2(\Delta x_2)}{dt^2}$ using Laplace transforms yields

$$s^2 \cdot \Delta X_2(s) = -\frac{K_{d1}}{M} \cdot s \cdot \Delta X_2(s) - \frac{K_{s1}}{M} \cdot s \cdot \Delta X_2(s) + \frac{1}{M} F_{x_3}(s) , \tag{2.293}$$

where $\Delta X_2(s)$ and $F_{x_3}(s)$ are the Laplace transforms of $\Delta x_2(t)$ and $f_{x_3}(t)$, respectiveley. We have shown how to determine the free-body system of mass M_1.

The rear-end suspension system, shown in Fig. 2.15, is much more complex as the simple free-body system, shown in Fig. 2.16. Using D'Alemberts law we have

$$\sum F = K_{d1} \cdot \left(\Gamma \cdot x_2' - x_1' \right) + K_{s1} \cdot \left(\Gamma \cdot x_2 - x_1 \right) - M \cdot g = M \cdot x_1'' , \tag{2.294}$$

where Γ is the ratio of $\dfrac{r_1}{r_2}$, where r_1 is the distance from pivot and r_2 is the distance between spring and damping. Consider $\Lambda_1 = x_1'$ we find

$$\Lambda_1 = \frac{1}{M} \left[K_{d1} \cdot \left(\Gamma \cdot x_2' - x_1' \right) + K_{s1} \cdot \left(\Gamma \cdot x_2 - x_1 \right) - M \cdot g \right]. \tag{2.295}$$

Determining the time response for systems described by state variables, we have to rearrange (2.295) by moving the derivative over to the left side and setting υ_1 equal to it, we obtain the first state equation

$$\Lambda_1' \cdot M - K_{d1} \cdot \left(\Gamma \cdot x_2' - x_1' \right) = K_{s1} \cdot \left(\Gamma \cdot x_2 - x_1 \right) - M \cdot g . \tag{2.296}$$

Let

$$\upsilon_1' = K_{s1} \cdot \left(\Gamma \cdot x_2 - x_1 \right) - M \cdot g . \tag{2.297}$$

Therefore,

$$v_1^{\cdot} = \Lambda_1' \cdot M - K_{d1} \cdot \left(\Gamma \cdot x_2' - x_1' \right). \tag{2.298}$$

Solving v_1' in terms of Λ_1' we have a set of second state equations

$$v_1 = \Lambda_1 \cdot M - K_{d1} \cdot \left(\Gamma \cdot x_2 - x_1 \right) \tag{2.299}$$

$$\Lambda_1 = \frac{1}{M} \cdot \left[v_1 + K_{d1} \cdot \left(\Gamma \cdot x_2 - x_1 \right) \right] \tag{2.300}$$

$$x_1' = \frac{1}{M} \cdot \left[v_1 + K_{d1} \cdot \left(\Gamma \cdot x_2 - x_1 \right) \right]. \tag{2.301}$$

Since we still have x_2 in the equations given above, we may develop an expression for x_2 in terms of x_3 and x_1. This can be achieved by writing equations for the forces of the spring damper pairs. Since the lever is assumed to be mass less we can use a simple ratio to determine the relationship between x_1, x_2, and x_3 as follows

$$F_1 = \Gamma \cdot F_2 = \frac{r_1}{r_2} \cdot F_2. \tag{2.302}$$

Obviously, if we write the equations to determine the forces, we find the compression between the lever and the mass and the lever and the road, respectively, yielding

$$F_1 = K_{d1} \cdot \left(\Gamma \cdot x_2' - x_1' \right) + K_{s1} \cdot \left(\Gamma \cdot x_2 - x_1 \right) \tag{2.303}$$

$$F_2 = K_{d2} \cdot \left(x_3' - x_2' \right) + K_{s2} \cdot \left(x_3 - x_2 \right). \tag{2.304}$$

As a result, the behavior of the automobile rear-end-suspension system considered here results in the following set of differential equations

$$-\Gamma \cdot K_{d2} \cdot x_3' + \Gamma \cdot x_2' \cdot \left(K_{d2} + K_{d1} \right) - x_1' \cdot K_{d1}$$
$$= K_{s1} \cdot x_1 - \Gamma \cdot x_2 \cdot \left(K_{s2} + K_{s1} \right) + \Gamma \cdot x_3 \cdot K_{s2}. \tag{2.305}$$

Rearranging the terms in (2.305) yields

$$K_{d1} \cdot \left(\Gamma \cdot \dot{x}_2 - \dot{x}_1 \right) + K_{s1} \cdot \left(\Gamma \cdot x_2 - x_1 \right)$$

$$= \Gamma \cdot \left[K_{d2} \left(\dot{x}_3 - \dot{x}_2 \right) + K_{d2} \left(x_3 - x_2 \right) \right]. \tag{2.306}$$

With the help of (2.306) we can define $\dot{\upsilon}_2$ and determine the second state equation as follows

$$\dot{\upsilon}_2 = -\Gamma \cdot \dot{x}_3 \cdot K_{d2} + \Gamma \cdot \dot{x}_2 \cdot \left(K_{d2} + K_{d1} \right) - \dot{x}_1 \cdot K_{d1} \tag{2.307}$$

$$\dot{\upsilon}_2 = -\Gamma \cdot x_2 \cdot \left(K_{s2} + K_{s1} \right) + \Gamma \cdot x_3 \cdot K_{s2} + \dot{x}_1 \cdot K_{s1}. \tag{2.308}$$

Solving υ_2 for x_2 and substituting the equation into the previous state equations yields

$$\upsilon_2 = -\Gamma \cdot x_3 \cdot K_{d2} + \Gamma \cdot x_2 \cdot \left(K_{d2} + K_{d1} \right) - x_1 \cdot K_{d1} \tag{2.309}$$

$$x_2 = \frac{\upsilon_0 + x_1 \cdot K_{d1} + \Gamma \cdot x_3 \cdot K_{d2}}{\Gamma \cdot \left(K_{d2} + K_{d1} \right)}, \tag{2.310}$$

and as a result of D'Alembert's law, we obtain

$$
x(t,\upsilon) := \begin{bmatrix} \dfrac{K_{s1} \cdot \upsilon_1 - \left(\upsilon_0 + \upsilon_1 \cdot K_{d1} + \Gamma \cdot x_3 \cdot K_{d2} \right) + \Gamma \cdot x_3 \cdot K_{s2}}{M} \\[4pt] \upsilon_2 + K_{d1} \left\{ \left[\dfrac{\upsilon_0 + \upsilon_1 \cdot K_{d1} + \Gamma \cdot x_3 \cdot K_{d2}}{K_{d2} + K_{d1}} \right] - \upsilon_1 \right\} \\ \hline M \\[6pt] K_{s1} \cdot \dfrac{\upsilon_2 - M \cdot \left\{ \upsilon_2 + K_{d1} \left[\left(\dfrac{\upsilon_0 + \upsilon_1 \cdot K_{d1} + \Gamma \cdot x_3 \cdot K_{d2}}{K_{d2} + K_{d1}} \right) - \upsilon_1 \right] \right\}}{K_{d1}} \end{bmatrix} \tag{2.311}
$$

We can now use these equations to approximate the behavior of the rear-end-suspension system using the Runge Kutta method that can be done, for example, with the rkfixed function of the software package Mathcad. Runge, a German mathematician, was born 1927 in Bremen, Kutta, a German mathematician, was born 1867 in Pitschem Upper Silesia, Germany.

2.10 Nonlinear State-Space Models*

State models of real-world system are often nonlinear. Consequently, the techniques indicated previously that are valuable in the analysis of linear state equations are not of direct applicability here. In fact, it is not possible to give general formulas or even procedures that are guaranteed to prove useful for non-linear state equations.

Assuming a first-order state model of the form

$$\frac{dx}{dt} = f(x,u),$$

$$y = g(x,u),$$

$$(2.312)$$

where x is the state variable, u is the input variable, and y is the output variable.

The state model is defined by the two functions $f(x, u)$ and $g(x, u)$. Since the analysis of the state equations depends most importantly on the differential equation

$$\frac{dx}{dt} = f(x,u),$$

$$(2.313)$$

most attention will be directed toward this equation. The essence is that in some special cases it is possible to integrate the differential equation; in particular consider the special case of a constant input function $u(t) = \bar{u}, t \geq 0$, as introduced by McClamroch, we obtain

$$\frac{dx}{dt} = f(x,\bar{u}).$$

$$(2.314)$$

This first-order differential equation can be determined using the method of separation of variables to obtain

$$\int_{x(0)}^{x(t)} \frac{dx}{f(x,\bar{u})} = \int_{0}^{t} dw.$$

$$(2.315)$$

In some cases, depending on the function $f(x,\bar{u})$, it is possible to describe the integral of (2.315) as an explicit function of $x(t), x(0)$, and $\bar{u}$, given by

$$\phi(x(t), x(0), \bar{u}) = t.$$

$$(2.316)$$

Finally, it can be possible using this implicit equation to determine $x(t)$ as an explicit function of $t, x(0)$ and $\bar{u}$, attempting to integrate the state equation by the separation of variables.

It is not possible to solve (2.316) for any initial state and any input function, which is usually rather easy to determine if there are constant solutions of the state equation. Consider the input is constant $u(t) = \bar{u}$ for $t \geq 0$, then if the condition

$$0 = f(\bar{x}, \bar{u}) \tag{2.317}$$

holds for $\bar{x}$ it follows that if $x(0) = \bar{x}$ then $x(t) = \bar{x}$ for all $t \geq 0$, i.e., $\bar{x}$ is an equilibrium state corresponding to the constant input $\bar{u}$; the corresponding output function is necessarily constant and given by $y(t) = g(\bar{x}, \bar{u})$ for all $t \geq 0$.

One of the most important methods in the analysis of nonlinear state equations is the approximation of the nonlinear equation by a suitable linear equation. Supposing that $\bar{x}$ is an equilibrium state corresponding to the constant input function $\bar{u}$; then the new approximated variables v, w, and z can be defined by

$$\begin{aligned} u &= \bar{u} + v, \\ y &= g(\bar{x}, \bar{u}) + w, \\ x &= \bar{x} + z. \end{aligned} \tag{2.318}$$

Hence we receive the equivalent nonlinear state model

$$\begin{aligned} \frac{dz}{dt} &= f(\bar{x} + z, \bar{u} + v), \\ w &= g(\bar{x} + z, \bar{u} + v) - g(\bar{x}, \bar{u}), \end{aligned} \tag{2.319}$$

where v and w are new input and output variables and z is a new state variable. If the input function u is close to the constant function $\bar{u}$ and if the state function x is close to the equilibrium state $\bar{x}$ then it is reasonable to approximate the functions $f(x, u)$ and $g(x, u)$ by linear relations in x and u, which results in

$$\begin{aligned} f(\bar{x} + z, \bar{u} + v) &\cong f(\bar{x}, \bar{u}) + \frac{\partial f}{\partial x}(\bar{x}, \bar{u})z + \frac{\partial f}{\partial u}(\bar{x}, \bar{u})v, \\ g(\bar{x} + z, \bar{u} + v) &\cong g(\bar{x}, \bar{u}) + \frac{\partial f}{\partial x}(\bar{x}, \bar{u})z + \frac{\partial g}{\partial u}(\bar{x}, \bar{u})v \end{aligned} \tag{2.320}$$

where the partial derivatives are evaluated at $\bar{x}$ and $\bar{u}$. Substituting these approximations into (2.320) results, after simplification, in

$$\frac{dz}{dt} = \left(\frac{\partial f}{\partial x}(\bar{x},\bar{u})\right)z + \left(\frac{\partial f}{\partial u}(\bar{x},\bar{u})\right)v, \tag{2.321}$$

$$w = \left(\frac{\partial f}{\partial x}(\bar{x},\bar{u})\right)z + \left(\frac{\partial g}{\partial u}(\bar{x},\bar{u})\right)v$$

which is a linearized state model. If the functions $u(t)$ and $x(t)$ are close to the constant functions $\bar{u}$ and $\bar{x}$, the variables v and z are small, and the error is less and the linearized state equations given above are valid.

2.11 References and Further Reading

Burghes DN, Borrie MS, (1981), Modelling with Differential Equations, John Wiley & Sons, New York

McClamroch NN, (1980), State Models of Dynamic Systems, Springer, New York, Heidelberg, Berlin

Dorf RC, (1986) Modern Control Systems, Addison-Wesley, Reading

Edwards CH, Penny DE, (2001), Differential Equations and Linear Algebra, Pearson Education/Prentice Hall

Möller DPF, (1992), Modeling, Simulation and Identification of Dynamic Systems (in German), Springer, Berlin, Heidelberg, New York

Möller DPF, Popovic'),D, Thiele G, (1983), Modeling, Simulation and Parameter-Estimation of the Human Cardiovascular System, Vieweg Publ., Braunschweig, Wiesbaden

Ogata K, (1967), Stata Space Analysis of Control Systems, Prentice-Hall, Inc. Englewood

Rowland JR, (1986), Linear Control Systems, John Wiley& Sons, New York

Savant CJ Jr, (1962 Fundamentals of the Laplace Transformation, McGraw-Hill Book Company, Inc., New York

Seely S, (1964), Dynamic Systems Analysis, Reinhold Publishing Corporation, New York

Seireg A, (1969), Mechanical Systems Analysis, International Textbook Company, Scranton, Pennsylvania

Strang G, (1998), Introduction to Linear Algebra, Wellesley-Cambridge-Press

2.12 Exercises

2.1   What is meant by saying that a dynamic system is continuous with time?
2.2   Give an example of an ODE.
2.3   Solve the ODE for an initial-value problem.
2.4   Solve the ODE for a boundary-value problem.
2.5   Give the vector-matrix notation for a linear continuous-time system.
2.6   What is meant by saying that a dynamic system is a linear system?
2.7   Describe a simple oscillatory system by using nonlinear differential equations of second order.
2.8   What is meant by saying that a dynamic system has the order n?
2.9   What is meant by saying that a dynamic system has the degree m?
2.10 What is meant by the term rank?
2.11 A dynamic system can be described by the equations

$$12x + 6y + 4z = 36 \qquad\qquad (2.322)$$
$$-12x + 6y + 8z = 24$$
$$12x + 6y + 8z = 48$$

Prove whether the system has a unique solution or not.
2.12 Define what is meant by the term controllability?
2.13 Define what is meant by the term observability?
2.14 Define what is meant by the term identifiability?
2.15 Define what is meant by the term convolution integral?
2.16 Define what is meant by the term Laplace transform?
2.17 Define what is meant by the term eigenvalues?
2.18 Describe the meaning of the phase plane shown in Fig. 2.2.
2.19 Give the differential equation system describing the phase plane for the state variables x_1 and x_2.t
2.20 Define what is meant by the terms stable, asymptotically stable, and unstable?
2.21 Define what is meant by the term Nyquist criterion?
2.22 Give a brief description for the loci of the Nyquist criteria as shown in Fig. 2.4.
2.23 Define what is meant by the term Routh Hurwitz criterion?
2.24 A dynamic system can be described by the equation

$$d(s) = s^3 + s^2 + 8 \cdot s + 60 \cdot s . \qquad\qquad (2.323)$$

Prove by using the Routh Hurwitz criterion whether the system is stable or not.
2.25 The shock absorber spring system of an automobile as shown in Fig. 2.10 can be described by the second order-differential equation system in (2.48). If the mass has been displayed from its rest position to a distance x_m and the exter-

nal force $F(t)$ is suddenly released, we get (2.49) which should be solved for a damped linear oscillation for position, velocity, and phase space.

3 Mathematical Description of Discrete-Time Systems

3.1 Introduction

Consider a discrete-time system as a system in which the state variables change only at a discrete set of points in time, as shown in Figs. 1.20 and 1.21. For a deeper understanding of discrete-time systems, a number of terms are helpful to use, such as:

- Entity, which is an object of interest
- Attribute, which characterizes the property of an entity
- Activity, which represents a time period of a specific length
- Event, which is defined as an instantaneous occurrence that can change the state of the system
- State, which characterizes that collection of variables necessary to describe the system at any time, relative to the objects of the study
- Status of elements, such as busy or idle

Let the formalisms describing discrete-time systems be of the same flexibility as general-purpose computer programming languages. Hence it is not accessible for any analytical method to calculate the trajectories for model quantities in its original form. However, with some restrictions in modeling such calculations can be done up to a certain grade. To succeed in this analytical way some assumptions of the system behavior and some restrictions in the design of the models are necessary. Restrictions in modeling will lead to restrictions in the significance of the results of a simulation study. On the other hand, a restricted set of facilities for the model description enables the designer of a simulation system embedding more support into the system itself during the modeling process by ready-made components with blocks of model code.

3.2 Statistical Models in Discrete-Time Systems

In real-world systems analysis modelers often have to state that real-world systems are of an imposing complexity and variety, where events more or less never repeat exactly. Hence real-world modeling requires skills in recognizing the statistical behavior of the various phenomena that must be incorporated into the model, based on a mathematical description that has a normalization in models as abstract representation, helping to understand the real-world phenomena. Therefore, part of the scientific work consists in formalization, which yields the respective mathematical description of models of the real-world systems studied. This task is scientifically oriented in order to gain a better understanding of real- world phenomena through an abstract representation. This can be based on experimentation and observation, to create representations and laws that formalize a verified hypothesis concerning the real-world phenomena. These formalizations are only useful if they succeed in seizing the essential features of the real-world systems. Then they permit extrapolation that allows generalization from past experience to future events from which one can learn how to manipulate the real- world system for ones purposes. The formalization used to mimic random processes of real-world systems are the basic concepts in probability and statistics as they relate to discrete-time or discrete-event systems modeling and simulation.

3.2.1 Random Variables

Let X be a variable that can assume any of several possible values over a range of such possible values. Assuming X is a variable in which the range of possible values is finite or countably infinite, the probability mass function of X is

$$p(x_i) = P(X = x_i \tag{3.1}$$

$$p(x_i) \geq 0$$
$$\sum_i p(x_1) = 1 \ . \tag{3.2}$$

Assuming X is a variable in which the range of possible values is the set or real numbers, the probability mass function of X is

$$P(a \leq X \leq b) = \int_a^b f(x)dx \tag{3.3}$$

$$f(x) \geq 0 \quad \text{for all } x \text{ in } \Re \tag{3.4}$$

$$\int_{\Re_x} f(x) = 1 .$$

Let $\dot{E}(X)$ be the expected value of the random variable X. The expectation function is given by

$$E(X) = \sum_i x_i p(x_i), \tag{3.5}$$

if X is discrete, and by

$$E(X) = \int_{-\infty}^{\infty} x f(x) dx, \tag{3.6}$$

if X is discrete. The expected value is also called the mean, denoted by μ. Defining the n-th moment of X results in the variance of the random variable X

$$V(X) = E\left[(X - E(X))^2\right] = E\left[(X - \mu)^2\right]. \tag{3.7}$$

Statistical models in discrete-event simulation include queuing systems, inventory systems, reliability and maintainability systems, as well as manufacturing systems.

3.2.2 Distributions

Random variables can be based on continuous distributions or discrete distributions that are used to describe random phenomena. For continuous distributions one can use the

- Erlanger distribution
- Exponential distribution
- Gamma distribution
- Normal distribution
- Uniform distribution
- Weibull distribution

For continuous distributions one can use the

- Bernoulli distribution
- Binomial distribution

- Geometric distribution
- Poisson distribution

Example 3.1
Assuming the number X of defective assemblies in the sample n of manufactured assemblies is binomially distributed. Let $n = 30$ and the probability of defective assembly $p = 0.02$ results in

$$P(X \leq 2) = \sum_{x=0}^{2} \binom{30}{x}(0.02)^x (0.98)^{30-x} = \tag{3.8}$$

$$0.5455 + 0.3340 + 0.0988 = 0.9783$$

The mean number of defectives in the sample is

$$E(X) = n \cdot p = 30 \cdot 0.02 = 0.6 . \tag{3.9}$$

The variance of defectives in the sample is

$$V(X) = n \cdot p \cdot q = 30 \cdot 0.02 \cdot 0.98 = 0.588 . \tag{3.10}$$

Example 3.2
Assuming a class of vacuum pumps has a time to failure that follows the Weibull distribution, named after the Swedish engineer, born 1887 in Schleswig Holstein, Germany, with $\alpha = 200$ h, $\beta = 0.333$, and $v = 0$. The mean time to failure yields for the mean Weibull distribution

$$E(X) = v + \alpha \Gamma\left(\frac{1}{\beta}+1\right) = 200\Gamma(3+1) = 200(3!) = 1200 \ h , \tag{3.11}$$

and for the variance Weibull distribution

$$V(X) = \alpha^2 \Gamma\left(\frac{2}{\beta}+1\right) - \left[\Gamma\left(\frac{1}{\beta}+1\right)\right]^2 . \tag{3.12}$$

The probability that a vacuum pump fails before 200 h can be calculated based on the cumulative distribution function of the Weibull distribution as follows

$$F(x) = 1 - e^{-\left(\frac{x-v}{\alpha}\right)^\beta} = 1 - e^{-\left(\frac{2000}{200}\right)^{0.333}} = 1 - e^{-2.15} = 0.884 . \tag{3.13}$$

3.3 Discrete-Event Simulation of Queuing systems

One of the most well known application areas in modeling and simulation of discrete-event systems are simulations of queuing systems that represent random-based processes. The key element of a queuing system are the customers and the servers. Hence a queuing system can be described by the following attributes:

- The calling population, which represents the population of potential customers
- The system capacity, which is the limit on the number of customers the discrete-event system can accommodate at any time
- The composition of the arrivals, which can occur at scheduled times or at random times
- The queuing discipline, which is the behavior of the queue in reaction to its current state
- The service mechanism, which means that the service times may be constant or of some random duration

The attributes represent the elements of discrete-event systems. Elements are necessary to describe real-world systems and can be classified as:

- Permanent elements, which are
  - Queues
  - Stations
  - Servers
  - etc.

- Temporary elements, which are
  - Jobs
  - etc.

- Times, which are
  - The inter arrival times between two jobs following each other
  - The service time needed in the server
  - etc.

These elements can be represented through graphical symbols, such as circles for an indication of the waiting line in a queue, or blocks, which represent the servers, etc. By combining elements complex queuing nets can easily be built up visualizing the way a job moves through a net of service stations. Moreover, the symbols of the elements describe the limits of the queuing models by sources and sinks for jobs, and the possibility to branch and to merge the flow of the jobs.

The intention of queuing models is to gain information about characteristic quantities that describe the workload of the servers, or the time the jobs need to pass through the discrete-event system. High workload of stations makes the discrete-event system highly efficient for the operator, but increases the waiting time for jobs. Hence modeling and simulation is necessary for the prediction of these values to be able to parameterize the system in an acceptable manner.

Example 3.3
Consider a simple single-channel telecommunication system with the following elements: a calling population, a waiting line, and a server. Let the calling population be infinite; that is, if a unit leaves the calling population and joins the waiting line or enters service, there is no change in the arrival rate of other units that may need service. Arrivals for service occur one at a time if we use a randomized schedule; once they join the waiting line, they are eventually served. In this simple single-channel telecommunication model service times are assumed to be of some random length according to a probability distribution that does not change over time. Assume that the system capacity has no limit, meaning that any number of units can wait in line. Furthermore, the units should be served in the order of their arrival by a single server, which results in the first-in, first-out (FIFO) service schedule.

Let arrivals and services be defined by the distribution of the time between arrivals and the distribution of the service times, respectively. For any simple single-channel telecommunication queue, such as the one of Example 3.3, the overall effective arrival time has to be less than the total service rate, otherwise the waiting line will grow without bound. If queues grow without bound, they are called explosive or unstable. In cases where the arrival time will be for short terms greater than the service rate, there is a need for queuing networks with routing capabilities.

Queuing systems can be represented by terms such as stable, event, simulation clock, etc. Hence, the state of the queuing system is represented by its number of units as well as the status of the server, which can be busy or idle. An event then represents a set of circumstances that cause an instantaneous change in the state of the system. In case study Example 3.2 there are only two possible events that can affect the state of the single-channel telecommunication system, the arrival event, which means the entry of a unit into the system, and the departure event, meaning the completion of service on a unit. Furthermore, a simulation clock is used to track simulated time.

If a unit enters the discrete-event system, the unit can find the server either busy or idle, which results in two cases:

1. The unit begins service immediately if the server is idle.
2. The unit enters the queue for the server immediately if the unit is busy.

It is not possible for the server to be idle and the queue to be not empty, which can be interpreted as a third case. The results of which can be expressed in a matrix form for the potential unit actions upon arrival, as shown in Table 3.1.

Table 3.1. Cases of unit actions upon arrival (For details see text)

		Queue status	
		Not empty	Empty
Server status	Busy	2	2
	Idle	3	1

After the completion of a service, as shown in Table 3.1, the server can become idle or remain busy with the next unit. The relationship of these two outcomes to the status of the queue is shown in Table 3.2. If the queue is not empty, another unit can enter the server keeping him busy, or if the queue is empty, the server will be idle after a service is completed, which is indicated by the disjunctive indication of case 1 or 2. Again, it is impossible for the server to become busy if the queue is empty when a service is completed, which is indicated by case 3.

Table 3.2. Server outcomes of Table 3.1 after service completion (For details see text)

		Queue status	
		Not empty	Empty
Server status	Busy	1 or 2	3
	Idle	3	1 or 2

Simulating queuing systems requires the stipulation of an event list for determining what will be next. This event list tracks the future times at which different types of events occur. Hence the simulation system is able to calculate the respective simulation clock time for arrivals and departures. If events occur at random times, the randomness needed can be realized through random numbers. Random numbers are distributed uniformly and independently on the interval $[0,1]$. Random numbers are uniformly distributed on the set $\{0, 1, 2, 3, ..., 8, 9\}$. They can be generated with the respective queuing systems simulation packages.

Example 3.4
For the simple single-channel telecommunication queuing systems in Example 3.1, the inter-arrival times and service times can be generated from the distribution of random variables. Consider having seven customers with the inter-arrival times 0, 2, 6, 4, 3, 1, 2. Based on the inter-arrival times the arrival times of the seven customers at the queuing systems results in 0, 2, 8, 12, 15, 16, 18. Due to these boundaries the first customer arrives at clock time 0, which sets the simulation clock in operation. The second customer arrives two time units later, at the clock time 2, the third customer arrives six time units later, at the clock time 8, etc. The second time values of interest in Example 3.2 are the service times that are generated at random from a distribution of service times. Let the possible service times be one, two, three, and four time units. Hence we are able to mesh the inter-arrival times and the service times, simulating the simple single-channel telecommunication queuing system, which results in the schedule, shown in Table 3.3.

Table 3.3. Simulating the single-channel queuing system (For details see text)

Customer no.	Arrival time	Service begins	Service time	Service ends
1	0	0	4	4
2	2	4	3	7
3	8	8	2	10
4	12	12	4	16
5	15	17	3	20
6	16	20	2	22
7	18	22	4	26

As shown in Table 3.3, the first customer arrives at clock time 0 and service starts immediately, which requires four time units. The second customer arrived at clock time 2, but service could not begin until clock time 4. This occurred because customer 1 did not finish service until clock time 4. The third customer arrives at clock time 8 and is finished at clock time 10, etc. The strategy that serves customers in Example 3.2 is based – again – on the first-in, first-out (FIFO) basis, which keeps track of the clock time at which each event occurs.

Furthermore, the chronological ordering of events can be determined from Table 3.3, as records of the clock times of each arrival event and of each departure event, depending on the customer number. The chronological ordering of events is needed as a basis concept for the realization of discrete-event simulation systems.

Further interesting parameters for discrete-event simulation systems are the:

- Workload, which represents the percentage of the simulation time a resource was working
- Throughput, which is the number of jobs per time unit that leave the system
- Mean waiting time
- Mean time in system
- Queue length
- Mean number of waiting jobs
- etc.

Moreover, knowledge of the layout of the queuing networks is of importance for the use of discrete-event simulation systems. The layout depends on:

- Open-queuing systems, which have sources and sinks. The job-pass through the queuing net and leave it when all demands are satisfied. Typical examples of open-queuing systems are production lines, where the jobs are the raw materials that have to be treated by certain operations and leave the system as ready-made products.
- Closed-queuing systems, which are identified by a closed loop in which the jobs move through the queuing net. The number of jobs are fixed for the whole simulation time. A typical example of a closed-queuing system

is a multi user system with n terminals and a single central processing unit (CPU). The jobs circle between the terminals and the CPU; their number stays constant all over the simulation time.

The way the jobs are processed through the queues is based on specific concepts that show how to organize queues. The most common concepts are:

- First-in, first-out (FIFO)
- Last-in, last-out (LILO)
- Shortest job first
- Round robin
- Shortest remaining processing time
- Multi level feed back
- Service in random number

Simulating queuing systems generally requires the maintenance specifying the dynamic behavior of the discrete-event system, which can be done using simulation tables, designed for the problem being investigated. Hence, the content of the simulation table depends from the observed system and can give answers such as:

- The average waiting time of a customer is determined by the total time the customers wait in queue, divided by the total numbers of customers.
- The average time a customer spends in the queuing system is determined by the total time the customers spend in the queuing system, divided by the total numbers of customers.
- The average service time is determined by the total service time, divided by the total number of customers.
- The average time between arrivals is determined by the sum of all times between arrivals, divided by the number of arrivals − 1.
- The probability a customer has to wait in the queue is determined by the number of customers who wait in queue, divided by the total number of customers-
- The fraction of idle time of the server is determined by the total idle time of the server, divided by the total run time of the simulation.
- etc.

Moreover, it has to be decided whether:

- It is possible to leave the queue without being served at all.
- The number of jobs in the queue is limited.
- There are priorities for the jobs (static and/or dynamic).
- It is possible for a job with high priority to interrupt the service for a low-priority one and to occupy the service station immediately when entering the queue.
- etc.

To standardize the description of queuing models, Kendall introduced notation for queuing systems, which includes information about the processes such as job arrivals and the distribution of the time that is needed in the server. This standard notation is based on a five-character code

$$A/B/c/N/k ,\qquad (3.14)$$

where A represents the interarrival-time distribution, B is the service-time distribution, c is the number of parallel servers – of a station ($c \geq 1$) –, N represents the system capacity and k is the size of the population.

The elements of queues and servers are represented in the term "station". Hence a station can be described, using Kendalls notation, as:

$$A/B/c\ -\ < strategy > \ [pre-emptive]\ [\max imal\ queue-length] .\qquad (3.15)$$

Kendall, a Swedish mathematician, was born 1907 in Conny Palm.

The short forms for the mostly used distributions of queuing systems are:

- G: general (no limitation concerning the distribution)
- D: deterministic
- M: exponential distribution
- etc.

Example 3.5
Kendalls notation can be used as follows:
$M/D/1$: which represents the simplest example, the FIFO principle.
$M/G/2$: which represents a so-called pre-emptive systems example, the LCFS princple.
$MM/1/\infty/\infty$: which indicates a single-server system with unlimited queue capacity and infinite calling population. Interarrival times and service times are exponentially distributed.

Performance measures for queuing systems are of importance for the validation of discrete-event simulation models, which are too complex to be modeled analytically. A queuing system typically has two stages of behavior, short-term or transient behavior, followed by long-term or steady-state behavior. If a queuing system is started it must operate for a period of time before reaching steady-state conditions. A discrete-event simulation model must run for a sufficiently long period of time to exceed the transient period before measures of steady-state performance can be determined, which results in a specific queuing notation that contains

- Steady-state probability of having n customers in system
- Probability of n customers in system at time t
- Arrival state

- Effective arrival state
- Effective rate of one server
- Server utilization
- Interarrival time between customers n-1 and n
- etc.

Based on this notation for the various classes of queuing system models a performance analysis can be introduced based on:

- Steady-state parameters for $M/M/1$ queues
- Steady-state parameters for $M/G/1$ queues
- Steady-state parameters for $M/E_k/1$ queues
- Steady-state parameters for $M/D/1$ queues
- Steady-state parameters for $M/M/1/N$ queues
- Steady-state parameters for $M/M/c$ queues

For the first three queues the service times are exponentially distributed for M, generally distributed for G and Erlangen distributed for E. For the fourth case, D, the service times are constant. For $M/M/1/N$ queues, the system capacity is limited to N, for $M/M/C$ queues the channels c operate in parallel.

While simulation of queuing systems often is done manually, based on simulation tables, one has to decide, comparing the difference between possible analytical and simulative solutions, which of the two methods should be used. This comparison can be restricted, reflecting limitations and advances.

Limitations of the analytical solutions are:

- Preconditions, concerning the distribution of the inter arrival times and time to be served
- Substantial problems, to handle queuing strategies
- Numerical efforts, to solve the state equations
- Results only for the steady state
- Only mean values, no predictions about the minimum and the maximum or the history of individual jobs

An advantage of analytical solutions is:

- Results, which are general for use of all possible parameterizations

A limitation of simulation-based solutions is:

- A single simulation run only corresponds to a single random sample, all simulation results are singular solutions for the given initial state, they are not general results for the whole model

Advantages of simulation-based solutions are:

- No preconditions concerning the distributions
- Any strategy can be reproduced
- Observation of the individual history for jobs and queue lengths possible

Another important class of simulation problems of queuing systems involves inventory systems. An inventory system has a periodic review of length at which time the inventory level is observed, and an order that is made to bring the inventory up to a specified level of amount in inventory. At the end of the review period, an order quantity is placed.

Example 3.6
An inventory problem deals with the purchase and sale of newspapers. The paper sellers may buy the papers for 30 cents each and sell them for 50 cents each. Newspapers not sold at the end of the day are sold as scrap for 5 cents each. The problem to be solved with this inventory system is to determine the optimal number of papers the newspaper seller should purchase, which can be done by simulating the demands for a month and recording the profits from sales each day. The profit P can easily be calculated as follows:

$$P = \begin{pmatrix} sales \\ revenue \end{pmatrix} - \begin{pmatrix} cost\ of \\ papers \end{pmatrix} - \begin{pmatrix} profit\ loss \\ excess\ demand \end{pmatrix} + \begin{pmatrix} salvage\ sale \\ scrap\ papers \end{pmatrix}. \quad (3.16)$$

Based on Example 3.6 the primary measure of the effectiveness of inventory systems, which are total system costs, can be extracted. Contributing to total inventory cost are the following:

- Item cost which represents the actual costs of the Q items acquired
- Order costs which are the costs of initiating a purchase or production setup
- Holding costs which are the costs for maintaining items in inventory
- Shortage costs represent the costs of failing to satisfy demand

In general, inventory problems of the type discussed above are often easier to solve then queuing problems.

Furthermore, discrete-event simulation of queuing models is based on simulation languages, which use programming languages. Assuming a model consists of two events: customer arrival and service completion. The events can be modeled with event subroutines, which are ARRIVE and DEPART, respectively. These subroutines include an INCLUDE statement, and can be described with generalized statements as follows

```
SUBROUTINE ARRIVE
INCLUDE 'mm1.dcl'
...
Schedule next arrival
....
```

```
IF (SERVER.EQ.BUSY) THEN
.....
END

SUBROUTINE DEPART
INCLUDE 'mm1.dc1'
...
Check whether the queue is empty or not
....
IF (NIQ.EQ.0) THEN
.....
SERVER = IDLE
....
ELSE
Queue is not empty
NIQ=NIQ+1
....
END
```

3.4 Petri-Nets

Another formalism for the analysis of the behavior of discrete-time systems are Petri nets, which offer a model specification paradigm that lies in between an analytical-solvable model description and a pure simulation-based solution. In their origin, Petri-nets have been used for the analysis of synchronization problems within a set of parallel processes. Nowadays, the general definition was extended in various directions and various types of nets are applied to mostly all application areas of discrete-time simulation problems.

At the very beginning, the German mathematician Petri, born 1926 in Leipzig, Germany, published in 1962 a paper in which the definition of a Petri net was given. Petri-nets represent systems by means of a net structure, a weighted bipartite directed graph that specifies the static part of the system, which contains two different types of knots: points and transitions.

Definition 3.1
A Petri-net is a triple that contains two kinds of nodes, points (P) and transitions (T), as well as flow relations (F), called edges, yields

$$N := (P,T,F), \tag{3.17}$$

with

$$P \cap T = \Phi \qquad (3.18)$$

$$F \subseteq (P \times T) \cup (T \times P). \qquad (3.19)$$

∎

Consider the node type point as a container for data or information that will be symbolized by a circle. A transition can be assumed as a data-processing unit and will be symbolized by a square. Points and transitions are connected by means of arcs. Arcs are only allowed to connect a point with a transition, or a transition with a point. The first type of point is regarded as an input point for a transition, while the latter is regarded as an output point. Each arc is labeled with a weight that is a positive integer value. The marking or state of the net corresponds to the assignment of one or more tokens to each point. Another state can be reached as a consequence of the firing of a transition. Firing of a transition is due to the consumption of tokens from all its input points and the production of tokens for all of its output points. A transition is only allowed to fire if there are sufficient tokens available with its inputs points, where the number of required tokens is determined by the weight of the arcs. Hence we obtain:

$$P = \{p_1, p_2, p_3\}, \qquad (3.20)$$

$$T = \{t_1, t_2\}, \qquad (3.21)$$

$$F = \{(p_1, t_1)\ (p_1, t_2)\ (p_2, t_2)\ (t_1, p_3)\ (t_2, p_3)\}, \qquad (3.22)$$

which can be represented in the general form of a Petri net, as shown in Fig. 3.1.

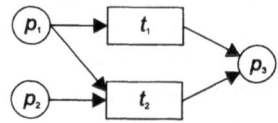

Fig. 3.1. Petri-Net representation of F

With this formalism typical aspects of processes become describable:

- Parallelism
- Generation and consuming of discrete objects during the process lifetime
- Conditions for executing operations or actions
- Causal dependencies between processes by using common resources

By their graphical representation a very descriptive model design is possible and the animation of the net is an obvious feature for representation.

Example 3.7
The points and transitions of the Petri-net, shown in Fig. 3.2, should be given.

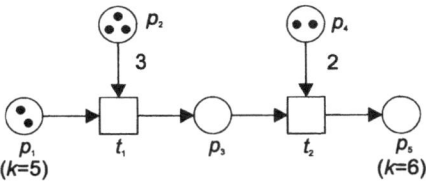

Fig. 3.2. Petri-net representation of points and transitions

The Petri-net shown in Fig. 3.2 has the following nodes and transition relations:

$$P = \{p_1, p_2, p_3, p_4, p_5\}, T = \{t_1, t_2\}, \tag{3.23}$$

$$F = \{(p_1, t_1)(p_2, t_1)(t_1, p_3)(p_3, t_2)(p_4, t_2)(t_2, p_5)\},$$

$$K(p_1) = 5, \ K(p_5) = 6 \quad \textit{for all other nodes the capacity} \to \infty$$

$$W(p_2, t_1) = 3, \ W(p_4, t_2) = 2 \quad \textit{all other edges are weighted } 1$$

$$M_0(p_1) = 2, \ M_0(p_2) = 3, \ M_0(p_3) = 0, \ M_0(p_4) = 2, \ M_0(p_5) = 0$$

which are the formal representation of a 6-tupel nodes-transition net

$$N := (P, T, F, K, W, M_0), \tag{3.24}$$

with

$$K : P \to N \cup \{\infty\}, \tag{3.25}$$

which represents the points-capacity function, and

$$W : F \to N, \tag{3.26}$$

which is the points-weighting function, and

$$M_0 : P \to N_0 \cup \{\infty\}, \tag{3.27}$$

which represents the initial marking of points with

$$\forall p \in P : M_0(p) \le K(p). \tag{3.28}$$

Petri nets fill many of the needs of systems modeling which is why they have extensions due to place/transistion nets with color, priority and time, which form the class of so-called high-level Petri-nets (HLPN). HLPN can be used for modeling more complex discrete-event systems models, while classical Petri-nets tend to be too large to handle. These extensions are

- Color, to describe tokens that can have one or more values, forming the so-called class of colored Petri-nets
- Time, which is included by associating time stamps with tokens, representing deterministic timed Petri-nets or random time as in stochastic Petri-nets
- Hierarchy, which enables the structuring of large systems

Compared with classical Petri-nets high-level Petri-nets necessitate renaming and the introduction of new symbols. Hence transactions are renamed as processors, and for points two types are defined, stores and channels. While channel represents a point, store is considered as a special point, which always contains precisely one token.

Definition 3.2
A high-level Petri-net is a n-tupel

$$HLPN := (P,T,F,K,R,B,M,V,W_0) \tag{3.29}$$

with

$$N := (P,T,F), \tag{3.30}$$

which is a Petri-net based on points, transitions, and flow relations and the functions K, Z, R, B, and M with

$$Z : T \times OT \to N , \tag{3.31}$$

which represents the switching-time function, and

$$R: T \rightarrow N, \tag{3.32}$$

which is the final switching-time function with $R(t) \leq Z(t, o.type)$ for $t \in T$, and $o \in O$,

$$B: (t, before(t)) \rightarrow BOOLEAN, \tag{3.33}$$

as switching condition for $t \in T$ with $Before(t) \subseteq O$, and

$$M: T \times O \rightarrow O \tag{3.34}$$

as the respective switching methods, and

$$V: F \rightarrow OT \tag{3.35}$$

as the edge-type function, and

$$\forall p \in P: 0 \leq |Wo(p)| \leq K(p) \tag{3.36}$$

as the initial condition of points P of type D with objects with the above given boundary.∎

Example 3.8
A motor assembly line, as part of a car production line, can be modeled based on a high-level Petri-net as shown in Fig. 3.3.

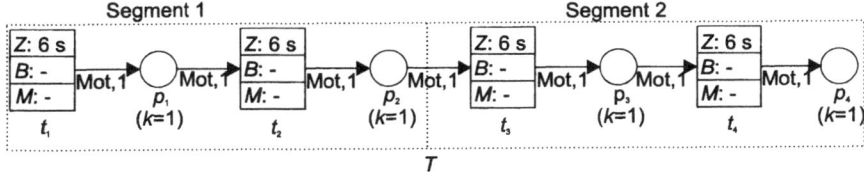

Fig. 3.3. High-level Petri-net representation of a conveyer belt of a motor assembly line

Figure 3.3 shows the model of the conveyer belt as part of the transportation processes of the assembly line. From Figure 3.3 one can conclude that the objects pass concurrent through the HLPN. While assembly processes need in some cases,

the take-over mechanism, the HLPN representation of a conveyer belt of the motor assembly line shown in Fig. 3.3 has to be expanded. The expansion of the HLPN has to allow the take-over in between stages 1 and 2 for changing the respective different assembled motor units, for ongoing assembling procedures. The expanded HLPN model, including the take-over mechanism, is shown in Fig. 3.4

From Fig. 3.4 one can conclude that in points p_1 and p_5 we have one object each and none in p and p_3, hence t_2 will be activated. In the switching state one object of type motor moves out of s_1 and one object of type WT moves out of p_5, both in the direction of transition t_2. After the switching time of 15 seconds from transition t_2 one object of type motor moves to point p_3 and one object of type WR moves to point p_2, which shows that the take-over mechanism is implemented in the correct manner.

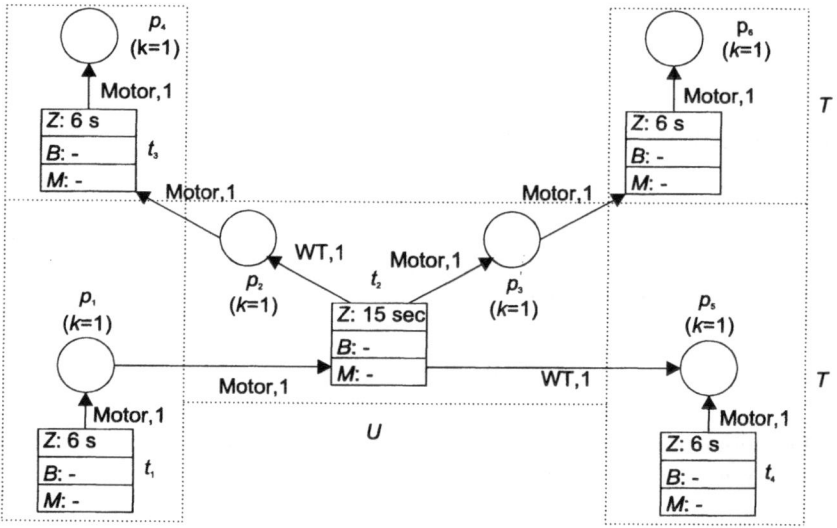

Fig. 3.4. High-level Petri-net representation of the take-over mechanism of a motor assembly line

Moreover, assembly processes need in some cases, the direct-access mechanism, which is another expansion of the HLPN representation of a conveyer belt of the motor assembly line shown in Fig. 3.3. The direct-access mechanism is of importance while implementing quality assurance within the HLPN model. The expanded HLPN model, including the direct-access mechanism, is shown in Fig. 3.5. In the direct access stage the object with the attribute $OK = true$ at point p_{12}, activates the transition t_{13}. When the value of the attribute is false, t_{14} will be activated and the defective motor moves during the switching state into the bypass loop. During transition t_{55} of a workplace M the attribute OK will be set true, which results via t_2 back to the assembly line.

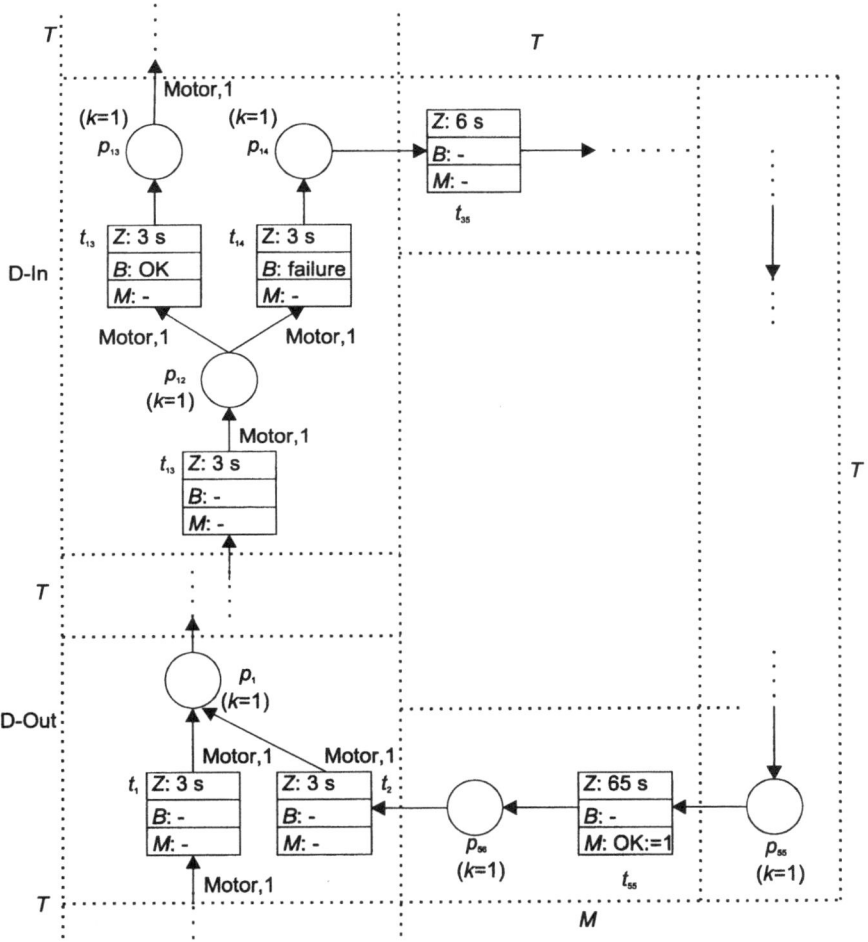

Fig. 3.5. High-level Petri-net representation of the direct-access mechanism of a motor assembly line

Developing a HLPN CAD-layout simulator for optimizing the order-entry-dependent assemblies, can be based using the HLPN models shown in Figs. 3.3, 3.4, and 3.5. The specific requirements of such types of simulation environment are specific commands, such as:

- Change
- Delete
- Expand

- Store
- Load
- Translate
- etc.

which can be implemented using programming languages such as C or Java.

Programming the application-specific simulator Layout can be done based on the above-mentioned control structure, as follows:

```
void PL_Editor(void);
void PL_Change(void);
void Copy_Ge(int in_nr, int out_nr);
void PL_Delete(void),
void PL_GE_Delete(int nr);
void Fetch_Car(void);
void PL_Expand(void);
void PL-Expand_Boundary(int nr, int*stages_nr)
....
```

The final graphic user interface screenshot of the developed HLPN simulator for assembly lines is shown in Fig. 3.6.

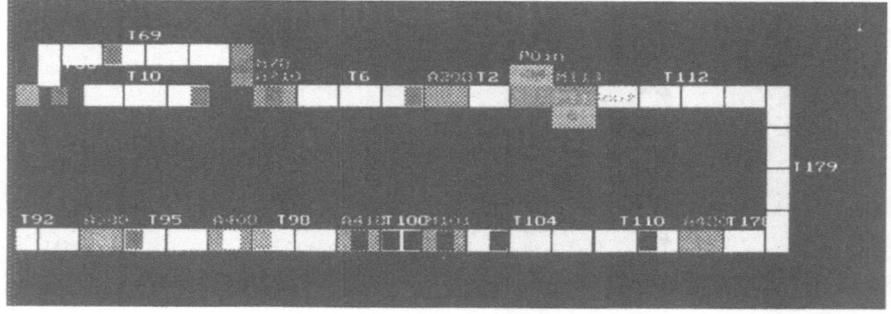

Fig. 3.6. Screenshot running an assembly line simulation based on high-level Petri-nets

While systems engineers often use models to investigate the properties of the system, like engineers responsible for motor assembling lines, as shown in Example 3.8, Petri-nets offer an easy way to describe concurrent systems for the inspection of the dynamic behavior of the system under test. An easy-to-use computer tool that supports the deveopment and excecution of object oriented Petri-nets, which include net instances, synchonous channels, and seamless Java integration for easy modeling – and simulation – is Renew. Renew, meaning reference net workshop, is available free of charge, including the Java source code (see Links in Sect. 3.5 and Apendix C). Renew can serve as a development environment and execution engine for workflow systems, such as the above-mentioned motor assembling line in the automotive industry, where the firing of transistions is coupled with the execution of the workflow tasks. Renew featured several extensions for the Petri-net formalism that includes clear arcs, flexible arcs, and inhibitor arcs. Also available are the expressiveness of timed Petri-nets, where time stamps are attached to tokens and to input and output arcs. The interactive debugging of complex net systems is supported by breakpoints and an inspection of Java token objects. The current release 1.6 contains several technical improvements suitable for production environmenst. These are remote simulation access, data base backing and a net-loading mechanism.

The main intention modeling real-world systems based on Petri-nets can be stated as follows:

- Users interest in certain systems properties.
- Simulation-based approach, which gives a singular predication for a single given initial state of the model. To deduce knowledge of the system behavior from the structure of the model (here: the given net) is possible, much more common predications can be made.
- Beneath its attractive graphical representation the main advantage of using net models to produce predications about the dynamics from structural analysis of the net can easily be done.

3.5 Discrete-Event Simulation of Parallel Systems

3.5.1 Introduction

For large-scale discrete-event models such as these used in manufacturing or traffic simulation the simulation run time can grow considerably large. Therefore, the simulation models outcome is too slow for real-time support of decision makers. Going deeper into the system behavior and model specification the reasons are obvious: All real-world activities that occur in the system are modeled by separately defined events. Hence the system has a wide extent and therefore many activities ongoing simultaneously at different locations within the system. However, these activities are treated successively by applying the next-event algorithm of the models run-time system. Therefore, the spatially distributed activities that occur simultaneously are serialized, which leads to the observed growth in model execution time.

The scope to shorten the execution time can be achieved by the following idea: Discrete-events that occur at the same time step are executed sequentially by two means:

- To treat interdependencies between the events correctly
- There is only one processing unit for the run-time system

In many applications events are independent of each other simply because of the spatial distribution of the system components within the plant or within a geographic region. Therefore the first reason loses ground and the second one can be removed by offering more than one processing unit.

However, the problem of interdependencies remains of interest at the spatial and/or logical interfaces between the independent units. For this reason it is important to know the subtasks that have to be solved before a correct discrete-event simulation run can be executed.

3.5.2 Basic Tasks

There are three subtasks that have to be taken into account:

- Partitioning, meaning the segmentation of the model into model parts
- Mapping and scheduling, meaning the assignment between the model components and the processing units
- Synchronization, meaning the appropriate design for the run-time system to avoid inconsistencies caused by retarded information flows between the model components

The explanations will show that these tasks depend on each other and that every solution depend on the model treated.

3.5.2.1 Partitioning of the Model Description

Two aspects are of importance while partitioning the model description: First the number of connections between two model components should be as small as possible, and secondly the information flow and/or the exchange of material, which can be expressed by such a connection, should be as low as possible.

This work can be done graphically by dividing a model into parts of code, representing each part by a node of a graph and plotting for every dependency between the model parts an edge between the corresponding nodes. With this data structure-partitioning algorithms can be imagined easily that fraction the model graph in a given number of sub graphs and minimize the traffic on the edges between these sub graphs.

Often, the partitioning can be done automatically by a graph algorithm that analyses the structure of the model. However, this is a problematic task because only static information can be evaluated. The model dynamics that influences the intensity of using connections between model parts is neglected completely and can be taken into account only during the run-time of the model itself. This results in repeating partitioning even during run-time to guarantee that dynamic information will be considered. This dynamically adapted model partitioning causes considerable costs in CPU time and can reduce the benefits received from parallel execution.

3.5.2.2 Mapping and Scheduling

The next step in parallelization deals with the mapping between the model parts and the processors available. Theoretically, two situations are possible:

- The number of processors is greater than or equal to the number of model parts: In this case it seems obvious how to map the model parts into the set of processors: Each processor should be responsible for the activities of one model component. But, in general, this mapping concept is not optimal at all. If the ratio of component communication and activities within a component tends to communication, the communication dominates run-time behavior, which means it can be possible to reach even negative speed-ups in simulation time.
- The number of processors is less than the number of model parts: This is the normal case of the mapping task. The optimal balance between the number of processors and the given number of model parts has to be found analytically or, – in most cases – through experiments. This mapping concept is the more practical one.

Independently of these deliberations the question arises, where to store the model data during parallel execution of a simulation run. There are also two possibilities:

- Common ("shared") memory: fast access but expensive
- Local ("distributed") memory for each of the processors: standard solution which causes consistency problems

As discussed, the concepts for mapping depends on the model dynamic which can even change during a simulation run. Hence, adapting mapping with the demands of the models dynamic behavior can be introduced as dynamical mapping, called scheduling. Due to the use in simulation, new criteria can be taken into account. The simplest and most obvious one is to use the current simulation time handled by a processor as a measure for its load. Hence, the scheduling algorithm should privilege those processors that have the slightest simulation time for the next mapping step.

It should be noted that all these thoughts are not specific to parallel simulation but the restriction in application gives some hints as to what special criteria can be established to adapt standard algorithms to the given special application, such as synchronization algorithms.

3.5.2.3 Synchronization

As long as events handling by the different processors are independent, each of them can proceed in simulation time without regard to the others. If the partitioning and the mapping steps are executed properly, this will be the standard situation. But another more complicated situation is possible: Assuming a partitioning that is motivated by a geographical division of space. Each processor executes the model description for such a region and proceeds in simulation time autonomously. Under these preconditions the following conflict may occur: At a given time step, t, one element moves from the region processed by processor P to the region processed by P'. However, the current simulation time in P' is already t' which is greater than the current time t in P. What the simulation run time system has to do is either to prevent such a situation or to ensure that the event will be treated correctly although it depends in the past for processor P'.

In general, the problem can be stated as follows: Each processor with its corresponding simulation run-time system holds its own current simulation time t and is able to receive messages and material from other processors. It has to be assured that these pieces of information are interpreted and processed correctly, especially concerning the time they arrive.

Solving this problem can be done by optimizing the speed up by reducing the overhead of the time control by deeper knowledge of the model specification and its implementation. The basic strategies behind this are called

- Conservative strategy
- Optimistic strategy

The conservative strategy is based on the idea that the processes mutually give guarantees that assure that the sending process will not announce any more events that lie before a point of time given in this guarantee. The active process only is allowed to proceed in time guaranteed of all the other processes for the intended time step. The idea is quite simple and its function is obvious.

Nevertheless, there are two problems: The process has to know enough about the future to give the adequate guarantee. This problem will be noticeable especially if there are complex interdependencies with feedback loops and can even cause a deadlock situation for the run-time system. The second problem is the low grade of parallelity, which normally is the outcome of this strategy: The processes wait for each other instead of proceeding in time in parallel, which causes a decrease in speed up for the entire set of processes.

The disadvantage of waiting for guarantees avoids the optimistic strategy. As the name says, every process is 'optimistic' and proceeds as fast as possible for its model components. If an event from a parallel working processor is sent with a time stamp less than the current time, the run-time system must be able to reset its own state to the time demands. After reset the incoming event can be considered correctly and the run can be continued as usual. Normally the calculations that the processor in advance became dispensable and must be repeated starting with the new initial state including the effects of the additional incoming event. Hence, this strategy has to assure that a process can be reset to any time in the past. This task is expensive, because information about already passed model states has to be stored. Standard algorithms for solving this problem are either to store all previous model states or to store a list of all the events executed and try an undo operation from the current state backwards to the time the incoming event arrives. The first possibility has quite a large demand for storage space, the second is algorithmically complicated.

It can be seen from the strategies that CPU time and storage can be substantial, which then reduces the intended speed-up through parallelism. The best methods avoiding these disadvantages are intelligent partitioning, mapping, and scheduling, which reduces the overhead for guarantees, resets, and undo operations.

3.6 References and Further Reading

Banks J, Carson JS, Nelson BL, Nicol DM, (2001), Discrete Event Simulation, Prentice Hall, New Jersey

Girault C, Valk R, (2002) Petri Nets for Systems Engineering, Springer, Berlin

Kheir NA, (1996), Systems Modeling and Computer Simulation, Marcel Dekker, Inc., New York

Mehl H, (1994), Methods of Distributed Simulation (in German), Vieweg Publ., Braunschweig

Zeigler BP, Praehofer H, Kim TG, (2000), Theory of Modeling and Simulation, Academic Press, San Diego

Links:
www.renew.de/
www.informatik.uni-hamburg.de/TGI/renew/bibliography.html

3.7 Exercises

3.1   What is meant by the term random variable?
3.2   What is meant by the term exponential distribution?
3.3   What is meant by the term Weibull distribution?
3.4   What is meant by the term FIFO?
3.5   Describe Kendalls notation by using a simple example.
3.6   Describe a simple queuing system by using a calling population, a waiting line and a server.
3.7   What is meant by the term average waiting time?
3.8   What is meant by the term probability of idle server?
3.9   What is meant by the term lead-time demand?

3.10 Give a simple example of the inventory problem.

3.11 What is meant by the term Petri-net?

3.12 Give the definiton of a Petri-net.

3.13 Give an example of a Petri-net.

3.14 What is meant by the term colored Petri-net?

3.15 Give the definiton of a colored Petri-net.

3.16 Give an example of a colored Petri-net.

3.17 What is meant by the term high-level Petri-net?

3.18 Give the definiton for a high-level Petri-net.

3.19 Give an example of a high-level Petri-net.

3.20 The algorithm for computing the state and output trajectories of a discrete-time system is based on the given input trajectory $x(n)$ and its initial state $q(0)$. With T_i as initial time, T_f as final time we have

$$T_i = 0, \ T_f = 9$$
$$x(0) = 1, \ x(9) = 0$$
$$q(0) = 0$$
$$t = T_i$$
$$\text{while } (t \leftarrow T_i) \ \{$$
$$y(t) = \lambda[q(t), x(t)]$$
$$q(t+1) = \delta[q(t), x(t)]$$
$$\}$$

Execute the algorithm by hand and fill in Table 3.4

Table 3.4

Time	0	1	2	3	4	5	6	7	8	9
Input trajectory	1	0	1	0	1	0	1	0	1	0
State trajectory	0									
Output trajectory	1									

4 Simulation Sofware for Computational Modeling and Simulation

4.1 Introduction

When discussing the fundamental concepts of modeling continuous-time and discrete- time real-world systems in Chap. 1 it was noted that an accurate mathematical model is necessary for the use of computer simulation, which is focused on a better and/or deeper understanding of the dynamic behavior of real-world systems. The complexity of man-made systems in engineering and science, as well as the complexity of systems in biology, medicine, and nature, mostly do not allow closed analytical solutions for all the sets of linear and/or nonlinear mathematical equations as they have been outlined in Chap. 2 to describe real-world systems. Assuming that the model has successfully been described, meaning the real-world system is represented in terms of differential equations, partial differential equations, state-space equations, difference equations, queues, Petri-nets, etc., a solution of which can be obtained based on computational simulation. Using computers for solving the equations that describe real-world systems in an effective and sufficient way, numerical integration methods are of importance. This is why, for a number of years, considerable effort has been devoted to the development of simulation software for continuous-time and discrete-time systems.

Definition 4.1
Simulation can be introduced as the process by which the understanding of the dynamic behavior of a real-world system can be obtained by observing the behavior of a mathematical model that represents the real-world system.■

From Definition 4.1 it can be regarded that simulation can be described as reproduction of the dynamic behavior of a real-world system, based on a model representation. The model of which contains the important attributes, relations, and objects of the dynamic system, which with the real-world system behavior can be approximated from simulation results by reasoning. The simulation process itself is said to be a computerized calculation of the outputs y of a mathematical model at the respective inputs u over the simulation time step t.

Due to the method of model description, the dynamic behavior of real-world systems can be studied as

- Physical similarity
- Isomorphism
- Mathematical reproduction

In consideration of physical similarity, e.g. a physical model of an airplane can be developed and certain characteristics, i.e. the strength and forces attaching the wings of the airplane due to turbulences can be observed, for example, in a wind tunnel, which is much easier to realize compared with the tests of the real airplane under normal and non-normal real-world conditions. In medicine, for example, a hydraulic circulatory system simulator related to the viscosity of the blood, can be built up by which the elasticity of the vascular bed and/or the pericardial vessels can be observed.

As a result of isomorphism, a real-world mechanical system, consisting of locally concentrated elements such as damper, spring, and mass, can be replaced by an electrical substituting system consisting of the respective elements like inductance, capacitor, and resistor, which, with the observation of the dynamic behavior of the mechanical real-world system, can be carried out much more easily. Hence a table of correspondences is helpful when developing isomorphic models of real-world systems, as shown in Table 4.1.

Table 4.1. Correspondences of isomorphism

Physically System	General Description	Electrical	Hydraulical	Pneumatical	Thermal	Translational	Rotational
Transversal Variable e(t)	Voltage, Pressure Velocity	U(t); Voltage	P(t); Pressure	P(t); Pressure	T(t); Temperature	V(t); Velocity	α(t); Angular Velocity
Transit Variable f(t)	Current, Flow Force, Momentum	i(t); Current	$\dot{V}$(t); Volume Flow	$\dot{m}$(t); Mass Flow	$\dot{q}$(t); Heat Flow	f(t); Force	M(t); Torque
e(t) Product	Power supplied to the element	p(t)=u(t)·i(t)	p(t)=P(t)·$\dot{V}$(t)	p(t)=P(t)·$\dot{m}$(t)	p(t)= $\dot{q}$(t)	p(t)= V(t)·f(t)	p(t)=α(t)×M(t)
e(t) Relation	Power Consumption e(t)=R·f(t)	R; Electrical Resistance	$R = \frac{8l\eta}{\pi r^4}$ Flow - resistance	identical to hydraulical	Thermal Resistance $R_\lambda = \frac{1}{\lambda x}$(Flow) $R_r = \frac{1}{\alpha}$(Transm.) $R_c = \frac{1}{\alpha}$(Convect.)	d^{-1}; Damping - factor	d$_r^{-1}$; Damping- factor
$\int$ e(t) dt	F(t)= 1/L · $\int$ e(t) dt	L; Inductor	$\frac{\rho l}{\pi r^2}$; Inertance	$\frac{\rho l}{\pi r^2}$; Inertance	—	c^{-1}; Spring - constant	c$_r^{-1}$; Spring - constant
$\int$ f(t) dt	e(t)=1/C · $\int$ f(t) dt	C; Capacitor	$\frac{A}{\rho g}$; Hydraulic Capacity	$\frac{V}{R \cdot T}$; Pneu- matic Capacity	m · c; Thermal Capacity	M; Mass	θ; Moving Mass
$\int$ e(t) f(t) dt	Energy done on system	E$_m$:Magnetic Energy of inductor E$_e$:Electric Energy of capacitor	E$_k$:Kinetic Energy of fluid flow E$_p$:Potential Energy of pressure head	E$_k$:Kinetic Energy of pneumatic flow E$_p$:Potential Energy of pressure	E$_r$:Thermal Poten- tial Energy of stored heat	E$_k$:Kinetic Energy of moving mass E$_p$:Potential Energy of compres - sed Spring	E$_k$:Kinetic Energy of rotating mass E$_r$:Potential Energie of twisted spring
Symbols		⊏⊐ R ▬ L ⊣⊢ c	⊃⊂ ▬ ⊣c⊢	⊃⊂ ⊣c⊢	T_1⊏R⊐T_2 T_1⊏c⊐T_2	⊔ R ⌁⌁ L ⊡ c	⊔⊃ R ⌁⌁⊃L ⊡⊃ c

Left margin labels: VARIABLES — PRIMARY; VARIABLES — INTEGRAL

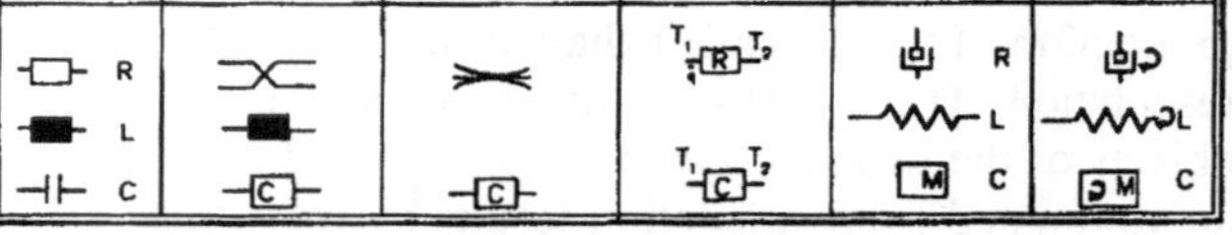

In general, the mathematical equations representing the mathematical model of a real-world system can be solved with the help of numerical-integration methods (see Appendix A). The mathematical models, which describe the intrinsic transient

system states, including the chosen physical system characteristics explicitly or implicitly, can be solved using different analytical methods, which can be separated into four groups:

1. Calculation of the characteristic polynomial after conversion of the differential equations of the time-continuous systems into a system of algebraic equations.
2. Eigenvalue calculation.
3. Numerical methods for the solution of the differential and/or difference equation systems.
4. Petri-nets, queuing theory, distribution theory, probability theory, etc. for modeling and simulation of time-discrete systems.

When digital computers became available in the late 1950s they replaced analog hardware and relays, which were the typical components of analog computers in the past. Today digital computers are used for the simulation of continuous-time systems as well as for discrete-time systems in the various application domains. As an example, a large refinery may have as many as 1000 feedback loops, in comparison a paper mill may have up to 5000 feedback loops, meaning that between 1000 and 5000 controller equations and additional plant-dependent equations have to be solved. Such types of dynamic systems could neither be solved with the analog computers in the past, nor with the so called hybrid computers, which combine analog and digital computer facilities. Hence there was (and is) a need for digital computer simulation techniques. The innovation sequence in digital computer simulation techniques from the early 1950s till today can be characterized as follows:

- 1955–1960: User programming, no user support, model building based on
  higher programming languages like FORTRAN, ALGOL, etc.
- 1960–1965: First generation of simulation software, very simple user support through automatically generated computational relations, graphical user interface.
- 1965–1970: Second generation of simulation software, better tools, interactivity.
- 1970–1980: Third generation of simulation software with extended and new possibilities of simulation tools like combined simulation, etc.
- 1980–1990: Fiurth generation of simulation software with domain specific and specialized simulators, animation possibilities, and easier model implementation
- 1990–2000: Fifth generation of simulation software, embedding artificial intelligence, model specification and experimental environments, expanding the possibilities of the tools of the fourth generation especially with much more sophisticated graphic tools that allowed 3D (spatial) and 4D (time) to became a standard.

- 2000–2010: Sixth generation of simulation software, embedding object oriented modeling frameworks, soft-computing methodology like fuzzy sets, neuronal nets, genetic algorithms, evolution theory, probabilistic methods, virtual and augmented reality environments in simulation, which allow tactile force-feedback interaction, and simulation at the internet.

From the very early beginning till today simulation software contains a simulation language and a set of commands for the control of the simulation process.

Although simulation software permits the description of the mathematical models to be implemented and its parameters, there is an obvious advantage when using special-purpose application simulation software that has been especially written for simulating continuous-time and discrete-time systems, from the viewpoints of the different application domains. These simulation software systems allow the user to implement and simulate models quickly and efficiently without being an expert in programming languages or numerical integration.

Moreover, the innovation in modern computer technology increases the possibilities of simulating complex systems. Hence we can consider computer simulation based on mathematical models as a third column apart from theory and experiments. In those many cases where experiments can not be realized, due to safety or security reasons, due to their time consuming nature, their costs, etc., it will be possible to obtain solutions for complex problems by means of mathematical and computational modeling and simulation, which results in a better understanding of real-world systems and the theory behind them. Examples are manifold such as the volume and time-dependent processes in engineering, chemical engineering problems involving transport phenomena, combustion analysis in an engine, 4D process analysis in geology, such as tunnel drilling with trenchless technology, structural mechanics application, applications in avionics such as vortex generation and shedding of vortices at the back hood of a planes wing, etc. Furthermore, modeling and simulation, by means of modern computer technology, provide information about the usability of hypotheses, e.g. for astrophysical problems such as the time course of matter distribution of a star formation, in molecular modeling due to the chemical structure for an optimal drug design, in physiology such as modeling and simulation of the nonlinear overall control mechanisms explained hypertension, or the nonlinear discrete modeling of tumor growth in humans and the respective test series for intra-individual sufficient tumor therapy, etc.

In order to extend the possibilities of applying mathematical and computational modeling and simulation, scientists worldwide are presently attempting the expansion of the classical descriptions by modern information processing based on the methods of artificial intelligence, soft computing, virtual and augmented reality, etc. In this sense artificial intelligence can be used twice in mathematical and computational modeling and simulation, as an advisory system for decision support, and for adaptive intelligent control where the expert system has to be linked to the simulation models. Artificial intelligence can be regarded as that part of computer science concerned with the development of intelligent computer programs. To make a program intelligent means providing it with specific knowledge on the respective problem domain. By contrast to this, soft-computing methods are preferable when modeling and simulation of real-world systems are based on vague data.

The task of knowledge-based advisory systems for mathematical and computational modeling and simulation purposes is

- User support in model synthesis, which focuses the knowledge on the modeling process, selecting and lumping domain primitives, defining models from schematic representations such as flow diagrams, symbolic notation, block diagrams, bond graphs, etc.
- Providing knowledge of the known mathematical properties of a model,
- user support in choosing the algorithms, numerical integration, parameter estimation, validation, layers for (neural-based) classifiers, definition of membership functions for fuzzy sets, etc.
- User assistance due to the integration of mathematical expressions, to overcome the fact that equations often are semantically disconnected from the application domain and less suggestive as schematic representations.

The knowledge representation itself can be based on rules, semantic nets, and frames, etc. In the case of

- Rule-based systems, IF-THEN rules are used to perform forward or backward chaining
- Frame-based systems, frame hierarchies for inheritance and procedural attachments are used
- Procedure-oriented systems, nested subroutines are used in order to organize and control program execution
- Object-oriented systems, objects are used, which communicate with one another via messages
- Logic-based systems, predicate calculus is used in order to structure the program and guide the execution
- Access-oriented systems, probes are used that trigger new computations when data is changed or read

Example 4.1
A rule-based system, which interprets waveforms from a scanning densitometer to distinguish between different causes of inflammatory conditions in patients should have the rule:

*IF the tracing pattern is asymmetric gamma **AND** the gamma quantity is normal **THEN** the concentration of gamma globulin is within the normal range*

The knowledge representation using semantic nets is a method that is based on network structures. A semantic net consists of points, which are called nodes, connected by links, called arcs, which describes the relations between the nodes. The nodes in semantic nets stands for objects, concepts, or events.

Example 4.2
Simple semantic nets can use an important type of arc, the **is a** relation, which can be expressed as follows:

The Queen Mary is a(n) ocean liner and Every ocean liner is a ship

When using the object-oriented method, a specified structure, consisting of objects (often called actors), is used that represent entities capable of exhibiting behavior.

Example 4.3
In an object-oriented air-battle simulation system the objects will be penetrators (offensive aircrafts), airborne radars (AWACS), ground radars, missile installations, missiles, service centers that interpret the radar reports, fighters (defensive aircrafts), fighter bases, command centers, and targets. Consider each object has distinct properties associated with each other that are embedded in a network hierarchy, showing inherited properties of higher-level objects, such as properties that are inherited in semantic nets and frames. For example the objects penetrators, fighters, missiles, and AWACS may all be linked to a higher-level object, which is called a moving object. In addition, the object AWACS can be linked to the higher-level object radar, and so on.

Furthermore, the knowledge-based advisory system manages simultaneously different kinds of information sources simultaneously like:

- Data bases and object bases: they consist of system data and/or objects, model data and/or objects, an experimental framework, experimental data obtained from the real-world system and/or objects, simulation data and/ or objects, etc. The data base can hold model data such as the exact model configuration to be simulated, or the outputs of the simulation run that should be stored in the data base. Moreover, the data base can be used for competitive representation of model entities, while these entities with their attributes and relations can be directly stored in an entity-attribute-set representation as a framework for organizing data.
- Algorithm bases: they contain numerical algorithms, parameter identification algorithms, soft-computing methods, formalized manipulation methods, etc .
- Knowledge bases: which comprise the general modeling knowledge, physical and mathematical knowledge about physical properties, simulation-interpretation knowledge, etc.

Simulation software systems (S) and expert systems (ES) can be embedded vice versa. This can be stated as a computer program using expert knowledge to obtain higher levels of performance in a narrow problem area, or soft-computing methods, such as neuronal nets, fuzzy sets, evolution theory, probability theory, which results in a more advanced man machine interface for the simulation model, as shown in Fig. 4.1.

Embedding, as shown in Fig. 4.1a, means that the simulation software system (S) is embedded in the knowledge-based system (ES) which makes domain-dependent knowledge explicit and separated from the rest of the system. Therefore, existing knowledge has to be embedded in the knowledge-based system. The communication is realized through the knowledge-based system, based on symbols. Hence the simulations run with obvious initial values and boundaries.

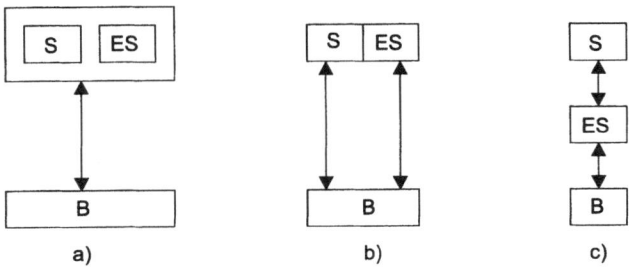

Fig. 4.1. Interaction of simulation software system (S), and expert system (ES) (see text)

Toothing the simulation software system (S) and the knowledge-based system (ES), as shown in Figure 4.1b, allows the simultaneous use of both systems. The simulation task decays in its specific sequences, which can be distributed. Toothing can be used in complex simulation studies where solutions can only be obtained in a sequence of different steps, but each step must be solvable.

For a man machine interface, as shown in Fig. 4.1c, the communication passes through the knowledge-based system, which enables an intelligent dialog.

The knowledge base in expert systems contains facts (data) as sources and rules (or other representations) which can use those facts as a basis for the decision-making processes. The inference mechanisms, or inference engines, contain an interpreter that decides how to apply the rules to infer new knowledge as well as a scheduler that decides the order (schedule) in which the rules can be applied. The resulting architectural structure of an expert system is shown in Fig. 4.2.

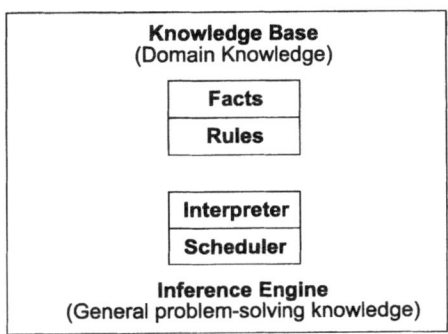

Fig. 4.2. Structure of an expert system

4.2 Digital Simulation Systems

Digital simulation software systems consist of the simulation language and the translator. The simulation languages for computational modeling and simulation of continuous and discrete systems are discussed in the following sections, which includes complete applications of the most popular simulation software systems. The simulation software translator generates the compiler code for the simulation program, which will be translated into machine code, and linked with standard elements from the system library. When the model shows an error, such as in the case of an implicit loop, the simulation will be terminated before the first simulation runs. Furthermore, the computer determines values for the continuous functions $x_i(t)$ of the system being simulated by producing a series of discrete values, such as $x(t_0)$, $x(t_1)$, $x(t_2)$,..., $x(t_n)$, which should be identical to its continuous equivalent at $t = t_k$. Due to the discretization, while using numerical-integration methods, the discretization accuracy is limited, which results in errors in numerical- integration.

Definition 4.2

A simulation run can be defined as a calculation of the state-variables transient behavior in discrete steps, based on a calculation sequence of the mathematical model, which starts with the initial condition $x(t_0)$, and finishes with the final state $x(t_e)$. The calculation of which can be based on fixed or variable step width.∎

Moreover, simulation can be regarded as experimentation with models, while using simulation software. This definition is particularly appropriate when considering interactive simulation.

The simulation software systems for computational modeling and simulation can be divided, as shown in Fig. 4.3, in software for

- Continuous-time systems
- Discrete-time systems

Simulation software systems for continuous-time systems are

- Block oriented simulation software, such as DARE, DORA, PROSIM, PSI, SIDAS, SIMULANT II, SIMNON, etc.
- Equation-based simulation software, such as ACSL, CADSIM, CSMD, Matlab-Simulink, SLCS-4, etc.

Simulation software systems for discrete-time systems are

- Transaction-based: time schedules are determined; elements of transaction-based simulation software are transactions, blocks, facilities, queues, logical switches, numerical and logical variables, functions, tables

- Event-based: the time schedules of which depend on the time characteristics of the event
- Activity based: the simulation run will be active if the previous specified conditions are realized
- Process oriented: the specified model element activates the next event

Examples of available simulation software systems for discrete-time systems are:

- ARENA, DEMOS, GPSS, GPSS/H MODSIM II, SIMAN, SIMPLE++, SIMPLEX, SIMULA, SIMSCRIPT 11.5, SLAM, SLX, etc.

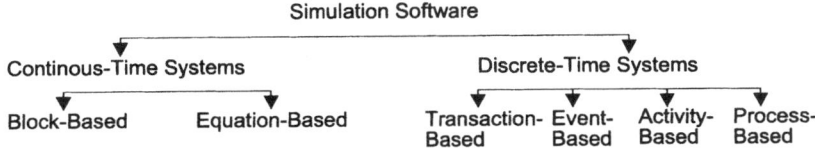

Fig. 4.3. Simulation software for continuous-time and discrete-time systems

4.3 Simulation Software for Continuous-Time Systems

A variety of continuous-time system-simulation software systems have been developed since Selfridge wrote the first paper on the subject in 1955. Cellier gave, in 1993, an excellent overview on the development of simulation software systems. In the last past few years, many simulation software systems have been developed, some of which will be discussed in the subsections of Sect. 4.3, which includes example applications. The widespread use of personal computers and workstations has had a profound impact on the use of simulation software. The trend is toward multipurpose, interactive simulation software systems that provide tools for mathematical modeling and simulation. Each approach has salient features, and familiarity with one of which can easily be transferred to another one.

Regardless of the simulation software selected, several requirements for simulation must be fulfilled by the user, such as the description of the real-world system to be modeled and simulated, which involves specifying the type of elements, or functional blocks, of which the system consists and describing how the elements are interconnected. This is accomplished by means of structure statements or commands that include all standard and special functional blocks used to build up the model of the real-world system, while using a so-called block oriented simulation software system. Moreover, the blocks had to be specified by parameters, inputs and outputs, the appropriate functions, initial conditions, arbitrary functions, run time, time interval, and other control commands, etc.

Most simulation software systems are very user-oriented and incorporate default conditions that enable a novice to obtain meaningful results immediately.

Moreover, simulation software can solve linear as well as the nonlinear differential equations of n-th order, describing the real-world system. In most cases the
higher order differential equations are reduced as sets of first-order differential
equations, which means for the block oriented simulation software the decomposition into a block oriented scheme of first-order blocks, as shown in Fig. 4.4,
which can easily be solved using numeric integration.

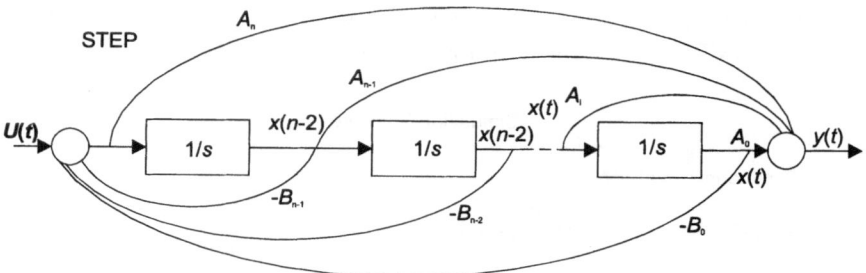

Fig. 4.4. Decomposition of a dynamic system of n-th order

4.3.1 Block Oriented Simulation Software

As an example of an interactive block oriented simulation software, the simulation
system PSI will be briefly introduced, developed at the TH Delft, Netherlands. PSI
can be used for studying the behavior of continuous-time – and discrete-time –
systems. The notation used in PSI is similar to the mostly-used ones in other block
oriented simulation systems. Block oriented simulation software systems use the
differential equations that represent the simulation model.

Example 4.4
Supposing the second-order differential equation

$$y''(t) = -y'(t) - y(t) + u(t),$$ (4.1)

which can be rewritten as set of first order integral equations, yields

$$y'(t) = y'(0) + \int_0^t y''(\tau)d\tau$$ (4.2)

$$y(t) = y(0) + \int_0^t y'(\tau)d\tau,$$ (4.3)

and modeled as a block oriented representation as shown in Fig. 4.5.

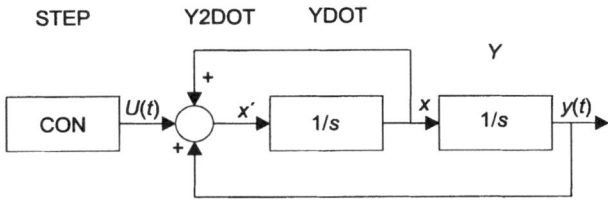

Fig. 4.5. Block diagram of a ssecond-order continuous system

In Fig. 4.5 Y2DOT represents the second derivative, YDOT represents the first derivative, Y is the original function, and STEP represents a constant $U(t)$.

The second-order system, shown in Fig. 4.5, needs specific facilities such as:

- Structure
- Parameters
- Numeric integration
- Output

The structure is given by defining the inputs of each block. If the inputs of all blocks are defined the structure of the simulation model is known, and the block-oriented simulation software is able to calculate the behavior of the system. The idea behind the block oriented simulation approach is that any simulation model can be built up from basic block elements such as integrators, gain-transfer functions, table lookup facilities, nonlinear function blocks, constants, etc. Each block is identified by its name, which determines the block and its output, as well as the block type and its inputs. In the example, shown in Fig. 4.5, we have

$$
\begin{aligned}
\text{STEP} \quad &= u(t) \\
Y \quad &= y(t) \\
\text{YDOT} \quad &= y'(t) \\
\text{Y2DOT} \quad &= y''(t)
\end{aligned}
$$

Together with the simulation software specific control commands the block structure can be defined. For PSI we use B as follows:

PSI·B
Configuration Specification
Block, Type, Input1, Input2, Input 3
B·STEP, CON   : STEP is a CON block
B·YDOT,INT,Y2DOT   : YDOT is an integrator with Y2DOT as input
B·Y,INT,YDOT   : Y is an integrator with YDOT as input
B·Y2DOT,SUM,STEP,Y,YDOT: Y2DOT is a summer to add all inputs
B·

The parameters of each block can be defined using specific control commands, which depends on the simulation software used. For PSI we use P, hence

```
PSI·P
Parameters
Blocks, Par1, Par2, Par3
P·STEP,1          : STEP has value 1
P·Y,0,1           : initial condition=0; input gain=1
P·YDOT,0,1        : initial condition=0; input gain=1
P·Y2DOT,1,-1,-1 : gains of the corresponding inputs
P·
```

The variables that determine the numeric integration, are the integration method, the integration interval, and the simulation time, which can be defined by specific control commands. For PSI we use T and obtain:

```
PSI·T
Integration interval=0.1
Integration time=10.0
```

Using 0.1 for the integration interval and 10.0 for the integration time, the simulation run, as shown in Fig. 4.6, will be calculated for 10 time units with an integration interval of 0.1 time unit.

All blocks are calculated during the simulation run, however, only some of which can be shown on the screen. Supposing that $y(t)$ is the output variable of interest which may be shown on the screen, some specific control commands are needed, which depends from the simulation software used. For PSI we use O, hence

```
PSI·O
Name of blocks to be shown=?
```

Now the required transient behavior can be calculated and the output variable(s) indicated are shown on the screen, as shown in Fig. 4.6. The simulation run will start using a specific control command. For PSI we use R

```
PSI·R
```

PSI allows the presentation of more than the one variable, meaning more complex figures can be shown representing the respective system variables, which include inputs as well as outputs.

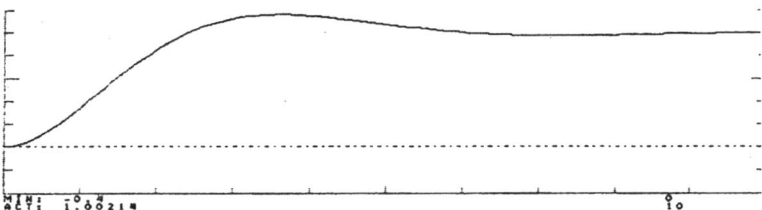

Fig. 4.6. Transient behavior of the output variable $y(t)$

As mentioned, simulation software systems for continuous-time systems have salient features but all are based on three types of instructions,

- Model instructions, which generate the algebraic block oriented structure of the mathematical equations of the real-world system
- Data instructions, which are used to assign the respective parameter values, initial conditions, etc.
- Control instructions, which determine the simulation run and select the output variables

The simulation run itself can be divided into four steps, which will be shown for the simulation software SIDAS, in detail,

- Interactive implementation of the block diagram of the mathematical model: SIDAS outlines a frame with a number of cross-points, at each of which a block can be inserted representing the respective mathematical function element. To insert a block, e.g. an integrator, one chooses a cross-point with the cursor and types "I" for insert, followed by the symbolic name of the block to be inserted, which is "INT" for the integrator block. Connecting or modifying blocks, a set of additional commands are available, such as change a block, delete a block, link a block to different blocks and end the input modus, etc. Moreover, special blocks are used to implement specific algorithms, transient behavior, etc.
- Set of block parameters: initial conditions, gain factors of integrators, simulation parameters such as the length of the simulation interval, the desired accuracy of the numerical integration, the numeric integration method, etc.
- Numerical solution of the sets of differential equations: based on successful sorting algorithms, which determine the calculation sequence in such a way that only theses elements are processed in the loop the values of which are updated in the respective sequence.
- Presentation of simulation results.

Sorting algorithms handle the algebraic loop problem of a simulation run. The simulation run has an algebraic loop problem if it is not possible to calculate all blocks $b \in B$ in such a way that the output of a block b_i is connected to the input of block b_j, then $i < j$. The value of algebraic blocks can be calculated if:

- The algebraic block is a constant block
- The output of the algebraic block at t_{n+1} is known either by the initial values or the output values one step behind the actual integration step
- The output value of the algebraic block has been already calculated in the actual time interval

Sorting algorithms are used to determine the calculation sequence of the functional elements used. The first step of a sorting procedure deals with the evaluation of the simulation configuration determining the matrix positions of the integrators with their inputs, the block numbers of the previous functional blocks in a counter-clock wise signal flow. This results in a list of sorted algebraic functions that allows the calculation for each block as a function of its argument in the determined sequence. If all integration steps in the sorted sequence are finished the next repetition of the sorted loop will be prepared setting

$$n := n + 1 \qquad (4.4)$$

and

$$t_n := t_{n+1} . \qquad (4.5)$$

The simulation runs through the sorted loop until the condition $t_n = t$ and/or another given condition is reached that terminated the procedure. For an algebraic loop one or more algebraic blocks are connected in a closed loop that contains no dynamic components such as integrators. Algebraic loops can be eliminated by opening the loop, which means manipulating the original equation.

Example 4.5
A model of the renal blood flow control system that allows elucidation how of the kidneys regulate the physiological variables that are affecting the renal function can be built for the long-term blood-pressure-regulation observation. The model is based on the Guyton overall renal function model, named after the American physiologist Guyton, born 1919 in Oxford, Mississippi, USA. A simplified version only uses a single feedback loop. The urine output (UO) as the most important renal variable has been considered as a nonlinear function of the arterial blood pressure (PAS), which yields

$$UO = f(PAS), \qquad (4.6)$$

with the implicit functionality that blood-pressure stabilization depends on the renal sodium and water output. From physiology it is known that sodium influences the renal characteristic, which is important for the arterial blood pressure control due to the ability of the kidney to maintain an appropriate balance between an increased sodium and water output in the case of an increase of perfusion pressure. After a malfunction of the renal characteristic, which is the case in high blood pressure, the threshold of the diuretic pressure is increased. Hence the extra cellular fluid volume (VECF) can be calculated as follows

$$VECF = f(UO, WS), \tag{4.7}$$

which influences the blood volume (VB), yields

$$VB = f(VECF), \tag{4.8}$$

where f is a nonlinear function. The blood volume effects the mean systemic pressure (PMS) as follows

$$PMS = f(VB), \tag{4.9}$$

where f is a nonlinear function. With the right atrium pressure (PRA) and the resistance of venous return (RVR) the venous return (VR) can be calculated as

$$VR = \frac{PMS - PRA}{RVR}, \tag{4.10}$$

assuming that an increased inflow of blood into the heart does not increase the right atrium pressure, which is based on the Frank Starling law of the heart, meaning the heart pumps, whatever amount of blood enters it, and does so without a significant rise in the right atrium pressure. Closing the loop, arterial blood pressure (PAS) can be obtained by multiplying venous return by the total peripheral resistance (RA), yields

$$PAS = VR \cdot RA, \tag{4.11}$$

the result of which is shown in Fig. 4.7.

In the block diagram, shown in Fig. 4.7, the notation NL indicates that the block represents a nonlinear function while the notation L represents a linear relationship of the respective block. The symbol • in a block indicates a multiplying function of the block, while the ÷ sign in a block indicates a division function of the respective block. The line from bottom left to top right within the block indicates that the block is operating as an Integrator, for which the appropriate numeric integration method can be chosen.

The model, shown in Fig. 4.7, can be implemented using the simulation system SIDAS, which results in Fig. 4.8.

The simulation results can be demonstrated in comparison with animal experimental results, as shown in Fig, 4.9. Dotted columns represent animal-experimental results, white columns are simulation results obtained with the mathematical model shown in Fig. 4.7.

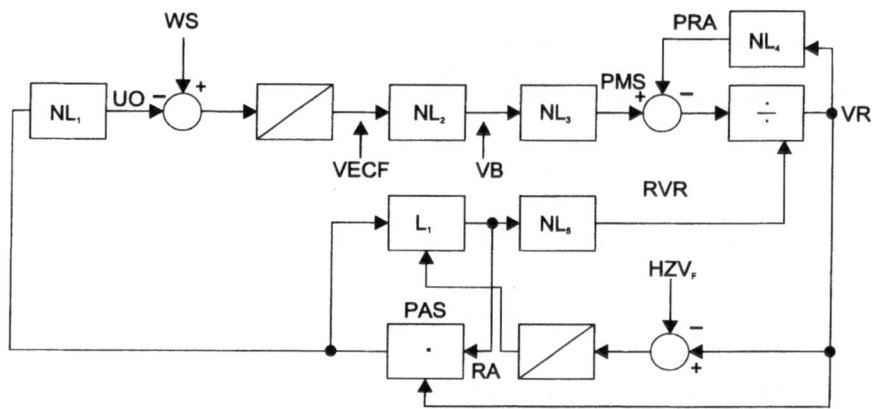

Fig. 4.7. Mathematical model of the renal influenced long-term arterial blood pressure control; WS means isotonic fluid uptake (water and salt)

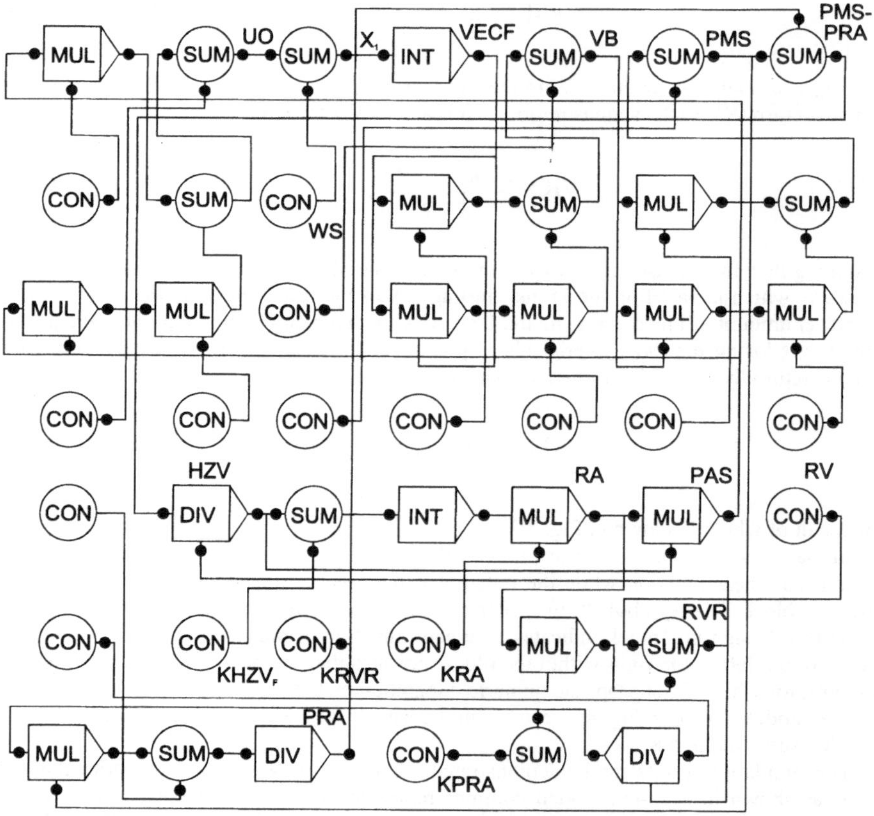

Fig. 4.8. SIDAS block diagram of the mathematical model of the renal influenced long term arterial blood pressure control, shown in Fig. 4.7.

Fig. 4.9.shows the steady state values of the system for the mean arterial blood pressure (PAS), the mean systemic Pressure (PMS), the total peripheral resistance (RA), the blood volume (VB), the cardiac output (VR), and the urine output (UO) for the normal case and experimental-induced hypertension based on the reduction of two-thirds of the renal mass – so-called Goldblatt clip – , for different isotonic water and sodium loads:

  (a) Normal case with isotonic water and sodium load WS = 1 mL/min 0.9 g/dl NaCL
  (b) Removal of two-thirds of the renal mass with isotonic water and sodium load WS = 1 mL/min 0.9 g/dl NaCl
  (c) (emoval of two-third of the renal mass with increased isotonic water and sodium load WS = 2.5 mL/min 0.9 g/dl NaCL

Fig. 4.9. one can conclude that the simulation results (white columns) are in good accordance with the experimental results (dotted columns). One expection can be seen for PAS and VB in case (c). The results can be understood as follows: The higher mean arterial pressure (PAS) is caused by the higher value of the total peripheral resistance (RA). The reason for the higher total peripheral resistance is physiologically based on the myogenic autoregulation, which has been implanted in the model. Hence, the model and the real-world system are in good accordance. Discussing the other haemodynamic parameters will result in similar good accordance between the model and the real-world system, Goldblatt hypertension.

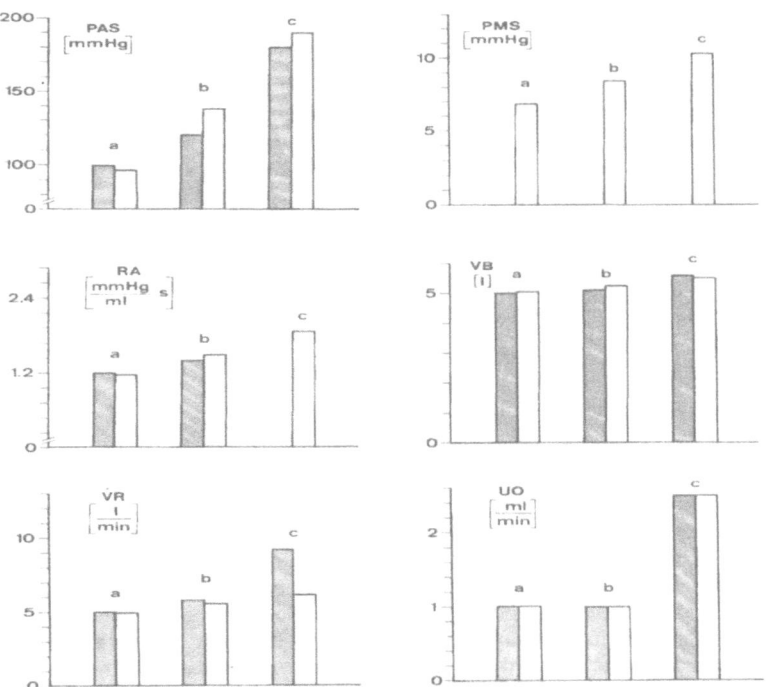

Fig. 4.9. Simulation and experimental results showing the influence of an isotonic water and sodium load for a normal kidney and for the removal of two-thirds of the renal mass

4.3.2 Equation-Oriented Simulation Software

It has been mentioned that the widespread use of personal computers and worksta-tions had a profound impact on the use of simulation software. Beside the interac-tive block oriented simulation software systems, described in Sect. 4.3.1, equa-tion-oriented simulation software systems had been developed for the direct im-plementation of the mathematical equation into the simulation software. Today a wide variety of different equation-oriented simulation software systems are avai-lable. As a typical example of an equation-oriented simulation software, the wide-ly known CSMP (continuous system modeling program) will be briefly intro-duced. The model-building process in CSMP is based on three steps:

- INITIAL
- DYNAMIC
- END

containing the respective structural, data, and control statements.

The structural statements transform the mathematical expressions of the real-world system through algebraic statements into functional blocks. The data state-ments connect the symbolic defined parameters, constants, and initial conditions, with the respective numerical values. The control statements schedule the time de-pendence of the output variables.

INITIAL contains the parameter values and initial values. DYNAMIC contains the model description. END terminates the simulation run, and contains the para-meters with its new values, which may be calculated in a second run.

Example 4.6
The ingestion and subsequent metabolism of a drug in a human individual should be simulated. The model used to study the ingestion, distribution and metabolism of the drug in the human individual is based on a two-compartment model, shown in Fig. 4.10.

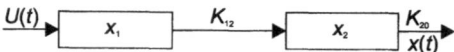

Fig. 4.10. Two-compartment model

Supposing that the drug is taken orally, it enters the gastrointestinal tract, is absorbed into the circulation and distributed throughout the body to be metabolized and finally elimi-nated. Compartment X_1 describes the gastrointestinal tract and the gastrointestinal vascular bed (circulation) of the individual; compartment X_2 stands for the bloodstream (between the distribution and elimination processes) of the individual, whereas K_{12} and K_{20} represent the distribution and elimination constants, respectively. Let us start at time zero, and let $X_1(t)$ denote the mass (concentration) of drug in compartment 1 and $X_2(t)$ be the mass (concen-tration) of drug in compartment 2. If the ingestion rate of the drug $U(t) > 0$, we find the plausible assumption for the two-compartment model that the rate of change of the mass of drug in the gastrointestinal tract is equal to the rate at which the drug is ingested minus the rate at which the drug is distributed from the gastrointestinal tract to the bloodstream:

$$\frac{dX_1(t)}{dt} = U(t) - \text{blood distribution rate compartment 1 to 2} . \tag{4.12}$$

(4.12) is commonly a mass-balance equation. In the case of first-order kinetics, the drug distribution rate from compartment 1 to 2 is assumed to be proportional to the mass (or concentration) of the drug in the first compartment. If $K_{12} > 0$ is the corresponding proportionality constant, then (4.12) becomes:.

$$\frac{dX_1}{dt} = U(t) - K_{12} \cdot X_1(t) , \tag{4.13}$$

where $K_{12} \cdot X_1$ is the inflow rate of drug distribution from the first compartment. Compartment 2 is described by a flow-rate equation that balances the inflow and outflow rates described by (4.14):

$$\frac{dX_1}{dt} = U(t) - K_{12} \cdot X_1(t) . \tag{4.14}$$

With respect to first-order kinetics, the outflow rate of compartment 2 is proportional to X_2. Thus (4.14) becomes:

$$\frac{dX_2}{dt} = \text{inflow rate} - \text{outflow rate} , \tag{4.15}$$

where K_{20} is the elimination constant.

(4.13) and (4.15) constitute the linear model of the pharmaceutical kinetics. In a matrix-vector format the constant coefficient linear differential equations yield (t > 0):

$$\begin{Bmatrix} \dfrac{dX_1}{dt} \\ \dfrac{dX_2}{dt} \end{Bmatrix} = \begin{Bmatrix} -K_{12} & 0 \\ K_{12} & -K_{20} \end{Bmatrix} \cdot \begin{Bmatrix} X_1(t) \\ X_2(t) \end{Bmatrix} + \begin{Bmatrix} U(t) \\ 0 \end{Bmatrix} , \tag{4.16}$$

with

$$U(t) = A \cdot e^{-K_{31} \cdot t} . \tag{4.17}$$

The model of the pharmaceutical kinetics described in (4.16) represents a second-order linear model. The first differential equation is uncoupled from the second differential equation, meaning there is no feedback from the second differential equation to the first differential equation, which means that the mathematical model not difficult for analytical studies. The second differential equation is coupled with the first differential equation.

The homogeneous differential equation of (4.13) is

$$\frac{dX_1}{dt} + K_{12} \cdot X_1(t) = 0 \,,$$
(4.18)

which can be solved using the approach

$$X_1(t) = e^{-\lambda \cdot t} \,.$$
(4.19)

Substituting (4.8) into (4.7) yields

$$-\lambda \cdot e^{-\lambda \cdot t} + K_{12} \cdot e^{-\lambda \cdot t} = 0 \,,$$
(4.20)

with the solution $\lambda = K_{12}$, which can be used in (4.19)

$$X_1(t) = e^{-K_{12} \cdot t} \,.$$
(4.21)

Therefore, the solution of (4.2) is

$$X_1(t) = c_1 \cdot e^{-K_{12} \cdot t} \,.$$
(4.22)

Consider the initial value as $X_1(0) = A$ we obtain with $c_1 = A$

$$X_1(t) = A \cdot e^{-K_{12} \cdot t} \,.$$
(4.23)

Substituting (4.23) into (4.15) results in

$$\frac{dX_2}{dt} = K_{12} \cdot A \cdot e^{-K_{12} \cdot t} - K_{20} \cdot X_2(t) \,,$$
(4.24)

which can be rewritten after multiplying by $e^{K_{12} \cdot t}$ as follows

$$\frac{dX_2}{dt} \cdot e^{K_{12} \cdot t} + X_2(t) \cdot K_{20} \cdot e^{K_{12} \cdot t} = K_{12} \cdot A \cdot e^{(K_{20} - K_{12}) \cdot t} \,.$$
(4.25)

The left side of (4.25) can be rewritten using the product rule for the calculus of differential equations, which has the form $\frac{d}{dt}\left(X_2(t) \cdot e^{K_{20} \cdot t}\right)$ hence, (4.25) can be integrated as follows:

$$X_2(t) \cdot e^{K_{12} \cdot t} = K_{12} \cdot A \cdot \int e^{(K_{20} - K_{12}) \cdot t} \cdot dt + c \ , \tag{4.26}$$

or

$$X_2(t) \cdot e^{K_{12} \cdot t} = \frac{K_{12} \cdot A}{K_{20} - K_{12}} \cdot e^{(K_{20} - K_{12}) \cdot t} + c \ , \tag{4.27}$$

which results in

$$X_2(t) = \frac{K_{12} \cdot A}{K_{20} - K_{12}} \cdot e^{-K_{12} \cdot t} + c \cdot e^{-k_{20} \cdot t} \ . \tag{4.28}$$

With the initial value $X_2(0) = 0$ we obtain

$$0 = \frac{K_{12} \cdot A}{K_{20} - K_{12}} + c \implies c = -\frac{K_{12} \cdot A}{K_{20} - K_{12}} \ , \tag{4.29}$$

and finally

$$X_2(t) = \frac{K_{12} \cdot A}{K_{20} - K_{12}} \cdot (e^{-K_{12} \cdot t} - e^{-k_{20} \cdot t}) \ . \tag{4.30}$$

The model of the pharmaceutical kinetics described in (4.16) can be implemented directly in common simulation systems. For the simulation software CSMP the model formalization is as follows:

```
INITIAL
        PARAMETERS K12=1.0, K20=0.5
        CONSTANT A=0., K31=0., X10=100., X20=100.
DYNAMIC
        X1DOT = - K12·X1 + U
        X2DOT = K12·X1 - K2·X2
        X1    = INTGRL (X10, X!DOT)
        X2    = INTGRL (X20, X2DOT)
        U     = A·EXP(-K31·TIME)
TIMER  DELT=0.1, OUTDEL=0.2, FINTIM=10.0
  PRTPLT U, X1, X2
  END
  PARAMETER A=100.0
  END
  STOP
```

The transient behavior of the derived two-compartment model can be investigated in some case study examples. The primary interest in studying the two-compartment model is to govern how input ingestion rate and/or the initial concentration of the drug in the body affect the subsequent amounts of drug in the bloodstream of the individual. The variable $X_2(t)$ – and hence compartment X_2 – is of great importance, because it is accessible for analysis by taking blood samples. For this purpose the behavior of $X_2(t)$ can be shown in Figs. 4.11, 4.12, and 4.13, using the two-compartment model described above.

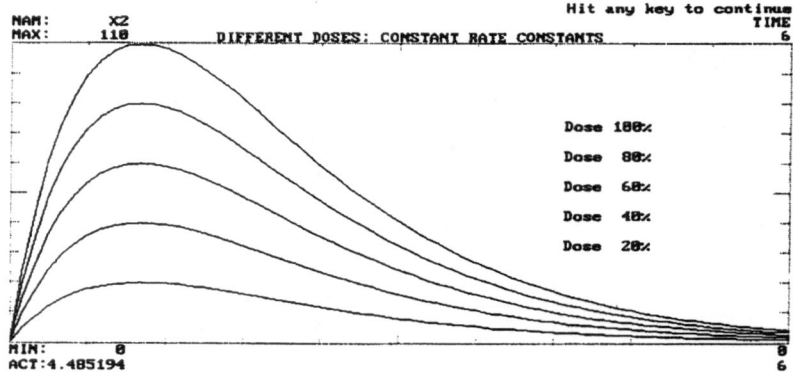

Fig. 4.11. Change of the dose with single oral intake

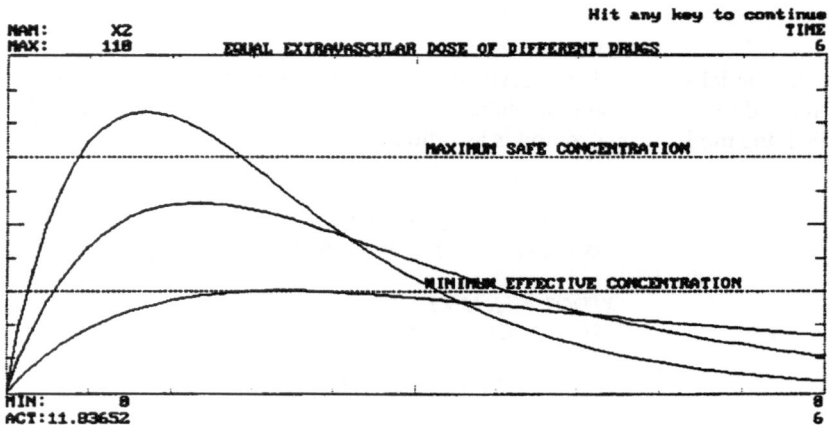

Fig. 4.12. Change the drug or its form with single oral intake to demonstrate the relation to the minimum effective and maximum safe concentration

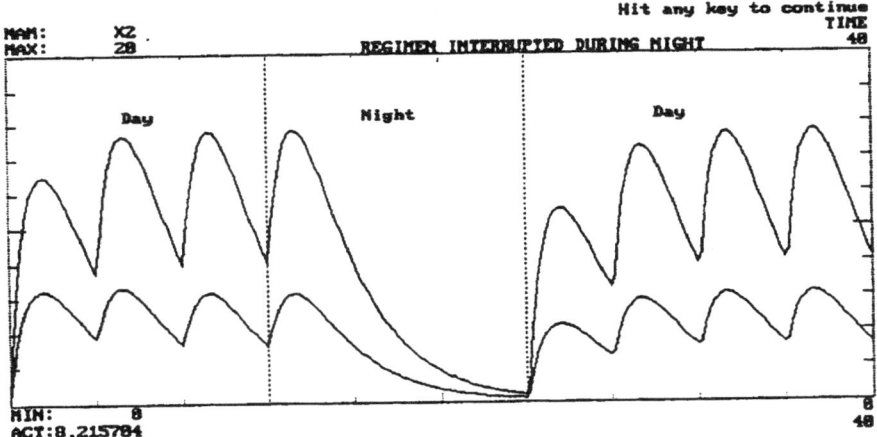

Fig. 4.13. The effect of interruption during night hours of a multiple-dose regimen

4.3.3 General-Purpose Simulation Software

The simulation models built are mostly either language or platform dependent, which can be block- or equation-oriented. Although they still provide valuable simulation results, these models lack the ability to be easily ported to another software, or easily capable of interoperation to generate larger and more complex models. Compared with the described bock-oriented and equation-oriented simulation systems in Sects. 4.3.1 and 4.3.2, general-purpose simulation systems are a group of simulation software, that are based on equations rather than on blocks allowing the user to handle much more complex models. A typical representative of general-purpose simulation software is the widely known simulation system ACSL (advanced continuous simulation language).

General-purpose simulation systems such as ACSL are compiler based. Hence a modification of the model structure requires the translation of the model, compilation, and linking with the required software tools. Consequently, this simulation software is less interactive and suited for the experienced user. Due to this ACSL has been developed for modeling of dynamic systems by time-dependent differential equations and/or transfer functions. Although the dynamic system being modeled are time dependent, the independent variable can be something other than time, such as distance or angle. Typical application areas where ACSL currently is being applied include: biomedical systems, chemical process representation, control systems design, heat transfer analysis, fluid flow, missile and aircraft simulations, plant and animal growth, power plant dynamics, robotics, toxicological models, vehicle handling, etc. The current ACSL information can be found in the respective handbooks and reference documentation as well as on the web, www.aegis.com.

While ACSL is a general-purpose simulation system, models can be developed from:

- Block diagrams
- Mathematical equations
- Other conceptualizations of systems

but probably the most common ones are free-body diagrams of physical systems, as shown for the pendulum example in Fig. 4.14.

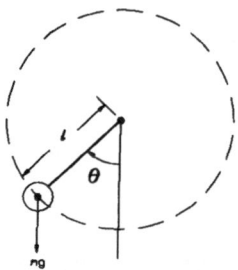

Fig. 4.14. Pendulum on a rigid rod

Example 4.7

Let us use the diagram in Fig. 4.14 to describe the motion of a nonlinear pendulum. The terms for this example are as follows:

m = mass [kg]
W = weight [N]
g = gravity [m/s²]
l = length of rod [m]
k = viscous damping [kg/(m/S)]
Θ = deflection of road from vertical [rad]
$\dot{\Theta}$ = rate of change of Θ [rad/s]
$\ddot{\Theta}$ = acceleration of Θ [rad/s²]
F = tangential force [N]

The tangential force, which acts to increase the angle Θ, can be written in two different forms. The first one results in:

$$F = -W \cdot \sin\Theta - k \cdot l \cdot \dot{\Theta} .$$ (4.31)

Secondly, with a constant mass, Newtons law can be expressed as:

$$F = m \cdot l \cdot \ddot{\Theta} . \tag{4.32}$$

Combining (4.31) and (4.32) results in the differential equation form

$$m \cdot l \cdot \ddot{\Theta} + k \cdot l \cdot \dot{\Theta} + W \cdot \sin \Theta = 0 . \tag{4.33}$$

The ACSL translator handles the differential equations and their numeric integration. Differential equations have to be written in terms of their highest-order derivative, while the integral will be calculated with the INTEG statement. In the pendulum example the acceleration of the angle $thdd \approx \Theta''$ is the highest derivative that can be expressed by rearranging (4.33), which results in

$$thdd = -\frac{(k \cdot l \cdot thd + W \cdot \sin(th))}{m \cdot l} . \tag{4.34}$$

The angular rate $thd \cong (\dot{\Theta})$ and position $th \cong (\Theta)$ are obtained by integration

$$thd = INTEG(thdd, thdic)$$

$$th = INTEG(thd, thic) ,$$

where $thdic$ and $thic$ are the initial conditions of thd and th, respectively. They form the model. Other statements in the program can support these equations or control the execution of the program. These equations can be written in ACSL in the order given even though th and thd are used in the $thdd$ equation.

In the ACSL language a program can be written without any structural statements other than the PROGRAM and the END statements, which are known as implicit structure, meaning all code is implied to be in a DERIVATIVE section. An explicit program describes the structure explicitly, meaning it allows to separate code, as shown in Fig. 4.15.

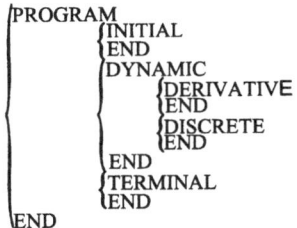

Fig. 4.15. Explicit structure in ACSL

The respective sections are:

- INITIAL: appears at the beginning of the program before time moves forward, and is evaluated only once per run. Calculations in this section involve initial conditions of state variables or the initialization of counters. The pendulum example needs INITIAL for the calculation of the conversion factor degrees per radian.
- DYNAMIC: moves forward in time. Within the DYNAMIC section, the DERIVATIVE section moves forward in a continuous manner, controlled by the integration algorithm. In the pendulum example, the angle *th* is converted to degrees in the DYNAMIC section.
- DERIVATIVE: contains differential equations and integrations. ACSL sorts the equations deciding what has to be calculated in what order.
- DISCRETE: are at the same level as DERIVATIVE, but are activated by INTERVAL or SCHEDULE statements, to describe discrete events.
- TERMINAL: is executed once after time has stopped. When program control transfers out of the DERIVATIVE or DYNAMIC section in response to the TERMINAL statement, it moves to the beginning of the TERMINAL section. TERMINAL is used for statistical calculations such as missile miss distance. Also parameters of the model can be changed in the TERMINAL section.

With this background and the explicit structure, shown in Fig. 4.15, the program listing for the pendulum example can be written in ACSL as follows:

PROGRAM Damped Nonlinear Pendulum

```
INITIAL
          !----Conversion factor, deg/rad
          DPR = 45.0 / (ATAN(1.0))
  END     ! of INITIAL

DYNAMIC
DERIVATIVE
          !----Integration algorithm and step size
          ALGORITHM   IALG = 4
          MAXTERVAL   MAXT = 0.0125
          NSTEPS        NSTP = 1

          !----Constants of model (units in kg m s)
          CONSTANT   mass   = 1.0
          CONSTANT   length = 0.5
          CONSTANT   kdamp = 0.3
          CONSTANT   thdic  = 0.0
          CONSTANT   thic   = 1.0
          CONSTANT   g      = 9.81

          !----Angular acceleration of mass
          thdd=(mass·g·sin(th)+kdamp·length·thd)/ (length·mass)
```

```
            !----Integrate for angular velocity and position
            thd    = INTEG (thdd, thdic)
            th     = INTEG (thd, thic)
END         ! of Derivative

            !----Communication interval
            CINTERVAL  CINT   = 0.025

            !----Termination condition
            CONSTANT  tstop = 4.99
            TERMT (T. GE. Tstop)

            !----Angle in degree for output
            xth    = DPR·th
END         ! of Dynamic

TERMINAL
            !----Call for debug dump
            LOGICAL dump
            CONSTANT dump = .FALSE
            IF (dump) CALL DEBUG

END         ! of Terminal
END         ! of Program
```

It should be noted, that the program controls, as shown in the program listing above, can be changed at runtime.

The communication interval CINTERVAL controls the frequency at which the DYNAMIC section is executed. This is where the values of the variables on the PREPARE list are logged to the intermediate data.

The integration algorithm IALG, maximum step size MAXT, and number of steps per communication interval NSTP have been placed in the DERIVATIVE section. IALG is an integer between 1 and 9. The default IALG is 4, which means Runge Kutta fourth-order numeric integration method (see Appendix A).

Output from ACSL is divided into the two categories high and low volume. This distinction is made while the screen should not be overwhelmed with data. Low-volume output consists of error messages and the result of DISPLAY, OUT-PUT, and RANGE commands. The results of all other commands are considered high-volume HVDPRN. PRINT /ALL asks for all variables on the PREPARE list to be printed in columns, and PRINT /NCIPRIN is the number of communication intervals per print line. If /NCIPRN is 10, every tenth line is printed. With SPARE commands before and after START, simulation execution CPU time is given. The PROCEDURE statement in the example runtime file contains several commands. With go, each command is executed in turn. The respective ACSL runtime command file for the pendulum example are shown in Fig. 4.16.

```
SET HVDPRN=.TRUE.
SET TITLE='Nonlinear Pendulum Example'
PREPARE t, th, xth, thd, thdd
PROCEDURE go
SPARE ; START ; SPARE
PRINT /ALL /NCIPRN=10
END
```

Fig. 4.16. Runtime command file for the pendulum example in ACSL

Results of using ACSLs runtime commands are shown in Fig. 4.17.

```
ACSL Runtime Exec Version 6  Level 10a  Page 1
SET HVDPRN=.TRUE.
SET TITLE='Nonlinear Pendulum Example'
PREPARE t, th, xth, thd, thdd
PROCEDURE go
SPARE ; START ; SPARE
PRINT /ALL /NCIPRN=10
END
END of file found on unit 4
Reverting to logical unit number 5
go
SPARE
Accumulated cp time  263.640000 , Elapsed cp time  0.
START
SPARE
Accumulated cp time 264.550000 . Elapsed cp time  0.91000400
PRINT /ALL /NCIPRN=10
```

Line	T	TH	XTH	THD	THDD
0	0.	1.0000000	57.2958000	0.	-16.50970000
10	0.25	0.52727300	30.2105000	-3.466430	-8.83224400
20	0.50	-0.42950000	-24.5942000	-3.403780	9.18676000
30	0.75	-0.88829600	-50.8956000	-0.0216652	15.23160000
40	1.00	-0.46029400	-26.3729000	3.1363300	7.74530000
50	1.25	0.39687100	22.7390000	3.0172500	-8.48898000
60	1.50	0.79093400	45.3172000	-0.08855000	-13.9235000

Fig. 4.17 ACSL pendulum log (PRN) file

It is trivial to say but plots are much more intuitive than data columns. Hence simulation software has plotting options embedded. Time plots for the ACSL pendulum example are using the following command:

ACSL> PLOT th, thd

Fig. 4.18 shows the plot calculated with the ACSL. All scales are chosen automatically.

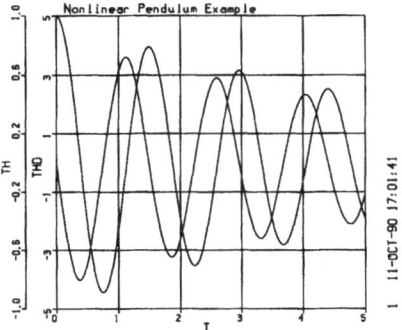

Fig. 4.18. Pendulum time plot

Using tags and symbols in a plot has the following commands in ACSL:

ACSL> SET TITLE(41) = İnitial angle 1.0 radians`
ACSL> SET SYMCPL=.T.,NPCCPL=20
ACSL> PLOT /XTAG='(sec)' th /TAG='(rad)' /CHAR=`t` & , thd
*   /TAG=`(rad/s)' /CHAR=1*

TITLE is Nonlinear Pendulum Example, but more information can be added. SYMCPL is a logical to determine whether symbols are to be drawn on the curve; NPCCPL is the number of data points between the symbols. PLOT /XTAG adds a string of characters to the X-axis label. PLOT /TAG adds a string of characters to the Y-axis label. PLOT /CHAR specifies the character to be drawn on the curve if SYMCPL is TRUE. The respective example is shown in Figure 4.19.

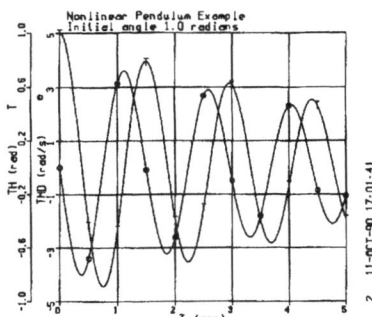

Fig. 4.19. Pendulum tags and symbols

Another possibility for plotting options are the phase plane plots, as shown in Figu. 4.20. A phase plane plot is a graph over state variables, i.e. X_1 over X_2. For a given set of initial conditions $X_1(0) = X_{10}$, and $X_2(0) = X_{20}$, the differential equation of the dynamic system, for the pendulum example of (4.10), have a unique solution $X_1 = \Phi(t)$, and $X_2 = \Psi(t)$, for $t > 0$. In this case it is helpful to represent the unique solution as a curve in the $X_1 X_2$ plane; the phase plane with t as parameter, marking the respective time stamps of the $X_1 X_2$ tuples.

For a phase plane plot of the velocity thd (X_1) versus the angle th (X_2) in ACSL we need the following commands:

```
ACSL> SET XINCPL=4, NPCCPL=40
ACSL>PLOT /XAXIS=th /XTAG='(rad)'& ,thd /HI=4  /TAG='(rad/s)'
        /CHAR=5
```

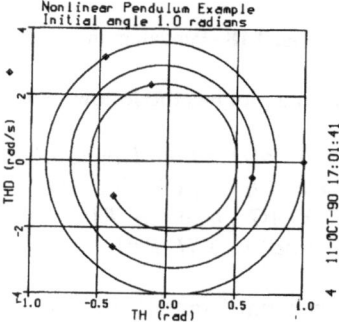

Fig. 4.20. Pendulum phase-plane plot

XINCPL sets the length of the X-axis (X_2), and Y-axis (X_1) equal, hence the result is a square plot with the zero-zero point in the middle. PLOT /XAXIS /XTAG specifies th as the X-axis variable. PLOT /CHAR, NPCCPL are set to produce a centered character.

4.3.4 Component-Based Simulation Software

Besides the previously introduced simulation systems in Sects. 4.3.1, 4.3.2, and 4.3.3, another trend in simulation software is towards component-based systems that provide a wide range of model parts – the components – which can be used for model building. A typical representative of a component based simulation software is ModelMaker. Model building with ModelMaker first involves constructing a diagram on the screen that represents the various model parts. This diagram is composed of a series of ModelMaker components, each of which are intended for a different type of mathematical operation. Each component has a definition that can be edited to insert its equation and any other information. ModelMaker can be applied to all areas of modeling in science, environmental science, mathematics, physics, chemistry, sociology, ecology, life science, etc. Using the component-based approach, first one can visualize the system and then built the model. ModelMaker's extensive range of functions allows to one model just about any system functionality: continuous and discontinuous functions, stiff systems and stochastic systems. The user can rationalize the results using ModelMakers analysis methods like optimization, minimization, Monte Carlo and sensitivity analysis. The Monte Carlo analysis enables model parameters to be specified as random distributions, while a model runs for a specified number of times, and during each run the value of the parameter is taken from a specified probability distribution. Hence, the effect of varying parameters of a component value can be observed. The current ModelMaker information can be found on the web, www. Modelkinetix.com.

The model building and analyzing process in ModelMaker contains several steps:

- Building a model with the respective components:

Compartments	Represent containers or integrators in the model
Flows	Signify transport of a quantity from one compartment to another
Variables	Values that are calculated as the model is run, according to the equations
Defined values	Values that are calculated at the start of a run or in response to the actions of an event
Influences	Indicate where the value of a component affects the value of another
Delays	Delay the value of another component for a defined period of time
DLL-functions	Add new functionality – design and integrate an own DLL
Sub-models	Create models within models
Lookups	Incorporate external numerical data into the model
Events	Adjust the values of other components and cater for discontinuous models
Parameters	Store constant values in the model
Text boxes	Add informative text or pictures to the model

- Running the model:

To run a model the user can choose for calculation one of the different numeric integration methods such as Runge Kutta, Mid Point, Euler, Bulirsch Stoer and Gear (see Appendix A). ModelMaker is also an appropriate solver for stiff systems simulation where processes happen on very different time scales. Other simulation features of the simulation language of ModelMaker are:

- User defined or adaptive output points
- Fixed or variable step length
- Error scaling

One also can use the repeated run facility to run a model several times. Moreover, ModelMaker provides the ability to analyze:

- Periodic models
- Stochastic models

Calculated values produced by running a model can be compared with experimental data by using optimization methods. During optimization, selected model parameters are systematically adjusted to find the best fit between the model and the experimental data. ModelMaker offers for optimization iterative numerical methods like the Marquardt or the Simplex methods. They are very powerful optimization methods but can be very time consuming. They may not always find the best parameter values fit and may be, in certain circumstances, simple. Simple analytical methods such as linear regression are not generally applicable. For optimization purposes ModelMaker offer a comprehensive statistical reporting.

- Reporting results:

Once the model has been simulated and analyzed the user can generate tables of results, graphs, and statistics. The respective features include:

- Fully customized graphs
- Comprehensive statistical output
- Tabulation of model values
- Cut and paste tools that allow to incorporate data, models, graphs, etc. into other applications.

Example 4.8
An important class of systems are those based on electrical networks, consisting of resistors (R), capacitors (C), and inductors (L). When building models based on electrical RCL networks we can use differential equations as the respective mathematical representation. For this class of systems we can apply the principle of physical isomorphism, while the real-world system, which is introduced as an electrical RCL network, can be described by an analogous structured second system, which is the one used for system analysis. The second system can be instead of the original electrical RCL network, a mass damper system. Both systems are shown in Fig. 4.21, which shows that an electrical system can also be modeled through a mechanical mass damper spring system and vice versa.
The mass damper spring system can be described by the second-order differential equation

$$M \cdot \ddot{x} + D \cdot \dot{x} + C \cdot x = F(t) , \qquad (4.35)$$

with M as mass, D as damping factor, C as spring constant, and X as elongation.

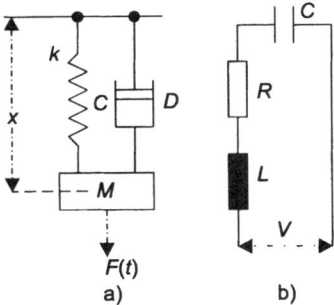

Fig. 4.21. Mass-Damper-Spring-System (a) and LRC-Network (b)

The differential equation of the isomorphic electrical RCL network can be described by the differential equation

$$L \cdot \ddot{q} \cdot R \cdot \dot{q} + \frac{q}{C} = \frac{V(t)}{C} \ ,$$ (4.36)

with L as inductance, R as resistance, C as capacitance, q as charge, and $V(t)$ as time dependent voltage source.

The comparison of both differential equations show that they have the same structure, due to the same mathematical notation concept behind, which can be rewritten in the general system description formula

$$A \cdot \ddot{x} + B \cdot \dot{x} + C \cdot x = D(t) \ ,$$ (4.37)

which results in the following correspondences, shown in Table 4.2.

Table 4.2. Correspondences between various dynamic systems

Mathematical model	Mechanical system	RCL network
x: state-space	x: oscillation	q: charge
x': derivative of x	x': velocity	i: current
A: system parameter	A: mass	L: inductance
B: system parameter	D: damper	R: resistance
C: system parameter	C: spring	C: capacitance
$D(t)$: input function	$F(t)$: force	$E(t)$: voltage

As Table 4.2 shows, the mechanical system notation and the electrical RCL network notation are physical systems for each other, which means that one system can be used for the other one describing the one system, but both notations can be described through the same mathematical model description formula. Hence we have a transform for system description, which can be expanded, if necessary.

(4.37) can be rewritten as

$$x'' + \frac{B}{A} \cdot x' + \frac{C}{A} \cdot x = \frac{D(t)}{A} .$$
(4.38)

Rewriting this differential equation, which is of second order, using n first-order differential equations we find:

$$x = x_1$$
$$x' = x_1' = x_2$$
$$x'' = x_2' = -\frac{B}{A} \cdot x' - \frac{C}{A} \cdot x + \frac{D(t)}{A}.$$
(4.39)

This results, due to the original second-order system, in two first-order differential equations:

$$x_1' = x_2$$
$$x_2' = -\frac{B}{A} \cdot x_2 - \frac{C}{A} \cdot x_1 + \frac{D(t)}{A}.$$
(4.40)

We can now introduce the first derivation $\dot{x}_1$ as Integrator1 and the second derivation $\dot{x}_2$ as Integrator2. These two integrators can be solved in ModelMaker by using compartments. They are defined by a symbol, a differential equation and initial values, and produce a series of values as output. The differential equation is solved as a function of the independent variable t by default.

Integrator 1 $x' = x_1' = x_2$
(4.41)

Integrator 2 $x'' = x_2' = -\frac{B}{A} \cdot x' - \frac{C}{A} \cdot x - \frac{D(t)}{A} .$
(4.42)

From these equations we see that Integrator1 and Integrator2 have an interrelationship which means each of the values of Integrator1 influences Integrator2, which also means that the values of Integrator2 affect the values of Integrator1. In ModelMaker this relationship is represented by arrows as follows:

Fig. 4.22. Bidirectional coupling of integrators

Note that the two arrows and the dotted line, as used in ModelMaker, represent the bidirectional influences.

The parameters that influence the integrators are obtained from the dynamic systems equations, which are given in (4.38). This equation is the equation of the Integrator2, which means it depends on the parameters A, B, C, and D. Changes in these parameters will influence the Integrator2. Due to this fact we can represent the model of the dynamic system in ModelMaker as shown in Fig. 4.23.

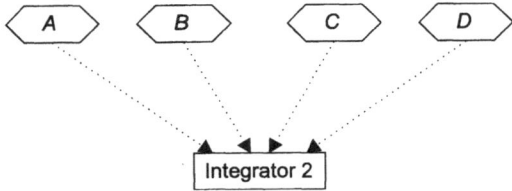

Fig. 4.23. ModelMaker implementation of the RCL network of Fig. 4.21b

Building the model in ModelMaker contains three categories, which are:

1. Define compartments:
   Integrator1: which is Integrator2 with equation $\dot{x} = \dot{x}_1 = x_2$
   Integrator2: which is $\ddot{x} = \dot{x}_2 = -\dfrac{B}{A} \cdot \dot{x} - \dfrac{C}{A} \cdot x + \dfrac{D(t)}{A}$,

   which can be represented by:

$$-\frac{B}{A} \cdot Integrator2 - \frac{C}{A} \cdot Intergrator1 + \frac{D}{A} \tag{4.43}$$

2. Define constants:
   A = inductance
   B = resistance
   C = capacitance

3. Define variable:
   D = voltage = $\cos(t)$

All elements are interconnected and we can study the dynamic system behavior for the several cases of interest, which can be done by changing the parameters A, B, and C, as well as the variables; assuming that the initial value can be 1 at both, Integrator1 and Integrator2. The model chosen to represent this system can be based on electrical RCL networks, as described above. The significant system variables are voltage, charge, and current.

With the system description of (4.37)

$$A \cdot \ddot{x} + B \cdot \dot{x} + C \cdot x = D(t), \tag{4.44}$$

which can be transformed into a series of first-order differential equations, for the implementation of ModelMaker's simulation model we use (4.20) and receive the resultant ModelMaker model, shown in Fig. 4.24

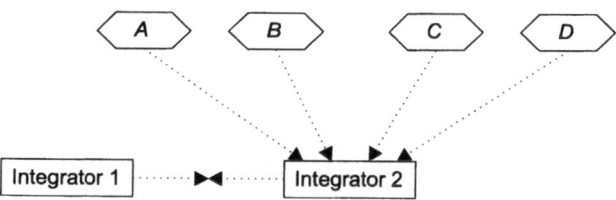

Fig.: 4.24. ModelMaker implementation model of the RCL network of Fig. 4.22 including the coupling with Integrator1

The following correspondences can be applied to the model:

- A = inductance
- B = resistance
- C = capacitance
- D = voltage
- Integrator2 = $\dot{x}_2$
- Integrator1 = x_1

A, B, and C are used as defined values, being the model parameters to be varied during simulation. The values of A, B and C do not vary during a single simulation run, their values are entered in the model as defined values. D is equal to $\cos(t)$ for all case study examples. The value of the cosine function is a variable over the course of a simulation run, it is placed inside a variable block. Integrator2 represents the current that flows through the system. Values A, B, C, D, and Integrator1 influence the calculation of Integrator2, and each of which must be connected to Integrator2 using an influence arrow type. Integrator1 represents the charge in the system that is influenced by the calculation performed in Integrator2, hence an influence arrow has to be inserted from Integrator2 to Integrator1. The model built up allows easy manipulation of the system parameters to be tested.

The ModelMaker implementation is ultimately determined by the mathematical model, which in any case is the basis for a simulation run.

Example 4.9
Case Study 1: $A = 1$; $B = 1$; $C = 1$
Represents a resistive circuit where capacitance and inductance are equal. Charge (Integrator1) and current (Integrator2) reach the state of harmonic oscillation at $t = 6$. The ampli-

tude of the oscillations of charge and current is near 2.0 (range of −1.0 to 1.0), as shown in Fig. 4.25.

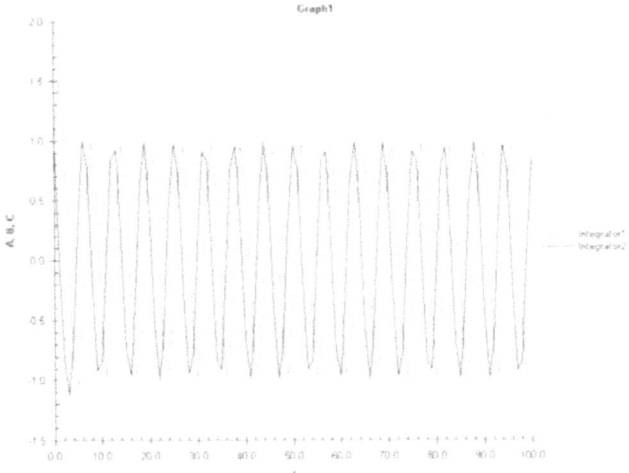

Fig. 4.25. Graph of case study 1 with $A = 1$, $B = 1$ and $C = 1$

The graph shows that charge and current reach a harmonic state very quickly, which is a result of the balance of resistance, capacitance, inductance, and the absence of a significant disturbance, which can be introduced by a large value of the resistance. The amplitude of the waves (i.e., current) is a result of the small resistance against the voltage.

Analytically, a motion is classified as being harmonic when the acceleration at any time is proportional in magnitude and opposite in sign to the displacement. Considering an object in a rectilinear motion with coordinate x, it will undergo a simple harmonic motion if

$$\frac{dx^2}{dt} = -\omega^2 \cdot x , \qquad (4.45)$$

where ω^2 is a constant of proportionality. It can be easily verified that the solution for the differential equation describing the simple harmonic motion is

$$x = a \cdot \sin \phi t + b \cdot \cos \omega t , \qquad (4.46)$$

from which

$$x = \Omega \cdot \sin(\omega t + \phi) , \qquad (4.47)$$

where a, b, and Ω, ϕ are inter-related integration constants representing the characteristics of the oscillation. These constants can be evaluated from the initial or boundary conditions of the motion. By examination of (4.47) it can be seen that ω is the frequency of the motion, while Ω and ϕ are its amplitude and phase angle respectively. Differentiating (4.47) with respect to time we obtain the expressions for the velocity and acceleration as

$$v = \frac{dx}{dt} = \omega \cdot \Omega \cdot \cos(\omega t + \phi) = \omega \cdot \Omega \cdot \sin(\omega t + \phi + \frac{\pi}{2}),$$ (4.48)

and

$$a = \frac{dx^2}{dt^2} = -\omega^2 \cdot \Omega \cdot \sin(\omega t + \phi) = \omega^2 \cdot \Omega \cdot \sin(\omega t + \phi + \pi).$$ (4.49)

These expressions show that velocity and acceleration can also be represented by vectors rotating with the angular frequency ω. These vectors have amplitudes $\omega \cdot \Omega$, and $\omega^2 \cdot \Omega$, and phase angles $\pi/2 + \Phi$, and $\pi + \Phi$ respectively.

From (4.48) and (4.49) we can conclude that for on harmonic motion, or any oscillation for that matter, the vectors representing the velocity and acceleration lead the displacement vector by $\pi/2$ and π radians, respectively. Therefore, the phase difference between the two is of importance.

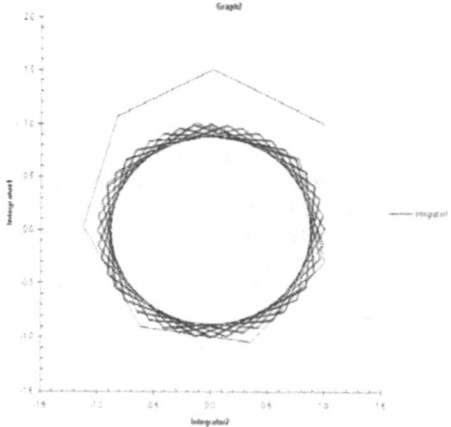

Fig. 4.26. Phase plane for case study 1 with $A = 1$, $B = 1$, and $C = 1$

The phase-plane representation in Fig. 4.26 shows that the system tends to zero (the resultant circle is centered at (0,0)), which means the system is stable. The diameter of the circle in Fig. 4.25 is 1.0, which is equal to the amplitude of the waves, shown in Fig. 4.25. The small resistance value results in a system reaching the stable state in a relatively short time period.

Example 4.10
Case Study 2: $A = 10$, $B = 10$, $C = 10$
Represents a resistive system where capacitance and inductance are equal. Charge (Integrator1) and current (Integrator2) reach the state of harmonic oscillation at $t = 19$. The amplitude of the oscillations of charge and current is 0.2 (range of -0.1 to 0.1), is shown in Fig. 4.27.

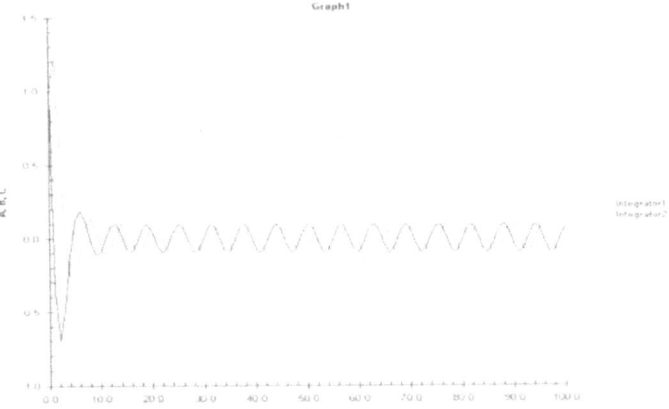

Fig.: 4.27. Graph I of case study 2 with $A = 10$, $B = 10$, and $C = 10$

The graph in Fig. 4.26 shows that charge and current reach a harmonic state in a relative shortly time. The amplitude of the waves (i.e. the strength of the current) in Fig. 4.27 is limited by the higher resistance values.

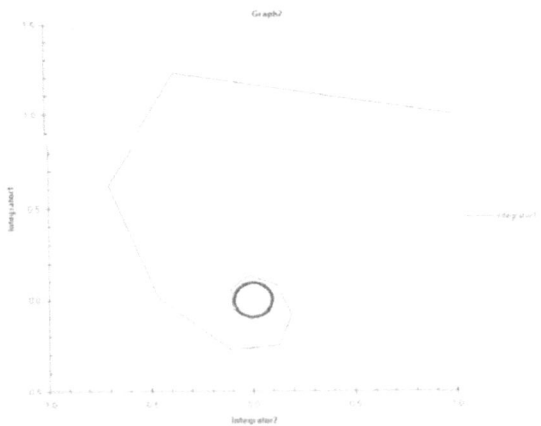

Fig. 4.28. Graph II of case study 2 with $A = 10$, $B = 10$, and $C = 10$

The graph in Fig. 4.28 shows that the system tends to zero (the resultant circle is centered at $(0,0)$), which means that the graph illustrates a stable system. The diameter of the circle in Fig. 4.28 is 0.5, which equals the amplitude of the waves as shown for graph I in Fig. 4.27. Compared with Fig. 4.26, it can be concluded that the system needs a longer time to reach the stable state.

Example 4.11
Case Study 3: $A = 1$, $B = 10$, $C = 1$

Represents a resistive circuit because capacitance equals inductance. Charge (Integrator1) and current (Integrator2) reach a state of harmonic oscillation at $t = 44$. The amplitude of the oscillations of charge and current is 0.2 (range of −0.1 to 0.1), as shown in Fig. 4.29.

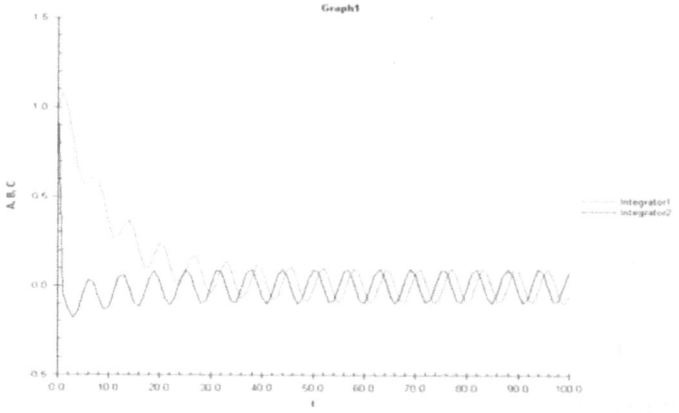

Fig. 4.29. Graph I of case study 3 with $A = 1$, $B = 10$, and $C = 1$

The graph in Fig. 4.29 shows that the system requires a long time to reach a harmonic state due of the dominance of the resistance in the system. The amplitude (i.e., the strength) of the current is the same as in Fig. 4.27, which results from the relationship between voltage, resistance, and current, while resistance is the same as in Fig. 4.27, and the resultant current is the same. The charge in the system, as shown for graph I, needs more time to reach the harmonic state relative to the current, as was true for the previous cases. Charge and current are out of phase initially, and their amplitude differs.

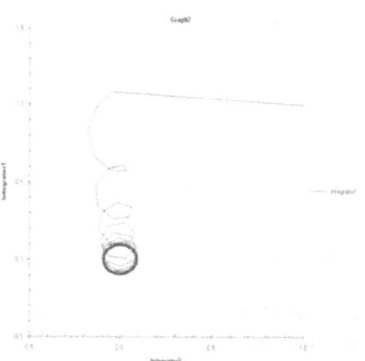

Fig. 4.30. Graph II of case study 2 with $A = 1$, $B = 10$, and $C = 1$

The graph in Fig. 4.30 shows that the system tends to zero, which means stability. In the case shown in Fig. 4.30, the system needs a relatively long time to reach the stable state, due to the imbalance of resistive, capacitive, and inductive forces in the system. The variance in amplitude initially between charge and current results in an irregular graph (the system ultimately reach an harmonic state).

Example 4.12
Case Study 4: $A = 1, B = 10, C = 5$
Represents a system that can be assumed as capacitive because capacitance (B) is greater than inductance (C). Charge (Integrator1) and current (Integrator2) reach a state of harmonic oscillation at $t = 12$. The amplitude of the oscillations of charge and current is 0.2 (range of -0.1 to 0.1), as shown in Fig. 4.31.

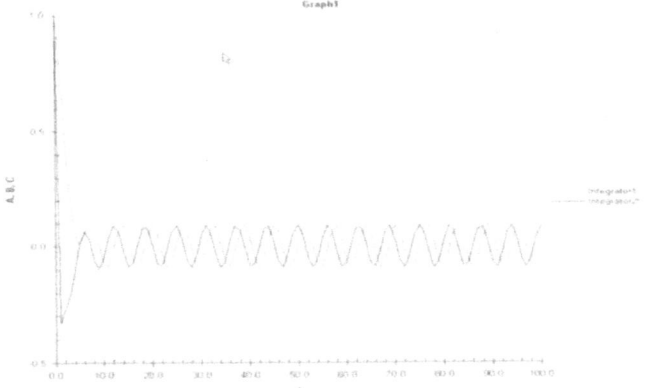

Fig. 4.31. Graph I of case study 4 with $A = 1$, $B = 10$, and $C = 5$

The graph in Fig. 4.31 shows that the charge and current reach an harmonic state in a short time as a result of the counterforce of an increased capacitance against the resistance that has not been altered from case study 3.

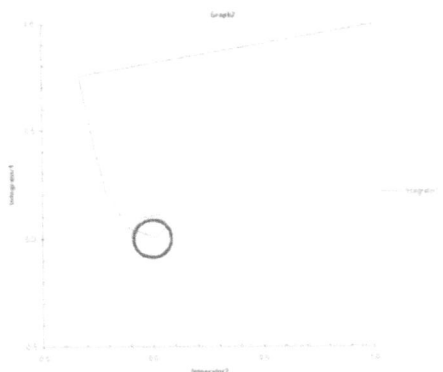

Fig. 4.32. Graph II of case study 4 with $A = 1$, $B = 10$, and $C = 5$

The graph in Fig. 4.32 shows that the system tends to zero, meaning a stable system. The center of the circle in Fig. 4.32 created graphically by the charge over the current plot is centered at zero with a diameter equal to the amplitude of the charge and current waves illustrated in Fig. 4.30. The progression of the system from its initial state to the stable state is relatively fast, as a result of the increase of capacitance against the larger resistance.

The simulation models are steady-state RCL networks enforced by an alternating current, represented by the function $D(t) = \cos(t)$, which generates a forced oscillation in the system with a predictable frequency. The results of the above case studies reveal the following:

- All cases are examples of damped harmonic oscillation; each set of variables results in harmonic oscillation of charge and current in electrical RCL networks, or oscillation and velocity in mechanical systems.
- All case studies result in the respective current of the electrical RCL network, or velocity in mechanical systems, which is in phase with voltage or force, respectively.
- The largest current is generated due to the impedance of the electrical RCL network $Z = (R^2 + (X_L - X_C)^2)^{1/2}$, in which $X_L = X_C$ has the smallest value of R. Case study 1 shows the largest current in this study.
- The electric current is demonstrated to be the electric charge per unit time; inversely, charge (represented by Integrator1) is shown to be a function of current (Integrator2) multiplied by time.
- The relation between current, voltage, and resistance, determined by Ohms law, is demonstrated in each case study, showing that voltage is the same for each case, the simulation proves that an increase in resistance results in a decrease in current.
- All case studies illustrate systems that reach a stable state.
- Increasing the value of parameter A, meaning increasing the mass (in a mechanical system) or inductance (in a RCL network system), the velocity can become smaller (in a mechanical system) or the current (in a RCL network system) will be lower. The Integrator2 (current) tends to the zero state if we the simulation runs for a longer time interval.
- Increasing the value of parameter B, meaning increasing the damper (in a mechanical system) or resistance (in a RCL network system), the time interval the system needs to reach a stable state or equilibrium states increase. The problem depends on the difference in amplitude and values between q (charge) and i (current).
- Increasing the value of parameter C, meaning increasing the spring (in a mechanical system) or capacitance (in a RCL network system), will make the system reach the equilibrium state.
- Using a different notation instead of using $\cos(t)$ for D, it might happen that the system can not fit to the stable state, and it can jump far away from the stable state. Using $D = \arctan(t)$ can be used as proof.

4.3.5 High-Performance Simulation Software for Technical Computing

Besides the previously introduced simulation systems in Sects. 4.3.1 till 4.3.4, another trend in simulation software is toward high-performance languages applicable for technical computing. These simulation software languages are mathematically and graphical-based combining the model parts using special scripts. Scripts are ASCI text files describing a sequence of statements and functions that can be saved and used as desired without having to recreate them each time they are needed. They are useful for automating series of simulation-language commands, such as computations that have to be performed repeatedly from the command line. Scripts operate on existing data in the workspace, or they can create new data on which to operate. Any variables that scripts create remain in the workspace after the script finishes so they can be used for further computations. A typical representative of a script-file-based simulation software is MATHLAB, which is a high-performance software for technical computing that integrates computation, visualization, and programming in an easy-to-use environment where problems and solutions are expressed in the mathematical notation. In MATLAB scripts are identified as M files, which contain the sequence of commands that are ordinarily processed following the command prompt. Several types of M files are used with MATLAB. The special toolboxes available for MATLAB are in fact made up of M files, developed for the particular application.

Example 4.13
The Collatz problem, namend after the German mathematician Collatz, born 1910 in Arnsberg, Germany, is to prove that the Collatz function will resolve to 1 for all positive integers. The M–files for this case study example have the filenames collatz.m and collatzplot.m. The file collatz.m generates the sequence of integers for any given n. The file collatzplot.m calculates the number of integers in the sequence for any given n and plots the results. For any given positive integer, n, the Collatz function produces a sequence of numbers that always resolves to 1. If n is even, divide it by 2 to get the next integer in the sequence. If n is odd, multiply it by 3 and add 1 to get the next integer in the sequence. Repeat the steps for the next integer in the sequence until the next integer is 1. The number of integers in the sequence varies, depending on the starting value, n.
TheMATLAB code for collatz.m (obtained from MATLAB M-File Section) is as follows:

```
function sequence=collatz(n)
% Collatz problem. Generate a sequence of integers resolving to 1
% For any positive integer, n:
%   Divide n by 2 if n is even
%   Multiply n by 3 and add 1 if n is odd
%   Repeat for the result
%   Continue until the result is 1%
sequence = n;
next_value = n;
while next_value > 1
    if rem(next_value,2)==0
        next_value = next_value/2;
```

```
else
    next_value = 3*next_value+1;
end
sequence = [sequence, next_value];
end
```

TheMATLAB code for collatzplot.m (obtained from MATLAB M-File Section) is as follows:

```
function collatzplot(n)
% Plot length of sequence for Collatz problem
% Prepare figure
clf
set(gcf,'DoubleBuffer','on')
set(gca,'XScale','linear')
%
% Determine and plot sequence and sequence length
for m = 1:n
    plot_seq = collatz(m);
    seq_length(m) = length(plot_seq);
    line(m,plot_seq,'Marker','.','MarkerSize',9,'Color','blue')
    drawnow
end
```

From Example 4.13 it can be seen that the name M file comes from the form of the filename, which is filename.m.

Statements in MATLAB have the form

$$>>\text{variable}=\text{expression},$$

where the two right arrows >> represent the command prompt, which precedes all commands, and variable identifies a variable, such as x, y, u, etc. The names for variables must begin with a letter and may include numbers and letters. For example, the expression $x = 9$ defines a variable x and assigns it a value of 9. Adding parentheses to the expression changes the order of processing. Modeling with MATLAB is initiated by opening an M–file window and typing in the respective commands necessary to describe the real-world system.

Example 4.14
Consider a first order RC network described by the voltage current relationship

$$V(t) = \left(1 - e^{-\left(\frac{1}{R \cdot C}\right)t}\right),$$

$$(4.50)$$

where the time constant is $T_C = R \cdot C$. The MATLAB command sequence is

$$>>V=(1-\exp(-t/T_C));$$

The MATLAB command sequence to plot the output of the first-order differential equation for an input of 1.0 and a time constant of 0.5 and the final time 3.0 is

>>T_C=0.5;
>>t=[0:0,1:3];
>>V=(1-exp(-t/T_C));
>>plot(t,V);
>>title('First Order System Response');
>>xlabel('Time');
>>ylabel('Output');
>>grid

where the x y plot is generated as the result of the command

>>plot(x,y);

where x and y have been previously defined or calculated by other command statements and expressions. Both the x-axis and the y-axis can be labeled with 'xlabel' and 'ylabel' commands, as shown above.

A plot can be given a descriptive title with the 'title' command

>>title('First Order System Response') .

The title command places the characters between the single quotation at the top of the plot.

MATLAB also provides several commands for manipulating, analyzing, and simulating systems in block-diagram form, which is helpful for control systems analysis and simulation. Very useful functions in MATLAB for the control systems domain are rlocus, which plots root locus from the translator function of the system, or Bode, Nichols, and Nyquist plots. MATLAB also includes functions for providing discrete-time systems. For example, the function c2d converts from continuous to discrete forms of system representation.

Furthermore, in MATLAB, functions exist in directories in the computer's file system. A directory may contain many functions (M–files). Function names are unique only within a single directory (e.g. more than one directory many contain a function called pie3). When typing a function name on the command line, MATLAB must search all the directories it is aware of to determine which function to call. This list of directories is called MATLABpath.

When looking for a function, MATLAB searches the directories in the order they are listed in the path, and calls the first function whose name matches the name of the specified function.

Consider writing an M–file, named pie3.m, and put it in a directory that is searched before the specgraph directory that contains MATLABs pie3 function, then MATLAB uses the pie3 function instead.

Object-oriented programming in MATLAB allows users to have many methods (MATLAB functions located in class directories) with the same name and enables MATLAB to determine which method to use based on the type or class of the variables passed to the function. For example, if p is a portfolio object, then pie3(p) calls@portfolio/pie3.m because the argument is a portfolio object.

Information about MATLAB is available on the web, at www.mathworks.org.

Together with SIMULINK, which has become the most widely used software language for modeling, simulating, and analyzing real-world systems, models can be easily built from scratch, or taken from existing models with the help of the tools offered by MATLAB-SIMULINK.

Simulations are interactive, which means that parameters can be changed during the simulation run and it can immediately be seen what happens. Furthermore, SIMULINK offers an instant access to all analysis tools of MATLAB, hence the results, obtained with SIMULINK, can be analyzed with MATLAB and visualized.

SIMULINK supports linear and nonlinear systems, modeled in continuous-time, sampled-time, or a mixture of both. Systems can also be multirate, i.e. have different parts that are sampled or updated at different rates.

For modeling, SIMULINK provides a graphical user interface (GUI) building models as block diagrams, using click-and-drag mouse operations. With this interface one can draw the models just as with pencil and paper, which is easier so as to formulate differential equations and difference equations in a language or program. For this purpose SIMULINK includes a comprehensive block library of:

- Sinks
- Sources
- Linear and nonlinear components
- Connectors
- etc.

and allows customizing and creating own blocks. Models in SIMULINK are hierarchical, which allows models be built top-down or bottom-up. Hence the user may view the system at a high level, then double-click on blocks to go down through the levels to see increasing levels of model details. This approach provides insight into how a model is organized and how its parts interact.

After a model has been built up it can be simulated, using a choice of integration methods, either from the SIMULINK menus or by entering commands in the MATLAB command window. The menus are particularly convenient for interactive work, while the command-line approach is very useful for running a batch of simulations (e.g. Monte Carlo simulations or sweeping a parameter across a range of values). Using scopes and the other display blocks users are able to observe the simulation results while the simulation is running. In addition, the interactivity of SIMULINK allows to be changed parameters and immediately shows what happens, for "what if" exploration. The simulation results obtained with SIMULINK can be transferred into the MATLAB workspace for post processing and visualization.

Model analysis tools include linearization and trimming tools, which can be accessed from the MATLAB command line, and the many tools in MATLAB and its application toolboxes. And because MATLAB and SIMULINK are integrated, the user can simulate, analyze, and revise his models in either environment at any point. The current MATLAB-SIMULINK information can be found on the web at www.mathworks.com.

Example 4.15
As an example we build a simple model in SIMULINK. The model integrates a sine wave and displays the result, along with the sine wave. The block diagram of the model is shown in Fig. 4.33.

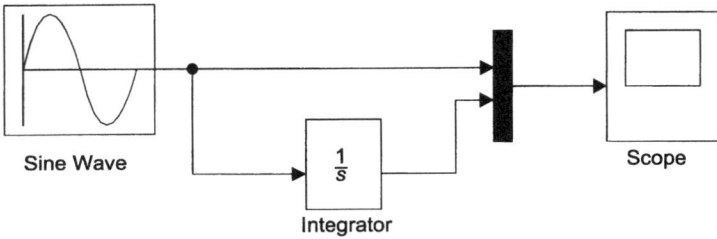

Fig. 4.33. Simple SIMULINK model

To create a model in SIMULINK, one has to type SIMULINK in the MATLAB command window. On Microsoft Windows, the SIMULINK Library Browser appears. Creating a new model on Windows, one has to select the **New Model** button on the Library Browsers toolbar. SIMULINK now opens a new model window. To create the model shown in Fig. 4.33, one needs to copy blocks into the model from the following SIMULINK block libraries:

- Sources library (the Sine Wave block)
- Sinks library (the Scope block)
- Continuous library (the Integrator block)
- Signals and systems library (the Mux block)

One can copy a Sine Wave block from the Sources library, using the Library Browser (Windows only) or the Sources library window (UNIX or Windows). To copy the Sine Wave block from the Library Browser, one has to expand the Library Browser tree to display the blocks in the Sources library. This can be done by clicking on the Sources node to display the Sources library blocks. Finally, one has to click on the Sine Wave node to select the Sine Wave block. Now drag the Sine Wave block from the browser and drop it in the model window. SIMULINK creates a copy of the Sine Wave block on the point where one dropped the node icon. To copy the Sine Wave block from the Sources library window, one has to open the Sources window by double-clicking on the Sources icon in the SIMU-LINK library window. SIMULINK displays the Sources library window. Now one has to drag the Sine Wave block from the Sources window to the model window. Thereafter one has to copy the rest of the blocks in a similar manner from their respective libraries into the model window. One can move a block from one place in the model window to another by dragging the block. Furthermore, one can move a block a short distance by selecting the block, then pressing the arrow keys. With all the blocks copied into the model window, the model designed looks as shown in Fig. 4.34.

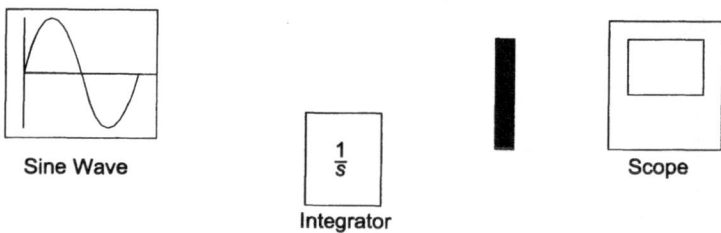

Sine Wave

$\dfrac{1}{s}$

Integrator

Scope

Fig. 4.34. SIMULINK block icons for build up of the model shown in Fig. 4.34

It can be seen from Fig. 4.34 that the block icons have an angle bracket on the right of the Sine Wave block and two on the left of the Mux block. The > symbol pointing out of a block is an output port; if the symbol points to a block, it is an input port. A signal moves out of an output port and into an input port of another block through a connecting line. When the blocks are interconnected, the port symbols disappear. The blocks can be connected. The Sine Wave block has to be connected to the top input port of the Mux block. For this reason the pointer has to be positioned over the output port on the right side of the Sine Wave block. It should be noted that the cursor shape changes to cross-hairs. The mouse button should be held down and the cursor should be moved to the top input port of the Mux block. It should be noted that the line is dashed while the mouse button is down and that the cursor shape changes to double-lined cross hairs as it approaches the Mux block. The mouse button should be released. The blocks are interconnected. It is also possible to connect the line to the block by releasing the mouse button while the pointer is inside the icon. If so, the line is connected to the input port closest to the cursors position. It should be noted that most of the lines connect output ports of blocks to input ports of other blocks. However, one line connects a line to the input port of another block. This line, called a branch line, connects the Sine Wave output to the Integrator block, and carries the same signal that passes from the Sine Wave block to the Mux block. Finishing the block connections, we obtain a model that looks like the one in Fig. 4.33.

The simulation parameters have to be specified. For this reason one can open the Scope block to view the simulation output. Keeping the Scope window open, set up SIMULINK to run the simulation for 10 s, which is the **Stop time** (its default value is set to 10.0). The simulation parameters can be set by choosing **Simulation Parameters** from the **Simulation menu**. If one close the **Simulation Parameters** dialog box by clicking on the **OK** button, SIMULINK applies the parameters and closes the dialog box. One can choose **Start** from the **Simulation menu** and watch the traces of the Scope block's input, as shown in Fig. 4.35.

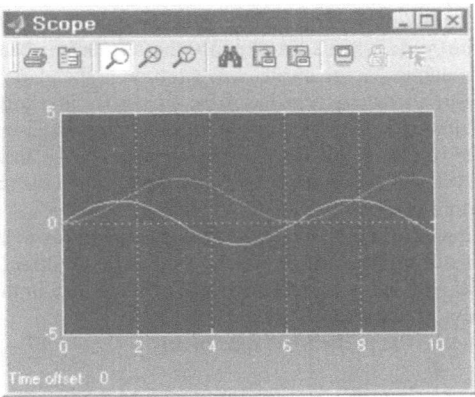

Fig. 4.35. Transient behavior of the SIMULINK model shown in Fig. 4.33

The simulation stops when it reaches the stop time specified in the Simulation Parameters dialog box or when one chooses **Stop** from the **Simulation** menu or presses the **Stop** button on the model windows toolbar (Windows only). To save the built up model, one can choose **Save** from the File menu and enter a filename and location. That file contains the description of the model. For termination of MATLAB and SIMULINK, we can use **Exit** MATLAB (on a Microsoft Windows system) or **Quit** MATLAB (on a UNIX system).

SIMULINK can be used for a wide range of applications in the different scientific domains. Hence we introduce a case study example showing its performance.

Example 4.16
The increasing demands in water quality result in the need for more effective process control of wastewater-treatment plants. One way of increasing the efficiency is using a model-based on-line process control, which requires a suitable model that describes the dynamic behavior of the process well enough and is as simple as possible at the same time.

Consider that the process of nitrification is the most important complex biochemical process in a wastewater-treatment plant, consisting of several enzymatic reactions. The first step of the reaction, the oxidation of ammonia to nitrite, is performed by bacteria of the type nitrosomonas, while the conversion of nitrite to nitrate is carried out by a bacteria of type nitrobacter. Both processes depend directly on each other since nitrite oxidation is a consecutive reaction to the ammonia oxidation and both reaction rates depend on the nitrite concentration. The nitrification reaction scheme is as follows:

$$NH_4^+ + 1.5O_2 \rightarrow NO_2^- + H_2O + 2H^+ \quad NO_2^- + 0.5O_2 \rightarrow NO_3^- . \tag{4.51}$$

The energy released in this process is used for the growth of the bacteria, which are slow growing as well as fast growing. Hence, the adapting ability of the bacteria for optimization of the wastewater-treatment plant is of importance, meaning that on a certain scale the mi-cro-organisms adapt to the changing conditions in the plant. The fact of the different bacte-ria growth results in different changes of the reaction rates, hence one of the reactions is preferred for the description of the plants dynamic. Consider a sudden increase of ammonia concentration in the inflow of the wastewater-treatment plant reactor, which results in an accumulation of nitrite, yields, some time before the nitrobacter bacteria have adapted to the increased load, a reduction in the nitrite concentration in the plant.

For simulation analysis a reactor type has to be chosen for investigating the nitrification, which takes place in this case study example in a so-called packed-bed reactor. Both types of bacteria are immobilized on a static mixer, meaning that the surface of the packed-bed is covered with a biofilm that is in contact with the wastewater and the air. Wastewater and air are injected at the bottom of the reactor, as shown in Fig. 4.36.

The reactor is equipped with recirculation, which is several times larger than the influent flow, to weaken the concentration profiles inside the reactor. Without recirculation an inhi-bition of the biological reaction is possible because of variations in the influent wastewater flow rates, concentration, and composition.

The main advantages of the packed-bed reactor are:

- Realization of high bacterial concentrations with high sludge ages independent of flow changes
- Robustness, because there is no wash-out of bacteria
- Limit of the maximum of the conversion rate, due to the diffusion of the biofilm

For a deeper understanding of the hydrodynamics inside the packed-bed reactor, which has an important influence on the distribution of substrate and hence on the reaction kinet-ics, we build up a reactor model that can be used for the simulation of the transient behav-ior, using MATLAB-SIMULINK. An intermediate state of mixing behavior can be mode-led using a cascade of reactors with recirculation, as shown in Fig. 4.37.

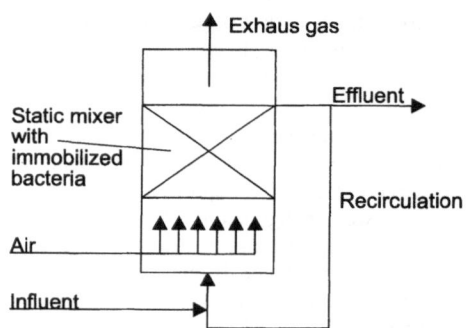

Fig. 4.36. Schematic diagram of a packed-bed reactor

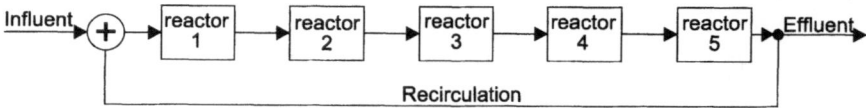

Fig. 4.37. Schematic diagram of a cascade of reactors with recirculation

Using a pulse input of a nonreacting tracer as a test signal (see Sect. 1.1), the transient behavior of the cascade of reactors can be simulated. Simulations with a cascade of five reactors, as shown in Fig. 4.38, show that this model behaves like a complete-mixing reactor if the recirculation flow exceeds a certain value, as shown in Fig. 4.38.

To gain a deeper insight into the dynamic behavior of the packed-bed reactor type, one has to formalize the mass balance for a specific substance as follows:

$$\begin{pmatrix} Rate \ of \ change \\ in \ reactor \end{pmatrix} = \big(flow \ in\big) - \big(flow \ out\big) \pm \begin{pmatrix} increase \ or \ decrease \\ by \ reaction \end{pmatrix} \quad (4.52)$$

The last term of the sum is added or subtracted, which depends on whether the concentration is increased or decreased by the reaction. For modeling the packed-bed reactor, flows (F) and concentrations c_i are the most important parameters, as shown in Fig. 4.39.

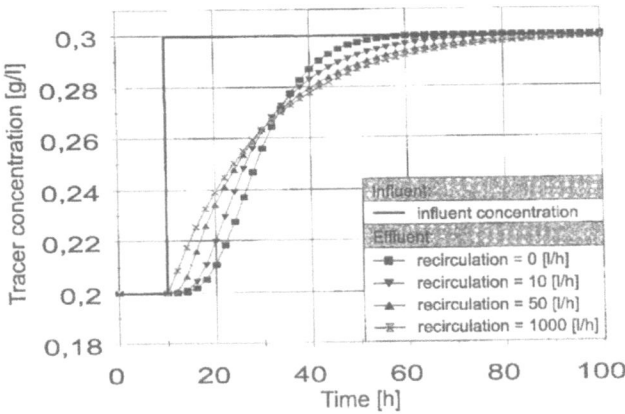

Fig. 4.38. Simulation results with five reactors cascade with different recirculation flows

Fig. 4.39. Flows and concentrations of the packed-bed reactor model

The mass balance equation of the packed-bed reactor, shown in Fig. 4.39, can be described as follows:

$$V \cdot \frac{dc_2}{dt} = F_1 \cdot c_1 - F_2 \cdot c_2 \pm V \cdot r \ . \quad (4.53)$$

Dividing (4.53) by volume V results in a differential equation notation which can be directly implemented in common simulation systems

$$\frac{dc_2}{dt} = \frac{F_1}{V} \cdot c_1 - \frac{F_2}{V} \cdot c_2 \pm r .$$ (4.54)

This first-order differential equation can be used for ammonia, nitrite, and nitrate which describe the change of the corresponding concentrations. The reaction rate r in (4.54) depends on the yield coefficient Y, the bacteria concentration c_B and the specific growth rate μ which describes the growth behavior of the micro-organisms, as follows:

$$r = \frac{\mu}{Y} \cdot c_B .$$ (4.55)

The yield coefficient is of constant value but changes caused by growth and death of the bacteria have to be taken into account, using a differential equation with a constant death rate k_d, yields

$$\frac{dc_B}{dt} = (\mu - k_d) \cdot c_B .$$ (4.56)

(4.55) and (4.56) are of importance while modeling the nitrification process depending on the specific growth rates. Hence μ has to be introduced due to its variations, which can be done using the Monod equation, which sets μ equal to a function of substrate concentration c_s, yielding

$$\mu = \mu_{max} \cdot \frac{c_s}{K_s + c_s} ,$$ (4.57)

where K_s is the so-called half-maximum saturation coefficient since μ equals $\mu_{max}/2$ if K_s equals c_s. Observation have shown that at high concentration the substrate can also act as a toxic growth inhibitors which can be taken into account using the Haldane equation, which is an extension of the Monod equation, given in (4.57), resulting in

$$\mu = \mu_{max} \cdot \frac{c_s}{K_s + c_s + \left(\dfrac{c_s^2}{K_1}\right)} ,$$ (4.58)

where $(c_s)^2/K_1$ is the so-called inhibition term, which is small in magnitude at low substrate concentrations, and increases at high values of c_s. There are other influences on the growth rate, which can be taken into consideration using additional terms in (4.58), expressing the respective kinetic influences,

$$\frac{dNH_4}{dt} = \frac{F_1}{V} \cdot NH_{4,1} - \frac{F_2}{V} \cdot NH_4 - r_{Ns} \qquad (4.59)$$

$$\frac{dNO_2}{dt} = \frac{F_1}{V} \cdot NO_{2,1} - \frac{F_2}{V} \cdot NO_2 + r_{Ns} - r_{Nb}$$

$$\frac{dNO_3}{dt} = \frac{F_1}{V} \cdot NO_{3,1} - \frac{F_2}{V} \cdot NO_3 + r_{Nb}$$

$$\frac{dN_s}{dt} = \left(\mu_{Ns} - k_{d,Ns}\right) \cdot N_s$$

$$\frac{dN_b}{dt} = \left(\mu_{Nb} - k_{d,Nb}\right) \cdot N_b$$

Consider the overall reactor kinetics is based on first-order differential equations, describing the respective concentration changes, we obtain for the concentration changes of ammonia (NH_4), nitrite (NO_2), nitrate (NO_3), nitrosomonas (NS) and nitrobacter (Nb) the respective set of differential equations where r_{Ns} and r_{Nb} are in relation to (4.55), and an expanded version of (4.58) is used for the calculation of μ_{Ns} and μ_{Nsb}.

The differential equations set (4.59) has been implemented in MATLAB-SIMULINK for simulation. The verification of the simulation study is based on measurements from a pilot plant, to show that the derived reactor model fits the reality of a real wastewater-treatment plant reactor, as shown in Fig. 4.40 for the concentration changes of ammonia (NH_4).

Fig. 4.40 shows the comparison of measured ammonia values of the pilot plant and the simulation results of ammonia for a period of six days, which shows a reasonable match between predicted (simulated) and measured data. Hence the model, shown in Fig. 4.39, which is described by (4.48) can be used for optimization as well as for prediction.

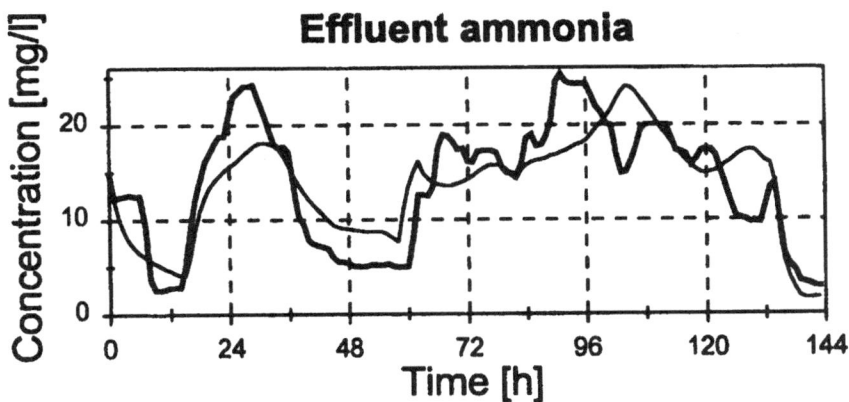

Fig. 4.40. Measured values (thick lines) and simulated results (thin lines) for the concentration changes of ammonia (NH_4) of the packed-bed reactor, shown in Fig. 4.39

4.4 Discrete-Time System Simulation Software*

Discrete-time systems simulation systems can be separated into general-purpose and special or application-purpose simulation systems. Their structure is similar to the continuous ones, but they contain an event-based control of time that allows classification of discrete-system simulation systems as follows:

- Transaction oriented simulation software; are based on a time-step control, determined through preprogrammed logical conditions related to the respective blocks. Language elements of transaction-oriented simulation systems are transactions, blocks, facilities, queues, pools and storages, logical switches, numerical and logical variables, functions, and tables.
- Event-oriented simulation software; are based on time-dependent and restricted event-handling language elements.
- Activity-oriented simulation software; are based on activity schedules that are started if specific constraints are fulfilled.
- Process-oriented simulation software; are based on the activation trigger of the following events as specified language elements.

The development of discrete-system simulation software results in several systems, which are the

- Universal or special-purpose discrete system simulation systems, such as ARENA, GASP, GPSS, MODSIM, SIMSCRIPT, SIMAN, SLAMSYSTEM, based on general-purpose programming languages
- Application-oriented discrete-system simulation systems, such as NETWORK, SIMFACTORY, SIMPLE++, SIMULAP, XCELL, etc., based on programming languages for special-application domains

The universal or general-purpose discrete system simulation languages are applicable in a wide range of applications, but, for example, the user has to have specific programming experience, such as TESS (the extended SLAM system), and SIMAN (simulation analysis language).

Application-based simulation systems are primarily more specific due to their application-domain dependent design that results in a better efficiency. But in contrast, their flexibility and their restriction to a specific application domain are less acceptable, from the general-purpose point of view. New software releases show two specific approaches:

- Parameterized, application-oriented systems, that are expanded by a user simulation system, like SIMFACTORY II.5 Rel.6, or an interface for a higher programming language is embedded, as in SIMAN V. Hence, the user can develop and implement his own control strategies, or his specific model components, which may be combined with the standard elements.

- Object-oriented simulation systems, based on specified basic components, which allow an object-oriented design of the application-specific components, such as the OS/2 metafiles for graphics in SLAMSYSTEM.

A typical representative of a special-purpose simulation programming language based on the process interaction-approach and oriented towards queueing systems is GPSS (general purpose simulation system). GPSS is a highly structured simulation software based on standard blocks, which represent events, delays, and other actions that affect the transaction flow. Hence, GPSS can be used to model situations where transactions, such as entities, customers, units of traffic, etc., are flowing through a system, which can be a network of queues, with the queues preceding scare resources. The GPSS block diagram is converted to block statements, and control statements are added, which results in a GPSS model. Furthermore, GPSS contains specific subroutines for coordinating the transactions:

- Coordination of transactions within one step
- Coordination of transactions within parallel steps
- Coordination of time-equal transactions

Its successor, GPSS/H, provides improvements of the fundamental concepts of an earlier version of GPSS, such as transaction flow view, facilities, queues, and storages. Its latest version includes

- Ampervariables that allow arithmetic combinations of values used in the simulation
- Animation
- Arithmetic expressions as block operands
- Built-in files
- Built-in mathematical functions
- Built-in random variate generators
- Expanded control statements
- Faster execution
- Floating-point clock
- Interactive debugging
- etc.

The animator for GPSS/H is proof animation, which provides a 2D animation, based on a scale drawing that can run in post-processed mode or concurrently.

Example 4.17

The kernel of a single-server queue-simulation program in GPSS/H contains the GENERATE statement which represents the arrival event, with the interarrival times given by the statements RVEXPO(1,&IAT). RVEXPO stands for random variable, exponentially distributed, while 1 indicates the random number to use, and &IAT indicates that the mean time for the exponential distribution comes from an ampervariable, indicated by the & character. The next statement is a QUEUE, named SYSTIME. The QUEUE statement works in conjunction with the DEPART statement to collect data on queues – r any other subsystem.

Another successor to GPSS/H, SLX, replaces many of the features of GPSS/H entirely and represents many of them with simpler, more general constructs, because SLX is a layered modeling system in which GPSS/H comprises only one of the five layers, which are as follows:

- Layer 0: C based kernel that supports a number of primitives required for simulation
- Layer 1: simulation and statistical primitives, consisting of data structures, subroutines, operators, and macros, written in SLX
- Layer 2: general GPSS/H features
- Layer 3: application-domain-specific packages
- Layer 4: special packages such as highly interactive front ends

SIMSCRIPT, a representative of another universal discrete simulation system that allows users to built models either process-oriented or event-oriented. SIMSCRIPT can be used to produce dynamic and static presentation graphics. Animation of the simulation output is realized using the SIMGRAPHICS software to produce interactive graphic front ends or forms for entering model input data.

A typical representative of the class of application-oriented discrete simulation software is SIMAN V, which incorporates the ARENA environment that includes

- Menu-driven point- and click procedures for modeling
- Animation of the model
- Input processor which assists in fitting distributions to data
- Output processor, which can be used to obtain confidence intervals, histograms, etc.

Example 4.18
Event handling should be introduced through the state event.
Whenever the value of the water level in a tank reaches its upper limit a quarter of its contents is taken away by the controller.

```
WHENEVER    tank_level >= level_max
BEGIN_BODY
          tank_level := tank_level · 0.75;
END_BODY
```

Independently from the way the tank is filled this event assures the level will not exceed the given limit (see Example 1.8).

4.5 Multi-Domain Simulation Software for Large-Scale Systems*

Real-world systems are often described through large-scale models, which can contain domain-dependent continuities as well as discontinuities, discrete events, changes of the structure, etc. Due to the different implications of real-world systems multi-domain simulation systems have to cover an efficient handling of the manifolds. Therefore, a specific design emphasis is necessary for synchronization and propagation of events and the possibility to find consistent restarting conditions after an event. Moreover, time-continuous parts or elements modeling real-world systems have to embedded into the overall large-scale system model. In recent decades, numerous simulation systems have been developed to assist modelers of large-scale simulation. Some are general-purpose simulation tools, others where developed for simulation in specific application domains, such as avionics (e.g. Easy5), electrical circuits (e.g. Spice), multibody systems (e.g. ADAMS), chemical processes (e.g. ASPEN), etc. A typical representative of a multi-domain simulation software usable for large-scale systems is Modelica, which is both a modeling language and a model-exchange specification. To accomplish this goal, the developers of previous object-oriented modeling languages such as Allan, Dymola, NMF, ObjectMath, Omola, SIDOPS and Smile worked together with experts from many engineering domains to create the specification for the Modelica language, based on their wide range of experiences. The current Modelica specification can be found on the web site at www.modelica.org. In addition to defining the specification for the Modelica language, the Modelica Association also publishes a standard library of Modelica models, called Modelica Standard Library. Modelica allows an equation-based, a component-based and a block-diagram approach. As an alternative to equation-based approaches, functions can be used in Modelica, which represent an algorithmic section, used when procedural semantics are required. Moreover, Modelica allows one to declare arrays of scalars, i.e. an array of floating-point numbers, as well as arrays of components, i.e. an array of resistor instances or an array of record instances. Arrays of scalars are useful for representing mathematical entities like vectors and matrices.

For the pendulum example in Sect. 4.3.3, based on the equations of motion, the model in the Modelica language specification is

```
model  SimplePendulum
   parameter  Real  L = 2;
   constant  Real  g = 9.81;
   Real  theta;
   Real  omega;
equation
   der (theta) = omega;
   der (omega) = -(g/L)·theta;
end  SimplePendulum;
```

Using the vocabulary of the Modelica simulation software the keyword **model** is followed by the name of the model. Thereafter the **parameters** and **constants** characterizing the model are defined, as well as the variables that appear in the equations. The parameters are quantities that remain constant during a simulation run but may have different values from one simulation run to another. The variables are those quantities that are functions of time. The constants are those quantities that are unlikely to change. The model will be completed by an **equation** section, which includes a built-in operator called **der,** which is used to represent the time derivative of a variable. The model equation includes first-order ordinary differential equation (ODEs), which describe the equations of motion of the simple pendulum. These equations look like

$$\begin{pmatrix} \dot{\Theta} \\ \dot{\omega} \end{pmatrix} = \begin{pmatrix} \omega \\ -\dfrac{g}{L} \cdot \Theta \end{pmatrix} \tag{4.60}$$

in the mathematical notation, where Θ is the angular position, L is the length of the pendulum, g is the acceleration due to earths gravity, and $\dot{\omega}$ is the angular velocity of the pendulum.

Modelica allows working with models combining continuous and discrete systems due to several formalisms:

- Ordinary differential equations (ODEs)
- Differential algebraic equations (DAEs)
- Bond graphs
- Finite-state automata
- Petri-nets
- etc.

The Modelica language has been designed to allow tools to generate very efficient code, this is because Modelica is an object-oriented modeling-language based simulation software. Moreover, Modelica allows, for example, hardware-in-the-loop simulation, and supports both high-level modeling by composition of predefined library units and detailed library component modeling by equations. In order that Modelica is also useful for model exchange, it is important that libraries of the most commonly used components are available, in fact, ready to use, and sharable between applications. Such a library includes

- Mathematical functions, i.e. sin, cos, ln, exp, int, etc.
- Type definitions, i.e. angle, voltage, etc.
- Interface definitions, i.e. pin, etc.
- Components.

The components library includes:

- Input/output blocks
- Electric and electronic elements
- Electric power systems
- Drive trains
- Gear boxes
- Multi-body systems
- Hydraulic systems
- Thermo-fluid flow
- Aircraft flight system dynamics
- Bond graphs
- Finite state machines
- Petri-nets
- etc.

Example 4.19

As an example for the components library we define a model for a resistor, created from interface model class **TwoPin**, that has two pins **p** and **n**, and add a parameter for resistance and Ohms law to define the behavior in Modelica. Ohms law describes the behavior of the resistance

$$v = i \cdot R , \tag{4.61}$$

which is, in Modelica,

model Resistor "Electrical Resistor"
parameter Resistance R=300 "Resistance";
ElectricalPin p, n;
equation
R*p.i = p.v − n.v;
p.i + n.i = 0:
End Resistor;

Example 4.19 shows how a model for the resistor can be written in Modelica. The "." in quantities such as p.v is a way of accessing the internal elements of a component. Since p is an instance of an **ElectricalPin** it contains the variable for voltage v. Therefore, the quantity p.v represents the voltage associated with pin p.

It is of importance when developing component models to use a consistent sign convention for the flow quantities. The normal sign convention for Modelica components is defined such that positive flow are into the component. Therefore, inside the **Resistor** model, the value of p.i refers to the current that flows into the resistor from pin p and the value of n.i refers to the current that flows into the resistor from pin n. From Example 4.19 one can see that a positive value for p.i. results when p.v is greater than n.v. That is consistent with the normal sign convention. Likewise, a positive value for n.i results when n.v is greater than p.v.

In Example 4.19, the current p.i is used to represent the current in Ohms law, given in (4.61). The choice between using p.i and n.i is arbitrary. However, if n.i have been used, the model equation has to be written as:

$$R \cdot n.i = n.v \; - \; p.v; \tag{4.62}$$

in order to satisfy the sign convention for the flow variables, i.e. a positive value represents a flow into the component. The value of 300 Ω for the resistance of the resistor is the default value of Modelica.

Modelica also supports multi-domain system models, characterized by the fact that their components belong to different engineering domains. As an example of a multi-domain system model an industrial robot is chosen. The model is composed of a control system and a plant model. The plant model contains electrical and mechanical components. Using the model editor, a model for an industrial robot can be defined by drawing a composition diagram as shown in a 3D view in Fig. 4.41.

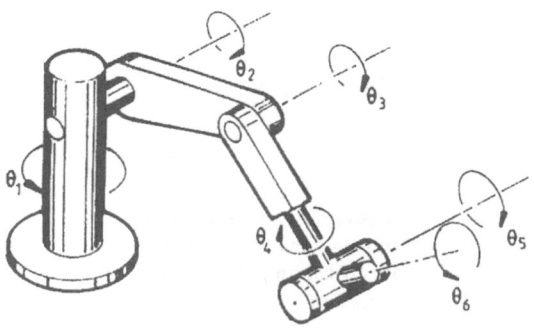

Fig. 4.41. 3D view of an industrial robot

Example 4.20

The industrial robot model shown in Fig. 4.41 can be decomposed such that the mechanical part of the robot consists of six revolute joints, six bodies, and finally the load. A body component describes the mass and inertia effects of the body. The joints of the robot are given by the axis. An axis is a key component that describes the motor and gear box that drive the joint, the control system, and the reference generation. A possible Modelica library representation, shown on an abstract level in Fig. 4.42, contains the reference acceleration of the axis qddRef as an input value of the connector, and a mechanical flange to drive a shaft on the output-side connector. The decomposition of the axis shows that the reference acceleration (qddRef) will be integrated twice in order to derive a reference velocity (qdRef) and a reference position (qRef). The reference values are fed into a controller (irControl), while the controller output is the reference current of the motor (irMotor), driving the gear box (irGear). The driving part of the gear box is a mechanical flange to which the axis of a shaft or of a robot joint can be connected.

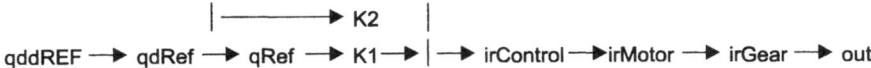

Fig. 4.42. Composition diagram of one axis of the industrial robot, shown in Fig. 4.42

Typical for the *axis* controllers (irControl) are the velocity and the position controller, the output of which is the desired reference current of the motor. The current of the motor is approximately proportional to the produced motor torque, the quantity to be controlled. The irMotor model of the electric motor consists of an analog current controller, which can be realized using operational amplifiers, and the DC motor with the components Ra, La, emf. The output current of the current controller represent the input signal of the motor. The DC motor produces a torque that drives a mechanical flange.

The composition diagram of the gearbox irGear of the driving system is modeled by the motor inertia, a rotational spring to model the gear elasticity, an ideal gear box representing the gear ration and a load inertia to model the rotational inertia of all parts at the driven side of the gear. A friction component connected between the motor shaft and the shaft bearings models the Coulomb friction of the bearings.

Describing how to model the details of a component, we can consider a simple motor drive system, as shown in Fig. 4.43. The system can be built up as a set of connected components: a controller, a motor, a gear box, and a load.

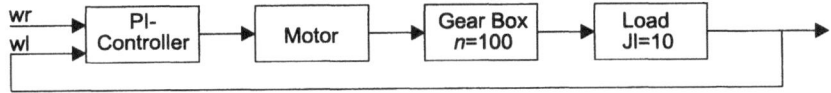

Fig. 4.43. Schematic diagram of the motor drive

Fig. 4.43 is a composite model that specifies the topology of the system to be modeled in terms of components and connections between the components. For example the statement "Gear Box ($n = 100$)" declares a component **gearbox** of class Gear Box and sets the default ratio, n, to 100.

The complete Modelica model of the system in Fig. 4.43 is shown in Fig. 4.44.

```
model MotorDrive
    PI            controller;
    Motor         motor;
    Gearbox       gearbox (n=100);
    Shaft         Jl (J=10)
    Tachometer    wl;
equation
    connect(controller.out, motor.inp);
    connect(motor.flange, gear box.a);
```

```
        connect(gearbox .b,    J1.a);
        connect(J1.b,          wl.a);
        connect(wl.w,          controller.inp);
    end  MotorDrive;
```

Fig. 4.44. Modelica model of the motor drive system in Fig. 4.43

The connections in Fig. 4.44 specify the interactions between the several components, as shown in Fig. 4.43. A connector contains all quantities needed to describe the interaction.

The Modelica model of the motor drive, shown in Fig. 4.43, represents a typical feedback loop for which the continuous time transfer function can be calculated using the computer algebra as an efficient simulation code. The Modelica model of a continuous-time transfer function is as follows:

```
partial block SISO
    input Real u;
    output Real y;
end SISO;
block TransferFunction
    extends SISO;
    parameter Real a[:]={1,1} "Denominator";
    parameter Real b[:]={1} "Numerator";
protected
    constant Integer na=size (a,1);
    constant Integer nb(max=na)=size (b,1);
    constant Integer na=na-1 "System order";
    Realb0[na]=cat(1, b, zeros(na −nb)) "Zero expanded vector";
    Real x[n] "State vector";
Equation
    //Controllable canonical form
    der (x[2:n])=x[1:n-1];
    a[na]·der (x[1]+a[1:n]·x=u;
    y=(b=[1:n]-b0[na]/a[na]·a[1:n])·x+b0[na]/a[na]·u
end  TransferFunction;
```

Beside Modelica other multi-domain simulation software systems had been recently developed, a typical representative of which is FEMLAB. FEMLAB is an interactive environment for modeling and simulation of physical phenomena, that can be described through partial differential equations (PDE). This finite element software combines ready-to-use applications with free-equation formulations. The formulations in the ready-to-use applications are open, which means that it is easy to change the existing equations and to add arbitrary couplings with other physical phenomena. Traditional modeling software include the most common multiphysics couplings as hard-wired elements. However, as computers have become more powerful and easy-to-use, modeling has penetrated into all disciplines of science and engineering, which has made it increasingly difficult for software developers to hard-wire all the possible couplings. The solution to this is given by equation-based simulation software, where these couplings are created dynamically through equation interpreters that transfer a given formulation to a numerical code. However, it is not necessary to have deep knowledge in mathematics or numerical analysis when using FEMLAB. Model building in FEMLAB will be simple by defining the relevant physical quantities rather than defining the equations directly. FEMLAB then internally compiles a set of PDEs representing the problem, but FEMLAB also offers an equation-based modeling application mode, which allows the user to define his own systems of partial differential equations. Beside providing these multiple modeling approaches, FEMLAB offers broad flexibility, either through a flexible self-contained graphical user interface or a user interface that integrates seamlessly with MATLAB, the package that provides the computational engine behind FEMLAB.

The underlying mathematical structure with which FEMLAB operates is a system of partial differential equations, which can be represented in the

- Coefficient form, suitable for linear or nearly linear problems
- General form, intended for nonlinear problems
- Weak form, usable as high-level finite element modeling software

Let u be a single dependent variable, meaning an unknown function on the domain that is to be determined from the partial differential equation (PDE) problem. A PDE problem in coefficient form yields:

$$d_a \frac{\partial u}{\partial t} + \nabla \cdot (-c\nabla u - \alpha u + \gamma) + \beta \cdot \nabla u + au = f \quad in \ \Omega \qquad (4.63)$$

$$n \cdot (c\nabla u + \alpha u - \gamma) + qu = g - h^T \mu \quad on \ \partial\Omega$$

$$hu = r \quad on \ \partial\Omega \ .$$

The domain of interest is usually called Ω, consisting of bounded subdomains and boundaries. The symbol Ω represents the union of all subdomains, and $\partial\Omega$ denotes the domain boundary. The outward unit normal vector on $\partial\Omega$ is denoted by n. The first equation is the PDE that has to be satisfied in Ω. The second and third equations are the boundary conditions that are satisfied due to the constraints in Ω. The second equation is referred to as a generalized Neumann boundary condition, and the third is referred to as the Dirichlet constraint. The symbol ∇ is the vector differential operator, defined as

$$\nabla = \left(\frac{\partial}{\partial x_1}, ..., \frac{\partial}{\partial x_n} \right), \qquad (4.64)$$

where the space conditions are denoted $x_1, \ldots, x_n$, where n represents the number of space dimensions.

When the coefficients depend only on the space coordinates, the PDE is called linear. If the coefficients depend on u, or the components of ∇u, the PDE is called nonlinear. When the coefficients are nonlinear the general or the weak solution form will be used.

Let u be a single dependent variable, meaning an unknown function on the domain that is to be determined from the partial differential equation (PDE) problem. Then the stationary problem in the general form reads

$$\nabla \cdot \Gamma = F \quad in \ \Omega \qquad\qquad (4.65)$$

$$-n \cdot \Gamma = G + \left(\frac{\partial R}{\partial u} \right)^T \mu \quad on \ \partial\Omega$$

$$0 = R \quad on \ \partial\Omega \ .$$

The first equation is the PDE. The second and third equations are the Neumann and Dirichlet boundary conditions, respectively. The terms Γ, F, G, and R are usually called coefficients, which can be functions of the space coordinates, the solution u, or the components of ∇u. It should be noted that the coefficients F, G, and R are scalar, while Γ is a vector called the flux vector. The T in the Neumann boundary condition denotes the transpose, and μ is the Lagrange multiplier.

The weak form comes from a certain equation involving integrals that can be derived from a PDE. The solution to a PDE is simply referred to as a solution. In FEMLAB, a solution computed using one of the three types of PDE, is always a weak solution. The coefficient or general form is used to enter the PDE coefficients. FEMLAB then derives the corresponding weak equation, the weak form of the PDE, which is then discretized by the finite element method, and solved. The weak solution allows direct access to the terms of the weak equation, which results in a maximum freedom defining finite element problems.

A PDE problem on coefficients form can be transformed into the corresponding weak form as follows. The PDE problem is known as

$$\nabla \cdot (-c\nabla u - \alpha u + \gamma) + \beta \cdot \nabla u + au = f \quad in \ \Omega \qquad\qquad (4.66)$$

$$n \cdot (c\nabla u + \alpha u - \gamma) + qu = g - h^T \mu \quad on \ \partial\Omega$$

$$hu = r \quad on \ \partial\Omega \ .$$

Let v be an arbitrary function on Ω, called test function. Multiplying the PDE by this function, rearranging terms, and integrating, we obtain

$$\int_{\Omega} v\nabla \cdot (-c\nabla u - \alpha u + \gamma)dA = \int_{\Omega} v(f - \beta \cdot \nabla u - au)dA , \qquad (4.67)$$

where dA is the area element, in 2D and volume element in 3D. Integration by parts yields

$$\int_{\partial\Omega} v(-c\nabla u - \alpha u + \gamma) \cdot nds - \int_{\Omega} \nabla v \cdot (-c\nabla u - \alpha u + \gamma)dA = \qquad (4.68)$$
$$\int_{\Omega} v(f - \beta \cdot \nabla u - au)dA$$

where ds is the length element. With the Neumann boundary condition

$$n \cdot (c\nabla u + \alpha u - \gamma) + qqu = g - h^T \mu , \qquad (4.69)$$

we obtain the following equation:

$$0 = \int_{\Omega} \nabla v \cdot (-c\nabla u - \alpha u + \gamma) + v(f - \beta \cdot \nabla u - au)dA + \qquad (4.70)$$
$$\int_{\partial\Omega} v \cdot (-qu + g - h^T \mu)ds$$

Together with the Dirichlet constraint, this is the weak form of the coefficients PDE problem. The mathematician Dirichlet born in 1805 in Düren, Germany.

Furthermore, it is possible to set up models in FEMLAB as stationary or time dependent, linear or nonlinear, scalar or multicomponent, as well as performing eigenfrequency or eigenmode analysis.

Solving the PDEs that describe the model, FEMLAB applies the finite element method (FEM), which means running that method in conjunction with adaptive meshing and error control as well as with a variety of numerical solvers. A detailed description can be found in the Reference Guide on the web site at www.femlab.com.

As mentioned PDEs are the fundamental basis to model scientific phenomena, which is why FEMLAB can model a large number of physical phenomena in many disciplines including

- Acoustics
- Bioscience
- Chemical reactions
- Diffusion
- Electromagnetics
- Fluid dynamics
- Fuell cells

- General physics
- Geophysics
- Heat transfer
- Micro-electromechanical systems (MEMS)
- Microwave engineering
- Optics
- Photonics
- Porous media flow
- Quantum mechanics
- Radio-frequency components
- Semiconductor devices
- Structural mechanics
- Transport phenomena
- Wave propagation
- etc.

The modeling software of FEMLAB is easy to learn, as is shown in the Getting Started Guide (see Sect. 4.9) which contains several ready-to-use applications for the application domains chemical engineering, electromagnetics and structural mechanics.

Example 4.21
A 3D current and heat balance problem will be investigated. The current heats the aluminum and the heat is dissipated to the adjacent air and to the silicon substrate, which results in a temperature gradient along the thickness of the aluminum plate, which seems to impede the model builder from reducing the problem down to two space dimensions. In this case study example, one can make use of the symmetry and asymmetry in the problem to reduce the geometry to one fourth of the original description.

The boundary conditions and the expression for the conductivity are identical in the 2D and 3D problems for current conduction, which implies electrical insulation everywhere except for the edges on the right and left.

The heat balance can be defined as follows: The temperature is fixed on the right and left edges of the silicon device. On the base of the silicon substrate and at the vertical boundaries on the substrate, one can assume thermal insulation. At all other surfaces one can define convective heat dissipation through film theory, using tabulated values for the heat transfer coefficient in the fictitious film, where all temperature differences are confirmed. At the boundary between the aluminum film and the substrate, FEMLAB automatically gives the continuity in temperature and heat flux.

The next step is to define the heat balance in the silicon device. In the silicon substrate, only conduction and accumulation of the heat occur. In the aluminum film, heat production is introduced also, by the conductor current. The production of heat is proportional to the square of the current density.

The thickness and length of the device geometry, shown in Fig. 4.45, differ by several orders of magnitude. This implies that a large number of elements would be required if one treated the problem without scaling, since the smallest length would set the edge size of the elements. Scaling can be used in different ways, such as scale the thickness of the deposit in order to get the same order of magnitude as the width of the strip. This can be achieved by introducing a new space coordinate along the thickness of the substrate and the deposit according to:

$$\tilde{z} = \frac{z}{L} . \qquad (4.71)$$

Setting L to 0.05 gives a scaled deposit thickness of 20 µm. The scaling is amounts for all properties and sources treated in the model.

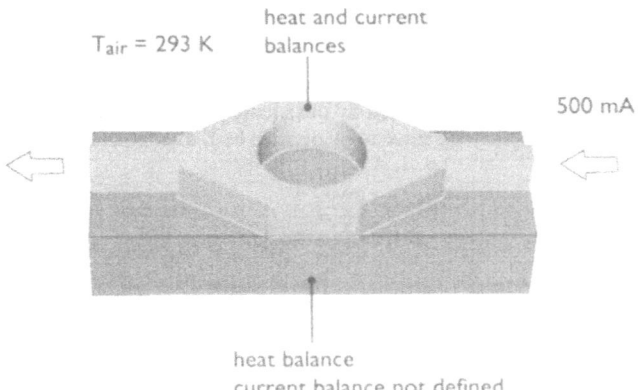

Fig. 4.45. Heat and current balance in a silicon device

Both the current and the heat balance are based on the change of flux per unit length. The change in current density per unit length has to be divided by the scaling length L to account for the scaling in geometry.

$$\frac{\Delta j}{\Delta z} = \frac{\Delta j}{L \Delta \tilde{z}} . \qquad (4.72)$$

In addition, the current density is proportional to the potential change per unit length, which implies that the flux balance has to be scaled by division by L^2.

$$\frac{\Delta(-\sigma \Delta V)}{(\Delta z)^2} = \frac{\Delta(-\sigma \Delta V)}{L^2 (\Delta \tilde{z})^2} . \qquad (4.73)$$

Introducing this in FEMLAB by using a scaled conductivity in the z-direction yields

$$\tilde{\sigma}_z = \frac{\sigma}{L^2} . \qquad (4.74)$$

The analogous calculation results in a scaled thermal conductivity:

$$\tilde{k}_z = \frac{k}{L^2}.$$ (4.75)

It should be noted that the electrical and thermal conductivities do not have to be scaled in the x- and y-directions.

The heat source is proportional to the square of the current density and has therefore to be divided by L^2 in the contribution in the z-direction,

$$\frac{(-\sigma \Delta V)^2}{(\Delta z)^2} = \frac{(-\sigma \Delta V)^2}{L^2 (\Delta \tilde{z})^2}.$$ (4.76)

In addition, all boundary conditions involving fluxes in the z-direction should be scaled by L. The scaled conductivity is divided by the square of L, which implies that the fluxes in the z-direction, at the boundaries, have to be divided by L to compensate, yielding

$$\frac{(-\sigma \Delta V)}{L^2 (\Delta \tilde{z})^2} = \frac{1}{L} j_z.$$ (4.77)

Based on the equations derived, the 3D geometry can be created in FEMLAB by extrusion a 2D projection of the film geometry, shown in Fig. 4.45.

- Select **Export to Workspace, Geometry as Objects**...in the **File** menu
- Press **OK** to export C01 – which is now available in the MATLAB command window –

To redefine the problem to 3D proceed as follows:

- Push the **New** button
- Press the **Dimension, 3D** radio button
- Select the **Physics Modes/Heat transfer** and press **OK**
- Open the **Multiphysics/Add_Edit_modes...** menu and highlight **Conductive Media DC,** press >> button
- Press **OK**

In the Draw Mode one can create the work plane:

- Select **Work plane I** in the **Draw** menu to work in the x y plane
- Select **Insert from workspace, Geometry object(s)...**
- Type C01 to export the cross-sectional 2D geometry
- Remote the left and the lower half of the geometry by performing Boolean operations
- Press the Zoom extents button

The cross-sectional geometry is now one fourth of the original. Extruding makes it 3D.

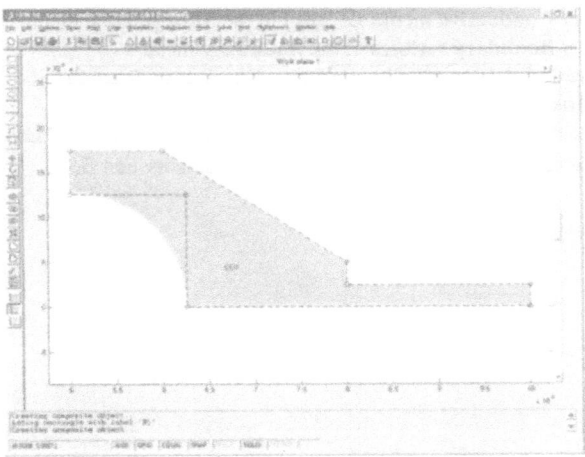

- Select **Extrude** in the **Draw** menu
- Press **OK**
- Type 20e-6 in the **Distance** edit field, press **OK**

Go back to work plane I and create the rectangular cross section of the substrate.

- Select **Work plane I** in the **Draw** menu
- Press **Rectangle** tool button
- Draw a rectangle by snapping to the upper left vertices of the existing geometry and drag it until the rectangle snaps to the lower right vertex of the deposit cross section
- Select **Extrude** in the **Draw** menu
- Type –40e-6 in the **Distance** edit field to create the 3D substrate
- Press **OK**

Options, such as input data of the model, are defined in the **Options** menu. The activation and deactivation of the current balance in the substrate can be defined in the **Subdomain** menu, as well as the expressions for the conductivity in the x, y, and z direction in the corresponding component position. Toggle to the heat transfer application mode to define the subdomain settings is as follows:

- Select **Heat transfer (ht)** in the **Multiphysics** menu
- Select subdomain 1 and define the **Subdomain settings** according to …
- Set the initial condition to T0
- Press **Apply**
- Select subdomain 2 and define the **Subdomain settings** according to …
- Set the initial condition to T0
- Press **OK**

With the Boundary mode one can define the boundaries of the 3D current and heat balance problem as follows:

- Select **Conductive media DC (dc)** in the **Multiphysisc** menu
- Select **Boundary settings** in the **Boundary** menu
- Press Ctrl-A to select all boundaries. Check **copy from 5**.
- Set **Insulation/symmetry** for all boundaries
- Select boundary **17** and press the **Inward current density** radio button
- Type in/side/thick in the **Inward current density** edit field

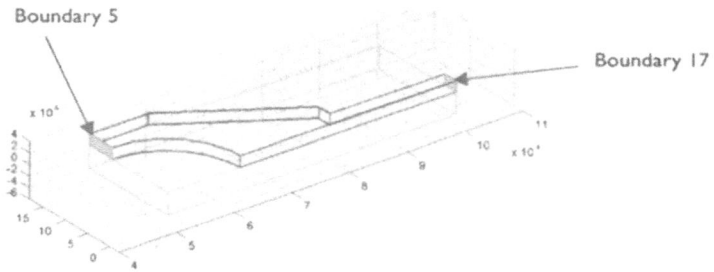

The inwards current density does not have to be scaled since it does not contain any component in the z-direction.

- Select boundary 5 and press the **Asymmetry/ground** radio button

Toggle to the heat transfer application.

- Select **Heat transfer (ht)** in the **Multiphysics** menu
- Select **Boundary settings** in the **Boundary** menu
- Press Ctrl-A to select all boundaries
- Set **Insulation/symmetry** for all boundaries
- Select boundaries **6, 10, 11, 14, and 15**

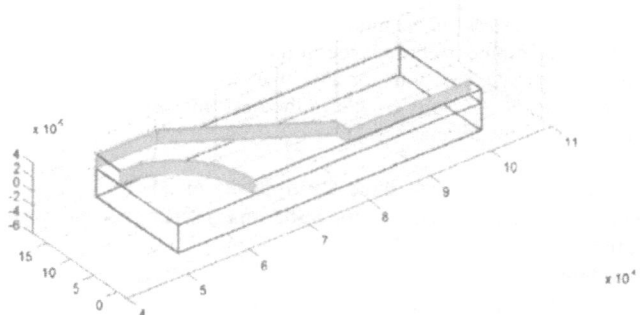

In the **Solve** mode one can compute the solution for the heat transfer. Once a good enough solution have been obtained, one can proceed to solve the time-dependent graph for the temperature field as follows:

- Select **Parameters** in the **Solve** menu
- Click on the **Multiphysics** tab and select both application models in the **Solve for variables** list
- Click on the **General** tab and press the **Time depends** radio button
- Click the **Timestepping** tab and type 0:0.001:0.03 in the **Output time** edit field
- Select the **fldaspk** solver in the time stepping algorithm pull-down menu
- Press **OK**
- Press the **Restart** button

In the **Post** mode the plot parameters are defined. The $T = f(t)$ plot is shown in Fig. 4.46.

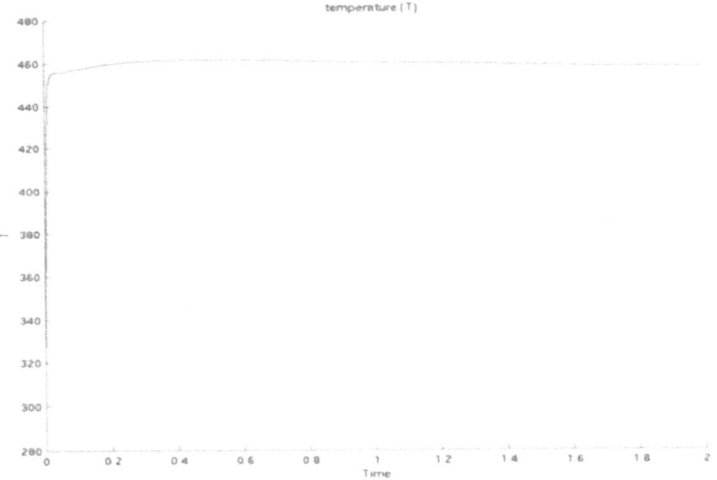

Fig. 4.46. Temperature plot of the silicon device, shown in Fig. 4.4.6

The plot in Fig. 4.46 compared with the 2D result shows that the time scale of the process is similar in the two- and three-dimensional cases but there is a different maximal temperature. This is due to the high heat capacity and thickness of the substrates, which is able to cool the deposit to a larger extent than air during the heating process.

4.6 Simulation Software for Mixed-Mode Circuits*

Traditionally, electronic circuit design was verified by building prototypes, subjecting the circuit to various stimuli, such as input signals, temperature changes, power-supply variations, etc., and then measuring its response using appropriate laboratory equipment. Prototype building is somewhat time consuming, but it produces practical experience from which to judge the manufacturability of the design. Computer programs that simulate the performance of an electronic circuit provide a simple, cost-effective means of confirming the intended operation prior to circuit construction, and of verifying new ideas that could lead to improved circuit performance. Berkeleys Spice and the Georgia Techs Xspice simulators are the classical ones in this domain. The B2SpiceA/DV4 circuit simulator helps to design analog, digital, and mixed-mode circuits. Rather than working on circuit design with physical components, which require expensive equipment and a lab, the B2SpiceA/DV4 allows one to perform realistic simulations on circuits without clipping wires or splashing solder. Editing and simulating circuits is a quick and easy procedure. The current B2SpiceA/DV4 information for the latest patches and more can be found on the web site at www.beigebag.com.

Example 4.22

The schematic B2SpiceA/DV4 editor allows one to enter a circuit design. When building a new circuit all parts will be added into the circuit by choosing them from menus and drawing wires to connect the devices, and setting the properties for the devices to customize their behavior. Building a simple resistor-transistor-logic (RTL) inverter circuit is as follows:

1. Place the devices
2. Customize the transistor
3. Set the device properties
4. Simulate the circuits AC and DC transfer curve
5. Simulate the circuits transient behavior

Step one: devices are chosen from the **Devices** menu or from the **Part Chooser** window, to place the device one has to click the mouse button. Step two: one has to choose the transistor type from the **Devices** menu and place the device in the circuit diagram. Step three: sets the device names and properties, i.e. the voltage source is named to **VC** and the Voltage to **5V**, the collector resistor is named **R** and its resistance set to **5k** etc. If all properties are set the circuit looks like Fig. 4.47

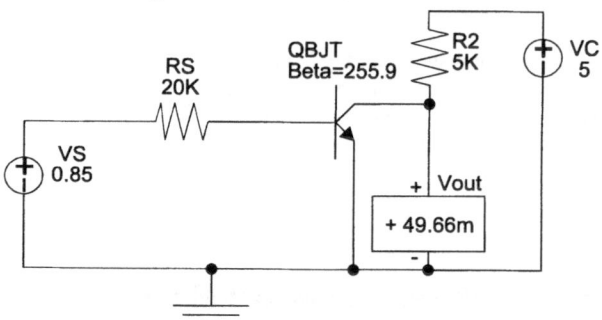

Fig. 4.47. B2SpiceA/DV4 model of a simple RTL inverter circuit

The analysis of AC and DC responses to a sine-wave input signal is to verify that the developed circuit behaves as an inverter. Hence step 4 deals with the simulation which is set up from the **Simulation** menu and the DC Transfer Curve simulation can be set up by clicking in the checkbox and on the button for **Single or Dual parameter DC sweep**. After entering the parameters

- Name 1 is VS (name of source to step)
- Start 1 is 0 (starting voltage)
- Stop 1 value is 5
- Step 1 value is 1E-1

the circuit can be run by clicking **Run Simulation**.

The fifth step performs a transient analysis of the circuit. The output response can be observed while the voltage source **VS** is modified to sinusoidal transient properties. Add a voltmeter in parallel with the voltage source for observation, named **IVin**, and an ampere meter in series to VC named **VIVC**. The modified circuit looks like Fig. 4.48, showing some values of the time-dependent quantities.

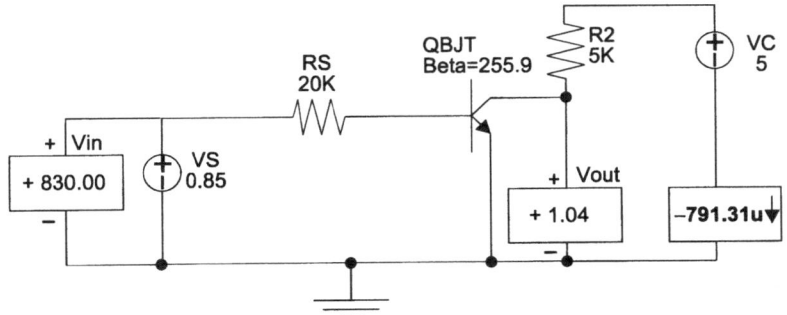

Fig. 4.48. B2SpiceA/DV4 model of the RTL inverter circuit in Fig. 4.47

The simulation run can now be started from the **Simulation** menu. The simulation results of the RTL inverter, shown in Fig. 4.48, are shown in Fig. 4.49.

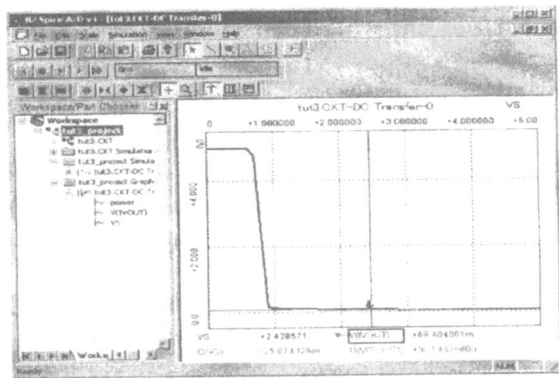

Fig. 4.49. B2SpiceA/DV4 simulation results of the RTL inverter circuit in Fig. 4.50

Example 4.2.3

For a digital circuit design we choose the components from the **Digital Devices menu** or from the **Part Choosers** digital subfolder. Building a three-input and function can be realized based on the same procedure as discussed for the analog circuit. First, one has to choose **Gates** from the **Categories** menu or from the **Part Chooser** window. Then scrolling to the 74LS10D_D and with a double click on that item the three-input nand gate follows the cursor to be placed. Next we choose and place an inverter from the **Digital Devices** menu and place it to the output of the nand gate. Now we can use the drawing tool and/or the selection cursor to draw the wires necessary. The circuit finally should look like that shown in Fig. 4.50.

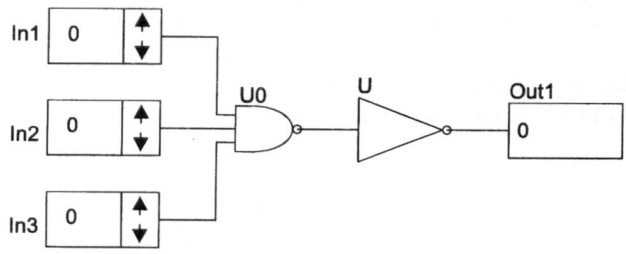

Fig. 4.50. B2SpiceA/DV4 model of the three-input and function

The simulation run can now be started from the **Simulation** menu. Following the simulation run on the screen it can be seen that the simulation time will increase to 20 ns, and the output value will change to 1. It should be noted, that all wires as well as the inputs and the output can be observed by using the probe tool.

4.7 Combined Simulation Software

Traditionally, simulation models that built are mostly either language or platform dependent. They are developed for continuous-time or discrete-time systems simulation approaches, normally using a single simulation software for the continuous-time world and a specific simulation software for the discrete-time world. In some situations real-world systems include diverse components that then requires different formalisms for modeling and simulation. This occur when the system components are continuous with concentrated parameters that show slow and fast parts, or for a system that contains a queuing part and a continuous part, which should be introduced as a combined system. Using an object-oriented approach one can simulate combined systems creating objects that simulate system submodels – queuing and continuous – running them concurrently. Hence, submodels of very different kinds can run and interact in the same simulation environment.

Typical representatives of combined simulation systems are PASION, and the recently launched AnyLogic. PASION is an object-oriented, Pascal-related simulation software that handles models specified in terms of **processes** and **events**. The PASION process is an object-type declaration that defines the object properties, such as **attributes** and **events**. At run time, objects are created according to the declared processes and activated. Then, the objects run by executing their events and interacting with each other, but the event queue and the clock mechanism are hidden from the user. The PASION models can be expressed in terms of the DEVS formalism (DEVS: discrete event simulation). Though PASION process declaration does not define formally model inputs and outputs, the relation between PASION processes and DEVS formalism are clear. Inputs and outputs and some model parameters are PASION **attributes** that can be PASCAL simple or structured variables of any type. This gives the model good versatility. This is why the modeler only has to know which of the process attributes are inputs, which are outputs, and which of them are model parameters. All other model variables are hidden. The internal and external transition functions are coded as the process events. Using the object-oriented terminology, process attributes are model data, and events are model methods. The difference between an event and a procedure is that events can be scheduled, and their execution occurs only through messages issued to the schedule, hidden for the user. Special scheduling algorithms are used to add the events on the event queue. There is no formal difference between PASION discrete-event and continuous process methods. Simply, a continuous object, described by a set of differential equations, has at least one event that integrates the system equations over a given time interval and schedules itself to be executed repeatedly. The **model type** is defined as a PASION process declaration. PASION links are available on the web at www.raczynski.com.

Compared with PASION, AnyLogic is a new-generation simulation software for combined simulation recently introduced. This simulation software is developed on the basis of the latest advances in complex system modeling theory and working standards in system design. It allows system exploration at any desired depth and any level of abstraction. The fundamental basis to model real-world systems with AnyLogic includes

- Arbitrary complex behavior logic, timing, topologies such as ring, chain mesh, etc., and routing
- Block based flowchart modeling
- Differential and algebraic equations
- Direct links to data bases and GIS
- HLA support
- MATLAB-SIMULINK-type library
- Modeling in Java, to run models on any Java-enabled platform, or even as applets on web browsers
- Messages passing, ports, custom routing
- Statechart modeling, to combine discrete and continuous behaviors
- UML for Real Time (UML-RT)

UML-RT in AnyLogic is specifically adapted for the development of complex, event-driven, real-time, real-world systems. Its modeling constructs have rigorous formal semantics that provide for model execution. UML-RT modeling means

- Explicit structural decomposition
- Clear separation of structure and behavior
- High degree of reusability

UML-RL is a complete working modeling standard. When developing a model in AnyLogic one develops classes of active objects representing the real-world objects. Active objects can encapsulate other active objects to any desired level. Running the AnyLogic model the object instances from their Active Object Class. There is one designated root class, which describes the model structure. Active objects interact with their surroundings solely through interface objects, which are ports and variables. Ports are used for discrete communication (message passing) optionally using a queue, while variables are used for continuous communication.

For using UML-RT AnyLogic provides a graphical user interface similar to classical visual development tools for programming languages. The included libraries give that development environment the extention to do discrete, continuous, and hybrid simulation modeling. If licensed by the user separately, NAG libraries or other numerical methods, not yet implemented in Java, can be added.

The notation of the interface makes active object classes highly reusable. Moreover, the modeler can define the object behavior as Java method and run it within the active object as a separate thread, the execution of which can be synchronized with other activities through

- Delay (timeout) method
- WaitEvent (static event) method
- WaitFORMessage(..) method of the PortQueuing class

The innovative core technology together with a remarkable set of features makes AnyLogic an advanced technology solution for a broad range of real-world application domains, such as

- Dynamic systems and control algorithms
- Education and training
- Enterprise modeling
- Mechanics and robotics
- Military
- Optimization
- Supply chain systems and manufacturing
- System dynamics and business simulation
- Traffic and pedestrian movement
- Telecom, networks and computer systems
- etc.

This core technology is continuously extended by different libraries, like MATLAB-SIMULINK, to provide the look and feel of the original SIMULINK simulation environment, or an Enterprise Library, which gives similar possibilities to build traditional discrete-event simulation models.

Providing Java classes, any AnyLogic Java class can be integrated in a Java program. On the other hand, external routines can be integrated, either by including external classes, or using the JNI interface to non-Java libraries.

As already mentioned, AnyLogic includes optionally OptQuest, the most common and widely used optimization software.

As long as there are no platform-dependend libraries needed for a specific simulation model. The precompiled Java-Applet can be put on a Webserver and used over the internet without any restrictions. This gives a main advantage in distributed development or project management.

From the vast number of examples, provided with AnyLogic, three selected models will be explicitly described.

Example 4.24
A fixed number of ants are passing over a 2D space. At the space boundaries, they rebound like billard balls. The ants must not collide, therefore a **visionDistance**, a **visionAngle**, a **turnAngle**, and a **stopTime** has been defined model-wide. An ant is an active object class, with four external parameters (x, y) for its current position in the space and its velocity (vx, vy). If two ants are about to collide (they are able to see each ither within the parameters **visionDistance** and **visionAngle**) they stop for the specified **stopTime**. Afterwards they change direction according to the **turnAngle**.

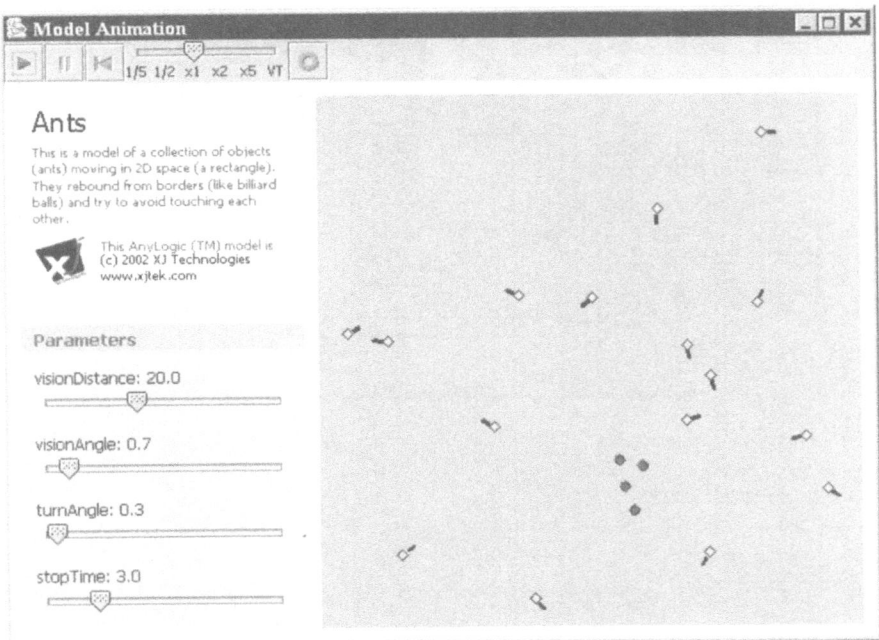

The schreenshot shows the model in an applet-window. It is remarkable that the parameters can be changed during the simulation run.

This example combines continuous processes like the distance check to each other performed by every ant itself with discrete state events (move-stop).

Example 4.25

Completely different from Example 4.24, Example 4.25 deals with very large simulation models such as they are used for System Dynamics – Urban Dynamics

Being responsible for the development of a city, different actions can be taken and their effects on the whole development are shown. The timesteps are set to one generation (30 years), but that value can be set individually by the user.

Compared to other System Dynamics specialized tools, this example shows the modularity, the model can be built on. The necessary equations are programmed in the Active Object Classes, where they are needed. The interactions between those classes are modeled by continuous external input and output variables. For this reason, the fairly large model is well structured and clearly arranged. There is no restriction in using differential equations or any other equations at the same time even within the same active object class.

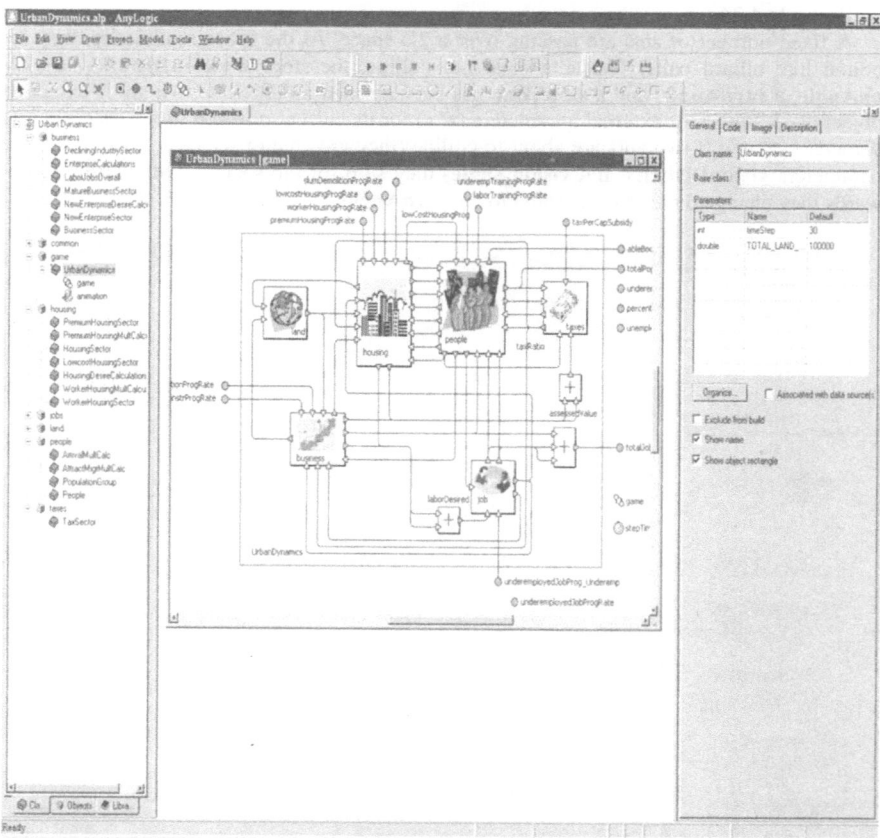

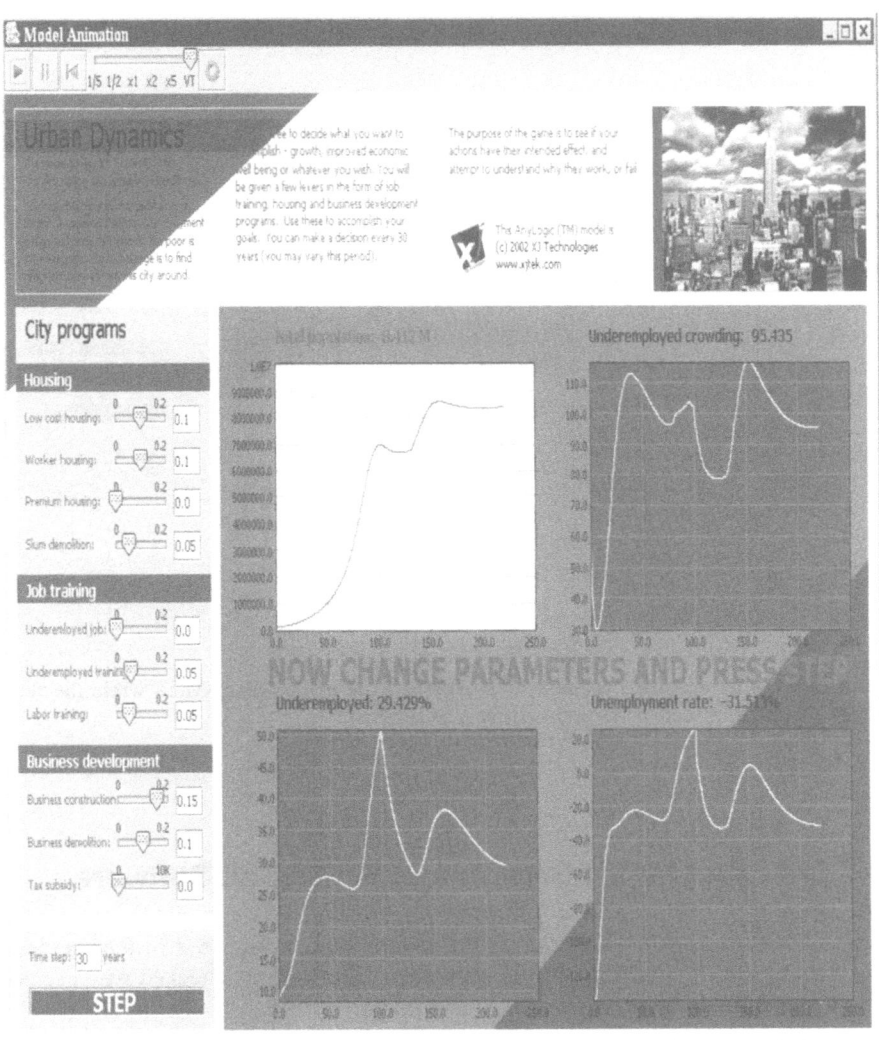

Example 4.26: Multiple-call centers

A network of similar equipped and staffed call centers will be optimized. Because of economical reasons, the individual call centers contain different numbers of staff, taking calls. Therefore, call centers are linked to others for dynamically routing of calls. This will avoid balked calls and therefore dissatisfied customers.

Starting with global parameters, like mean call duration and various cost factors, the call centers can be parameterized individually by specifying the average calls per hour, the queue capacity, and the number of operators.

Finally, the capacities of the links between call centers can be changed dynamically, to optimize the call-center capacities and additionally the network in between.

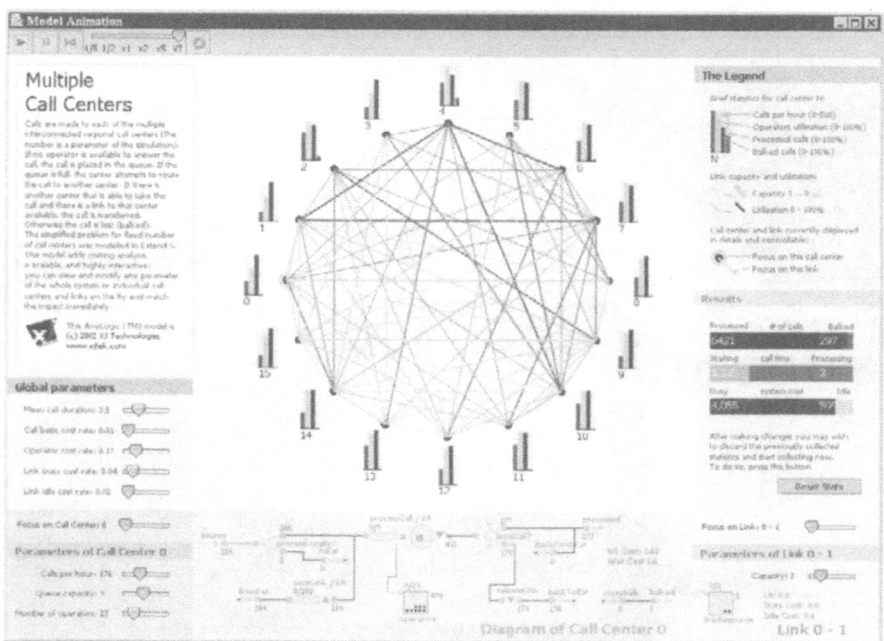

Various other examples are also of interest and are explained in detail wihin the fully functional time-limited evaluation version of AnyLogic.

4.8 Checklist for the Selection of Simulation Software

The following checklist summarizes the several topics discussed in the previous chapters to give an overview of the important features to be checked when selecting a simulation system.

☐ **Model building**

☐ discrete simulation	☐ continuous simulation	☐ combined simulation
☐ simulation language	☐ simulation software	☐ programming language
☐ general purpose	☐ application specific	☐ object oriented
☐ process oriented	☐ activity oriented	☐ event oriented

☐ **Application Domain**

☐ business dynamics	☐ biology/medicine	☐ engineering
☐ control systems	☐ multiphysics	☐ chemistry
☐ structural mechanics	☐ robotics	☐ telecom

☐ networking   ☐ traffic   ☐ manufacturing
☐ computer systems   ☐ cost modeling   ☐ supply chain
☐ workflow   ☐ material flow   ☐ …..

☐ Interface
☐ programming language☐ database/SQL   ☐ ASCII
☐ CAD/CAM/CIM   ☐ calculation   ☐ optimization
☐ …..

☐ Animation
☐ online   ☐ 2D/3D   ☐ VR/AR
☐ playback   ☐ zoom/scroll   ☐ pixel
☐ snapshots   ☐ windows   ☐ …..

☐ USER Interface
☐ menus   ☐ Help   ☐ GUI
☐ Windows   ☐ editor   ☐ keyboard
☐ communication   ☐ user guidance   ☐ ……

☐ Price
☐ list price   ☐ discount   ☐ licence agreements
☐ update costs   ☐ training   ☐ documentation
☐ service/hotline   ☐ hardware costs   ☐ expansion costs
☐ libraries   ☐ personnel costs   ☐ ………

☐ Results, presentation
☐ GUI   ☐ user defined   ☐ statistical module
☐ interval analysis   ☐ confidence analysis ☐ ……

☐ Service/References
☐ documentation   ☐ training   ☐ practical examples
☐ hotline   ☐ maintenance   ☐ test installation
☐ updates   ☐ user groups   ☐ installed base
☐ references   ☐ adaptation/extension ☐ …..

☐ User
☐ simulation language ☐ object-oriented   ☐ higher programming lan-
                                               guage
☐ training period   ☐ decision table   ☐ ……

4.9 References and Further Reading

Aström K, Albertos P, Blanke M, Isidori A, Schaufelberger W, Sanz R (Eds.), (2001), Control of Complex Systems, Springer, London, Berlin, Heidelberg

Banks J, Carson JS, Nelson BL, Nicol DM, (2001), Discrete Event System Simulation, Prentice Hall, New Jersey

Breitenecker F, Ecker H, Bausch-Gall I, (1993), Simulation with ACSL (in German), Vieweg Publ., Wiesbaden

Cellier FE, (1993), Integrated Continuous System Modeling and Simulation Environments, pp. 1–29, In CAD for Control Systems, Linkes D, (Ed.), Marcel Dekker Publ. New York

Engelbert J, Nguyen T, Thurston C, (2002), B^2 Spice A/D Version 4, Beige Bag Software Inc.

Hendiksen J, (1993), SLX, the Successor to GPSS/H, Proc. of the SCS Winter Simulation Conference

Jungblut J, Sievers M, Vogelpohl A, Bracio BR, Möller DPF, (1997), Dynamic Simulation of Wastewater Treatment: The Process of Nitrification, Simulation Practice and Theory Vol. 5, pp. 689–700

Kheir NA, (1996), Systems Modeling and Computer Simulation, Marcel Dekker, Inc., New York, Basel, Hong Kong

Leonard NE, Levine WS, (1992), Using MATLAB to Analyze and Design Control Systems, Benjamin & Cummings Inc., Redwood

Möller DPF, Popovic' D, Thiele G (1983), Modeling, Simulation and Parameter-Estimation of the Human Cardiovascular System, Vieweg Publ., Braunschweig, Wiesbaden

Möller DPF (1992), Modeling, Simulation and Identification of Dynamic Systems (in German), Springer, Berlin, Heidelberg, New York

Möller DPF (1997), Combined Simulation, Proc. UK Simulation Conference

Raczynski S, (1999), Combined Simulation: PASION Approach, Proceed. of the SCS Summer Simulation Conference, pp. 48–52

Russel EC, (1983), SIMSCRIPT II.5 and SIMGRAPHIS, Proc. of the SCS Winter Simulation Conference

Sandige RS, (2002), Digital Design Essentials, Prentice Hall, New Jersey

Schriber T, (1993), Perspectives on Simulation using GPSS, Proc. of the SCS Summer Simulation Conference

Selfridge RG, (1955), Coding a General Purpose Digital Computer to Operate as a Differential Analyzer, Proc. IRE Western Joint Computer Conference

Tiller MM, (2001), Introduction to Physical Modeling with Modelica, Kluwer Academic Publishers, Boston, Series Engineering and Computer Science Vol. 615

van Wyk van Brievingh RP, Möller DPF, (Eds.), (1993), Biomedical Modeling and Simulation on a PC, Springer, New York

Waterman DA, (1986), A Guide to Expert Systems, Addison-Wesley Publ. Company, Reading

Ziegler BP, Praehofer H, Kim TG (2000), Theory of Modeling and Simulation, Academic Press, San Diego

Technical Manuals
Advanced Continuous Simulation Language, Beginners Guide, Mitchell and Gauthier Associates, 1997, U.S.A.
Getting Started using Simulink, The Mathworks Inc. 2002
FEMLAB 2.3 Getting Started Guide, Comsol AB, Stockholm, 2002

Links
http://www.aegis.com
http://www.beigebag.com
http:// www.femlab.com
http://www.mathworks.com
http://www.modelica.org
http://www.modelkinetix.com
http://prozessoptimierung.arcs.ac.at
http://www.raczynski.com.
http://www.xjtek.com

4.10 Exercises

4.1 What is meant by the term simulation?
4.2 What is meant by the term physical similarity?
4.3 What is meant by the term isomorphism?
4.4 What is meant by the term rule-base system?
4.5 What is meant by the term semantic net?
4.6 What is meant by the term expert system?
4.7 What are the components of an expert system?
4.8 What is meant by the term block oriented simulation system?
4.9 Give an example of a block oriented simulation system.
4.10 What is meant by the term equation-oriented simulation system?
4.11 Give an example of an equation-oriented simulation system.
4.12 What is meant by the term transaction-oriented simulation system?
4.13 Give an example of a transaction-oriented simulation system.
4.14 What is meant by the term event-oriented simulation system?
4.15 Give an example of an event-oriented simulation system.
4.16 What is meant by the term activity-oriented simulation system?
4.17 Give an example of an activity-oriented simulation system.
4.18 What is meant by the term process-oriented simulation system?
4.19 Give an example of a process-oriented simulation system.
4.20 What is meant by the term mixed-mode simulation system?
4.21 Give an example of a mixed-mode simulation system.

4.11 Case Study Examples

The case study examples are embedded in this book to learn to build more advanced models and to get a step-by-step introduction into the selected simulators FEMLAB and ModelMaker.

4.11.1 FEMLAB

It will be shown how to create a geometry using FEMLABs CAD tool, how to set up model equations, and finally how to post-process the solution[1].

After starting a tutorial you can pause, fast forward or reverse the movie by using the buttons in the bottom of the tutorial window.

To be able to view the movies you need to have the Macromedias Flash 5 plug-in installed and ensure that you have at least Internet Explorer 4.x or Netscape 4.06+. It is also recommended to use at least 56 Kbit/s transfer rate.

Case Study 4.1: Forced and free convection heat transfer
This case study example describes a fluid flow problem with heat transfer in the fluid. An array of heating tubes is submerged in a vessel with fluid flow entering at the bottom. The figure below depicts the setup

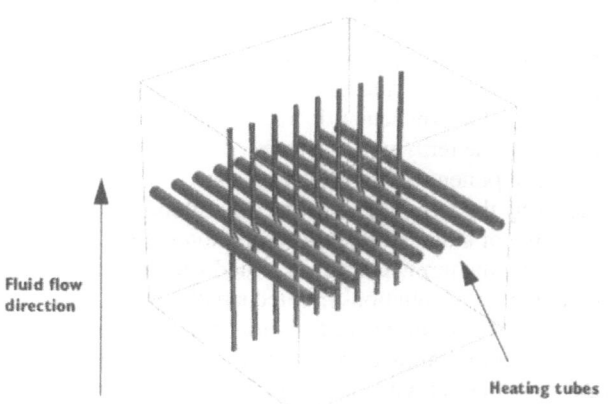

Fluid flow direction

Heating tubes

A first consideration when modeling should always be the true dimension of the problem. Many problems do not show variations in three dimensions and can be extrapolated from the solution of a related 2D case. Assuming any end effects from the walls of the vessel can be neglected, the solution can be assumed constant in the direction of the heating tubes, and the model is therefore reduced to a 2D domain (below).

[1] I would like to thank PD Dr. Stefan Funken, FEMLAB GmbH, Göttingen, Germany, for his support.

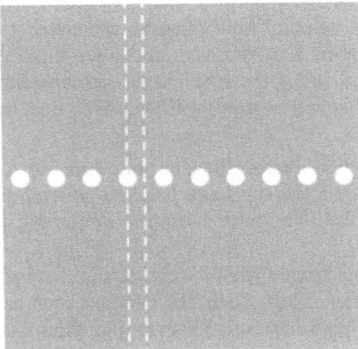

The next consideration is finding symmetries. In this case, inclusion of symmetry planes allows you to model only the thin domain indicated in the figure.

Governing Equations

This is a multiphysics model, meaning that it involves more than one kind of physics. In this case, you have incompressible Navier Stokes equations from fluid dynamics, together with a heat transfer equation, that is, essentially a convection-diffusion equation. There are four unknown field variables: the velocity field components u and v; the pressure, p, and the temperature, T. They are all inter-related through bidirectional multiphysics couplings.

The pure incompressible Navier Stokes equations consist of a momentum balance (a vector equation) and a mass conservation and incompressibility condition. The equations are

$$\rho \frac{\partial u}{\partial t} + \rho(u \cdot \nabla)u = -\nabla p + \eta \nabla^2 u + F \tag{4.78}$$
$$\nabla \cdot u = 0$$

where F is a volume force, ρ the fluid density and η the dynamic viscosity. We denote the vector differential operator. See further "Overview of PDE Modles" on pages 1–336 of the FEMLAB Getting Started Guide.

The heat equation is an energy conservation equation that says only that the change in energy is equal to the heat source minus the divergence of the diffusive heat flux

$$\rho c \frac{\partial T}{\partial t} + \nabla \cdot (-k\nabla T + p c_p T u) = Q \tag{4.79}$$

where c_p is the heat capacity of the fluid, and ρ is fluid density as before. The expression within the brackets is the heat flux vector, and Q represents a source term. The heat flux vector contains a diffusive and a convective term where the latter is proportional to the velocity field u.

In this model, the above equations are coupled through the F and Q terms. First, add free convection to the momentum balance with the Boussinesq approximation. In this approximation, variations in density with temperature are ignored, except insofar as they give rise to a buoyancy force lifting the fluid. This force will be put in the F-term in the Navier

Stokes equations. See further "Marangoni Convection" on pages 2–321 in the Model Library.

At the same time, the velocity field must be accounted for in the heat equation. Instead of applying the heat equation as it stands, you can put the divergence of the convective heat flux into the Q coefficient, using the fact that the velocity field is divergence free. This puts the equation on a form that can be used in FEMLAB's heat transfer application mode.

Case Study 4.2: Thermo electric heating in a bus bar

The resistive heating in a bus bar leads to a rise in temperature which, in turn, increases the resistance of the bus bar. The time evolution of current, temperature, and lumped resistance is studied.

This case study example examines the relationship between current throughput and temperature inside a solid copper bar. Such bus bars are used as conductors in industrial environments requiring very high currents, for example, aluminum smelters and certain chemical plants.

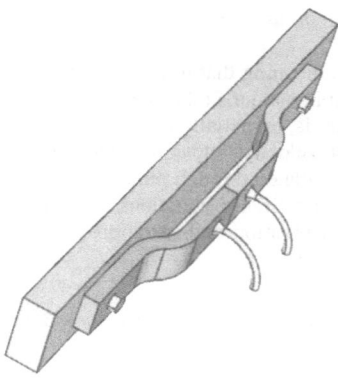

The thermo electric coupling is two-way: volume currents inside the bus bar, which are proportional to the conductivity, act as a distributed heat source, while at the same time the temperature affects the metal's conductivity. These kinds of dependencies make it necessary to create a multiphysics model.

As a first step, the stationary-current distribution at constant temperature will be modeled. It is highly recommended that you follow along with the description of the modeling process in this example even if you are not an electromagnetics expert. The discussion focuses on how to use FEMLABs graphical interface rather than on the underlying physics.

The second part of the example adds a heat transfer equation, bidirectionally coupled to the stationary-current model. This part is primarily about multiphysics modeling, showing how to connect phenomena from different fields of physics. It also shows how you define coupled variables, which can be, for example, an integral of the solution, evaluated on some part of the model geometry.

Geometry and boundary conditions

The bus bar is mounted as an intermediate step between two high voltage cables and a load. Because of the symmetry, it is only necessary to model one half of the true geometry.

The copper bus bar is electrically insulated everywhere, except for two contact plates around the mounting holes. The temperature is constant on the contact plates, and on the parts in contact with the load or mounting screw. All free faces have heat transfer coefficients corresponding to free-air convection, but the symmetry face is isolated in all respects.

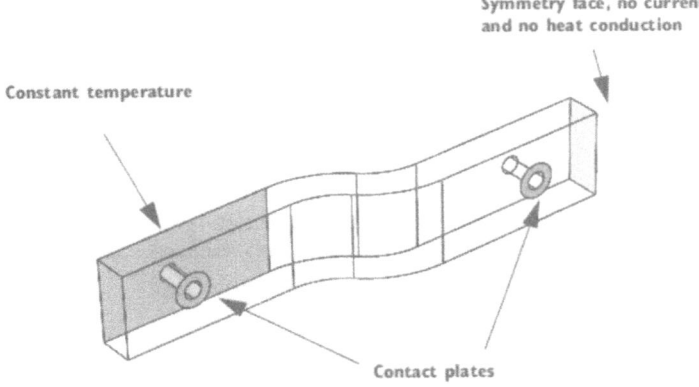

Case Study 3: 2D radiator
This model is an example of the 2D-physics-mode working environment. We will model heat conduction in a solid and therefore make use of the heat transfer application mode.

Case Study 4.4: 3D Radiator
This model is an example of the 3D physics mode working environment. We will model heat conduction in a solid and therefore make use of the heat transfer application mode.

Case Study 4.5: The catalytic burner

A mixture of air and hydrogen enters a catalytic reactor. Hydrogen reacts in the catalyst and the influence of convection and diffusion coupled to reactions is studied.

FEMLAB Chemical Engineering Module

Case Study 4.6: Coupled free and porous-media flow

The coupling between flow in an open channel and the flow induced in the porous walls of the channels is studied. The model couples the Navier Stokes equations in the open channel, with the Brinkman equations for porous-media flow. The results show that there is a substantial flow in the porous media and this flow is induced both by pressure and viscous effects.

Case Study 4.7: Simulation of a fixed-bed reactor for catalytic hydrocarbon
                        oxidation

This model treats the process for production of phthalic anhydride in a multitube fixed-bed reactor. The process is highly exothermic and cooling at the surface of the reactor tubes is accounted for in the model. The model includes mass balances for the involved species and an energy balance.

FEMLAB Electromagnetics Module

Case Study 4.8: Model of a cold crucible for molten metals

A cold crucible is modeled, which is used for elaboration of alloys that require a high degree of purity. The cold crucible is surrounded by an inductor, which makes the crucible act like a field concentrator due to the induced currents.

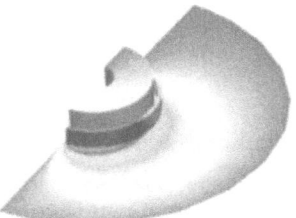

Case Study 4.9: Electromagnetic brake, exporting to SIMULINK

A metal disk is rotating in the air gap of a magnet. Currents are induced, and the forces on the current lines will slow down the disk. FEMLAB models the static problem of computing the induced currents, SIMULINK computes the time-evolution of the angular veloc-

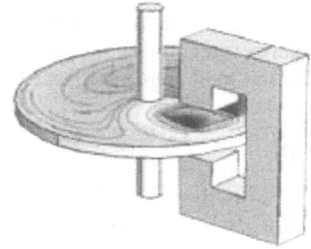

ity.

Case Study 4.10: Model of a monoconical RF antenna
The antenna impedance and radiation pattern are studied as a function of frequency of a monoconical antenna, with a finite ground plane and a 50-Ohm coaxial feed.

FEMLAB Structural Mechanics Module

Case Study 4.11: Stress-optical effects in a photonic waveguide
The stress-optical effect causes unwanted birefringence in a planar photonic waveguide. Plane-strain analysis followed by an optical mode analysis show the resulting split of the fundamental modes.

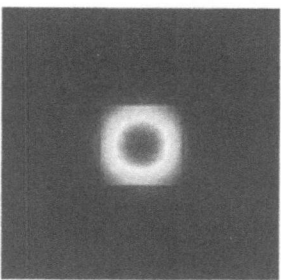

Case Study 4.12: Model of pressure vessel
A pressure vessel modeled with shell elements is subjected to an internal pressure higher than the surrounding atmospheric pressure. The deformation of the vessel is exaggerated in

the figure. The displacement around the pipe connections is especially important since rupture at the pipe joints can occur.

Case Study 4.13: Model of a tank filled with water
A tank is built up by shell elements and beam elements. The parametric solver is used to visualize the tank being gradually filled with water.

4.11.2 ModelMaker

ModelMaker is designed to mimic the process of conceptual model building. Hence the first step involves constructing a diagram on the screen that represents the various model parts. This diagram is composed of a series of ModelMaker components, each of which is intended for a different type of mathematical operation. Each component has a definition that can be edited to insert its equation and any other appropriate information. Once your model has been implemented, it can be run; the equations are solved, generating results that can be interpreted in graphs or tables.

It will be shown how to create a model of a system using ModelMaker's tool bar, how to set up model equations, and finally how to post-process the solution.

After starting the ModelMaker power-point tutorial you get a step-by-step insight into the modelling process.

Case Study 4.14: Model of a mass damper spring system
The mass damper spring system can be described by the differential equation

$$M \cdot x'' + D \cdot x' + C \cdot x = F(t), \tag{4.80}$$

with M = mass, D = damping factor, C = spring constant, X = elongation. With the transform $A \cdot x'' + B \cdot x' + C \cdot x = D(t)$ the equation above can be rewritten as follows

$$x'' + \left(\frac{B}{A}\right) x' + \left(\frac{C}{A}\right) x = \frac{D(t)}{A}. \tag{4.81}$$

Rewriting this differential equation, which is of second order, using n first-order differential equations we find:

$$x = x_1$$
$$x' = x_1' = x_2 \tag{4.82}$$
$$x'' = x_2' = -(B/A) \cdot x' - (C/A) \cdot x + D(t)/A .$$

This results, due to the original second-order system, in two first-order ODEs:

$$x_1' = x_2$$
$$x_2' = -(B/A) \cdot x_2' - (C/A) \cdot x_1 + D(t)/A . \tag{4.83}$$

which can be solved using ModelMaker. You are required to build the model of the second-order system given above and implement it with ModelMaker according to the following specifications with initial conditions of integrators being 1:

- $A = 1, B = 1, \ C = 1, D = \cos(t)$
- $A = 10, \ B = 10, C = 10, D = \cos(t)$
- $A = 1, B = 10, C = 1, D = \cos(t)$
- $A = 1, B = 10, C = 5, D = \cos(t)$

Case Study 4.15: Model of ingestion and subsequent metabolism of a drug
In project 2 ingestion and subsequent metabolism of a drug in a given individual are examined on the bases of a combined simulation containing continuous-time and discrete-tiem elements.
Background:
A two-compartment model is used to study ingestion, distribution, and metabolism of a drug in the individual. It provides the background information of the mechanism of action of drugs in general pharmacological terms and the significance of pharmacokinetic parameters in determining the efficacy of drugs. In particular, the drug is ingested, e.g. orally as medication, the drug enters the gastrointestinal tract from where it is then distributed throughout the bloodstream of the individual to be metabolized and eliminated. The primary interest of studying compartment models is to govern how input ingestion rate and/or the initial concentration of the drug in the body affects the individual.

The pharmacokinetic model described in the equation above is a second-order linear model. The first differential equation is uncoupled from the second differential equation, the second differential equation is coupled with the first differential equation. It should be noted that this observation is important, since mathematical models should not be excessively difficult for analytical studies.

Limitations of the model:
The main purpose of the model is to demonstrate the basic time courses of drugs in different fluids, tissues and/or organs and excreta of the body. Hence the model is kept as simple as possible while remaining accurate and realistic. The main limitations are:

- The model is linear, i.e. non linearities are not considered
- The model equations used neglect feedback influences
- Real biological systems don not have set points
- We can not instruct the model to adjust to output variables to certain values of the state, we only change its parameters

The model is not suitable for studying the effects of physical workload because the mechanisms that maintain the oxygen utilization during workload are neglected.

In this model, combined simulation is realized, the event has to be introduced using the block-independent event. Moreover, the ModelMaker realization should use a source compartment as the respective flux_in variable, and a drain compartment as the respective flux_Compartment2 uptake

Requirements:
You are required to give the analytical mathematical solution of the two-compartment model (for this purpose see Chap. 2).

You are required to build the model of the second-order system given above and implement it with ModelMaker 3.0 according to the following initial specifications of compartment 1 is 0 and compartment 2 being 1.

Single dose-injection
1. $k_{12} = 0.1, k_2 = 0.1$
2. $k_{12} = 0.1, \ k_2 = 0.05$
3. $k_{12} = 0.01, k_2 = 0.01$

Event triggered multiple dose-injection
4. $k_{12} = 0.1, k_2 = 0.1$
5. $k_{12} = 0.1, \ k_2 = 0.05$
6. $k_{12} = 0.01, k_2 = 0.01$

Case Study 4.16: Modeling and simulation of a single and a double pendulum
Modeling and simulation of dynamic systems with different degrees of freedom are part of this third project, based on the physical system of a pendulum on a rigid rod.

Background:
The physical model of the pendulum can be described as

$$F_a = -F_d - F_g = -D \cdot v - M \cdot g \cdot \sin(\Phi) = -r \cdot \Phi' - M \cdot g \cdot \sin(\Phi) \qquad (4.84)$$

with F_a: acceleration force, F_d: damping force $= D \cdot v$, F_g: gravidity force, M: mass, g: gravitation constant, v: velocity, r: radius, and the angle Φ. Assuming

$$a = \frac{F_a}{M}, \quad \Phi' = \frac{v}{r}, \text{and } \Phi'' = \frac{a}{r}, \tag{4.85}$$

we obtain

$$F_a = M \cdot a = M \cdot \Phi'' \cdot r \tag{4.86}$$

and due to this we find the second order vector differential equation:

$$M \cdot r \cdot \Phi \cdot \Phi'' + d\Phi \cdot r \cdot \Phi \cdot \Phi' + M \cdot \Phi \cdot g \cdot \Phi' \sin(\Phi) = 0, \tag{4.87}$$

which can be rewritten as

$$\Phi'' + \left(\frac{D}{M}\right) \cdot \Phi \cdot \Phi' + \left(\frac{g}{r}\right) \Phi' \sin(\Phi) = 0, \tag{4.88}$$

This second-order differential equation is a nonlinear differential equation due to the term $\sin(\Phi)$. With $\Phi = x$ we find the state equations $x = x_1$, $x_1' = x_2$, and $x_2' = a/r$ may be of interest when modeling the pendulum system. with the initial conditions $x_{10} = x(0)$, and $x_{20} = x'(0)$.

Requirements:
You are required to build the model of the single pendulum on a rigid rod as given above and implement it with ModelMaker 3.0 according to the following specifications (D: damping factor, M: mass, g: gravitation constant, r: radius) with initial conditions of the integrators being 1:

1. $D = 1, \ M = 1, g = 9.81, r = 1$
2. $D = 0.4, M = 2, g = 9.81, r = 5$
3. $D = 1, \ M = 10, g = 9.81, r = 10$
4. $D = 1, \ M = 10, g = 9.81, r = 1$
5. $D = 10, M = 10, g = 9.81, r = 10$

Background:
The physical model of the double pendulum can be described by the masses M_1 and M_2 that are connected by massless rods of length r_1 and r_2. The equations of motion of the two masses, expressed in terms of the angles Φ_1 and Φ_2 as indicated, are

$$(M_1 + M_2) \cdot r^2_1 \cdot d^2\Phi_1/dt^2 + M_2 \cdot r_1 \cdot r_2 \cdot d^2\Phi_2/dt^2 \cdot \cos(\Phi_1 - \Phi_2) \tag{4.89}$$

$$+ M_2 \cdot r_1 \cdot r_2 \cdot (d\Phi_2/dt)^2 \cdot \sin(\Phi_1 - \Phi_2) = - (M_1 + M_2) \cdot g \cdot r_1 \cdot \sin(\Phi_1)$$

$$\cdot M_2 \cdot r^2_2 \cdot d^2\Phi_2/dt^2 + m_2 \cdot r_1 \cdot r_2 \cdot d^2\Phi_1/dt^2 \cdot \cos(\Phi_1 - \Phi_2)$$

$$- M_2 \cdot r_1 \cdot r_2 \cdot (d\Phi_1/dt)^2 \cdot \sin(\Phi_1 - \Phi_2) = - M_2 \cdot g \cdot r_2 \cdot \sin(\Phi_2)$$

Requirements:
You are required to show for Part 2 that $\Phi_1 = 0$, $d\Phi_1/dt = 0$, $\Phi_2 = 0$, $d\Phi_2/dt = 0$ defines equilibrium states, theoretically, and choose the respective parameter and give an explanation.

You are required to obtain linearized state equations for the double pendulum that are valid if the pendulum system is near its equilibrium state, and explain it.

You are required to document the time dependent behavior and the phase-plane behavior of the double pendulum for the five cases of the double pendulum and explain the meaning of the graphs.

1. $M_1 = 1, M_2 = 1, g = 9.81, r_1 = 1, r_2 = 1$
2. $M_1 = 10, M_2 = 1, g = 9.81, r_1 = 5, r_2 = 1$
3. $M_1 = 1, M_2 = 10, g = 9.81, r_1 = 1, r_2 = 5$
4. $M_1 = 10, M_2 = 10, g = 9.81, r_1 = 5, r_2 = 5$
5. $M_1 = 2, M_2 = 4, g = 9.81, r_1 = 2, r_2 = 4$

Case Study 4.17: Modeling of the population growth and balance
Population models predict either population growth without bound or inevitable extinction. The difference is based on whether the growth rate is positive or negative.

The population can be modeled at the beginning of time period t based on the logistic population model

$$x(t+1) - x(t) = r \cdot x(t) \cdot (1 - \frac{x(t)}{k}), \tag{4.90}$$

where r is the growth rate, and k represents the carrying capacity, which is the population level at which the birth and death rates of a species precisely match, resulting in a stable population over time.

Requirements:
You are required to determine the equilibrium populations.

You are required to simulate the model with parameter values $r = 0.007$ and $k = 1000$, and using an initial population of 250, running the model for 100 years.

You are required to try other values of r, k and initial populations.

Case Study 4.18: Modeling of the birth and death rates of a population
Instead of simply computing the net change in population the model has to be rearranged to keep track of the birth and death rates, both of which are likely to be non-negative.

The newly developed model will still have a container containing the population level. However, there will be now two flows, one associated with births and a second associated with deaths.

The birth and death rates can be modeled at the beginning of time period t based on the population model as follows

$$x(t+1) - x(t) = r \cdot x(t) \cdot (1 - \frac{x(t)}{k}) \tag{4.91}$$

$$x(t+1) - x(t) = r \cdot x(t) - r \cdot x(t) \cdot \frac{x(t)}{k},$$

where the first term $r \cdot x(t)$ can be interpreted as the birth rate in this model, while the second term, $r \cdot x(t) \cdot x(t)/k$, can be interpreted as the death rate.

Requirements:
You are required to determine the equilibrium of birth and death of the population.

You are required to simulate the model if birth rate is faster or slower than death rate.

You are required to expand the model assuming a new predator comes to the area.

You are required to expand the model due to food supplies and simulate the new model for the decrease and increase of food supply.

Case Study 4.19: Modeling the Lotka Volterra equations

The Lotka Volterra model is a classical ecological model that explains the oscillatory levels of certain fish catches in the Atlantic.

To obtain a more realistic model of two interacting species, we include the effects of competition of the prey x_1 among themselves for their limited amount of resources, and the competition among the predators x_2 for the limited amount of prey. The model yields

$$\dot{x_1} = x_1(G - B \cdot x_2)$$
$$\dot{x_2} = x_2(B \cdot x_1 - S) \tag{4.94}$$

where G is the growth rate, and B represents the carrying capacity which is the population level at which the birth and death rates of a species precisely match, resulting in a stable population over time, and S is the death rate.

Requirements:

You are required to determine the dynamic behavior when the initial conditions are: $x_1(0) = 10.000$, and $x_2(0) = 1.000$, and $B = 6 \cdot 10^{-6}$, $G = 0.005$, and $S = 0.5$.

Case Study 4.20: Reference nets for habour-based workflow analysis of ship transportation

5 Parameter Identification of Dynamic Systems

5.1 Introduction

As shown in Chap. 1, there are two different approaches building a model of a real-world system, the theoretical one, based on the derivation of the essential physical relationships of the real-world system, and the empirical one, based on experiments with the real-world system. Practical approaches combine both. The difficulty in implementing a mathematical model, developed theoretically, is that not all important system parameters, appearing in the model equations as some coefficients, are known a priori. Unknown parameter values can be determined through experiments with the real-world system. This can, in principle, be done through evaluation of the data measured at the system input and output by the use of parameter-identification methods, which will work either in a direct manner, as shown in Fig. 5.1, or in an indirect way, by using an adjustable-parameter vector, which is part of the mathematical model, as shown in Fig. 5.2. Hence the parameter-identification method can be stated as a link between data and models.

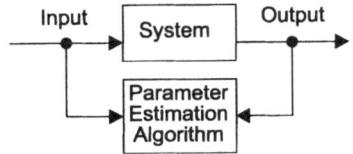

Fig. 5.1. Direct parameter identification

From Fig. 5.2 one can conclude that parameter identification of unknown system parameters can be done using a mathematical model of the real-world system and adapting its parameters. In fact only the model structure has to be known to build the adaptive model or to implement it as part of a software package. The initial parameter values themselves are guessed, for instance, on some preliminary knowledge about the real-world system. The type of model structure and the parameter identification method chosen for estimation purposes are of essential importance for the accuracy of the estimates.

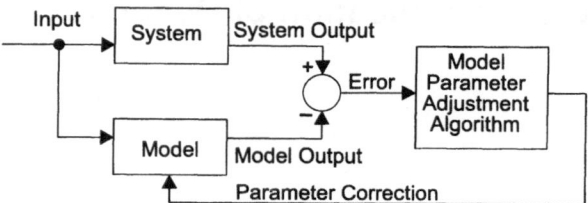

Fig. 5.2. Parameter identification using an adaptive model

From Fig. 5.2, the parameter-identification methodology can be considered synonymous for statistical and numerical procedures to obtain reasonable values for model parameters or data. The classical method is the linear regression technique, which goes back to the 18th century. It is easy to handle, hence linear regression has long been established as a convenient tool for analyzing data of dynamic systems. However, the diversion of using linear regression methods has led to an overemphasis on linear relationships. But, mostly, the relationships of real-world systems – or data – are nonlinear and linearization is nothing more than an approximation with a limited scope. Nowadays, statistical program packages are available with routines for solving nonlinear regression problems. But in recent years, highly efficient parameter-identification techniques for dynamic systems have been developed. Today, parameter-identification provides the link between data and models, in other words, between statistics and simulation.

It is the goal of any identification procedure to obtain the unknown model parameters for the real-world system. In practice the identification procedures are based on discrete measurements. The resulting model can be described in a continuous-time or a discrete-time representation and it can be linear or nonlinear in its parameters. System models are mainly used for the following purposes:

- Getting deeper insight into the physical structure of real-world dynamic systems
- Developing efficient adaptive controllers for real-world systems applications described by the identified model

Hence, the term identification specifies the determination of a model as an element of a given class of system models, based on input and output measures, to which the real-world system is equivalent. The meaning of an identification task therefore is as follows:

- Data and measurements of the time-dependent input and output signals of the real-world dynamic system are known. Unknowns are the structure and the parameters of the suitable mathematical model.
- Solving the identification problem, one has to arrange, that the mathematical model, which is connected in parallel to the real-world system, has the same static and dynamic behavior. Therefore, the parameters of

the mathematical model have to be optimized in such a way that the model outputs fit with the experimental data.

Let a time-continuous mathematical model of a real-world dynamic system, be described by a set of n ordinary first-order nonlinear differential equations, parameterized in an n-dimensional parameter vector $\underline{\Theta}$. Consider that this model fit the transfer behavior of the real-world system for a parameter vector $\underline{\Theta} = \underline{\Theta}_{RS}$ with sufficient accuracy, meaning that the output of the model and the real-world system coincide. If this assumption is true this model is said to be the true model of the real-world system and the corresponding parameter vector $\underline{\Theta}_{RS} = \underline{\Theta}_{TM}$ exists, with $\underline{\Theta}_{TM}$ as the true parameter vector.

Remark 5.1
Let **MM** be the set of mathematical models, and the chosen mathematical model MM be the best fit of a model for the real-world system RS. The chosen model MM is the best fit if the sum of squares is minimal.

$$\text{MM} \in \textbf{MM}: \ e[\underline{Y}_{RS}(t), \underline{Y}_{MM}(t)] \rightarrow \text{Min.} \tag{5.1}$$

Remark 5.2
Let e be an error functional that is based on the difference between the measures of the real-world system output $\underline{Y}_{RS}$ and the calculated numbers of the output of the mathematical model $\underline{Y}_{MM}$

$$e := e[\underline{Y}_{RS}(t), \underline{Y}_{MM}(t)]. \tag{5.2}$$

Hence the identification problem of the unknown parameters then can be expressed as an optimization problem minimizing a chosen performance criterion

$$J = \int_0^t (e)^2 \, dt \rightarrow \text{Min.} \tag{5.3}$$

Remark 5.3
Let the n-dimensional identification task be restricted to a p-dimensional model with the p-dimensional parameter vector $\underline{\Theta}$. Hence the identification task can be reduced to a parameter-estimation problem. The goal of the parameter estimation problem is the adaptation of the p-dimensional parameter vector $\underline{\Theta}$ of the mathematical model – which can be introduced as an identification model – such that its output $\underline{Y}_{MM}(\underline{\Theta}, t)$ will coincide with the output of the real-world system $\underline{Y}_{RS}(\underline{\Theta}, t)$.

Remark 5.4
Let an identification task be defined as a task to adjust the parameter vector $\underline{\Theta}$ of an identification model in such a way that its output sequence $\{\underline{Y}_{MM}(\underline{\Theta})\}$ coin-

cides with the output sequence of the real-world system $\{\underline{Y}_{RS}(\underline{\Theta})\}$. A real-world system is called identifiable if

$$\underline{\Theta}_{RS} = \underline{\Theta}_{MM}, \tag{5.4}$$

which is only the case if

$$\underline{Y}_{RS}(\underline{\Theta}) = \underline{Y}_{MM}(\underline{\Theta}). \tag{5.5}$$

Remark 5.4 takes into account cases in which the system outputs $\underline{Y}_{RS}$ of the real-world system are imprecisely, corrupted by measurement noise, introduced as $\underline{n}_{RS}$, which yields

$$\underline{Y}_{RS,Meas} = \underline{Y}_{RS} + \underline{n}_{RS}. \tag{5.6}$$

The outputs of the mathematical model $\underline{Y}_{MM}$ and of the real-world system $\underline{Y}_{RS}$ have to be compared. If $\underline{Y}_{RS}$ coincides with $\underline{Y}_{MM}$ the difference between the measured system outputs and the model outputs can be expressed as

$$\underline{e}_{MM}(\underline{\Theta}_{MM}) := \underline{Y}_{RS,Meas}(\underline{\Theta}_{RS}) - \underline{Y}_{MM}(\underline{\Theta}_{MM}), \tag{5.7}$$

with the output measurement error vector $\underline{e}_{MM}$ i.e.

$$\underline{e}_{MM}(\underline{\Theta}_{MM} = \underline{\Theta}_{RS}) := \underline{e}_{RS}. \tag{5.8}$$

The identification task is solved if $\underline{\Theta}_{MM}$ can be adapted in such a way that the estimated output error $\underline{e}_{MM}(\underline{\Theta}_{MM})$ is minimal, i.e. the outputs of the mathematical model and the real-world system coincide due to the equality of input and initial conditions. If this assumption is true the model is said to be the true model of the real-world system and the corresponding parameter vector $\underline{\Theta}_{RS}$ is called the true parameter vector $\underline{\Theta}_{TM}$, or true output error $\underline{e}_{RS}(\underline{\Theta}_{RS})$ vector.

5.2 Mathematical Notation of the Identification Task

Definition 5.1
Let a real-world system mathematically be described by a set of first-order linear or non-linear differential equations, with r input and m output variables

$$\frac{dx}{dt} = \underline{f}(\underline{x}(t),\underline{u}(t),t,\underline{a}), \tag{5.9}$$

and

$$\underline{x}(0) = \left[\frac{\underline{x}_M}{\underline{b}}\right] \tag{5.10}$$

$$\underline{y}(t) = \underline{g}(\underline{x}(t),\underline{u}(t),t,\underline{c}), \tag{5.11}$$

where $\underline{u}(t) \in \mathfrak{R}^r$ are the input variables, $\underline{y}(t) \in \mathfrak{R}^m$ are the output variables, $\underline{x}(t) \in \mathfrak{R}^n$ are the state variables, $\underline{x}(0)$ as the known initial conditions, and $\underline{x}_M$ as known measurable initial conditions; $\underline{a} \in \mathfrak{R}^A$, $\underline{b} \in \mathfrak{R}^B$, and $\underline{c} \in \mathfrak{R}^C$ are the parameters to be identified. ■

Remark 5.5
If f and g are the linear or nonlinear functions the initial value problem in (5.9) and (5.10) has a unique solution. It is, therefore, not excluded that the function f may be discontinuous at certain points of t, – this is due to time discrete events during simulation – and at certain points of x – due to state events during simulation – or at certain points of $\underline{u}$. Consider that the output variables $\underline{Y}_{RS}(t)$ and the input variables $\underline{U}_{RS}(t)$ of the real-world system are measured at discrete – not necessarily equidistant – times, then

$$t = t_j, t_j \in [0,T], j = 1,...,k. \tag{5.12}$$

The output measurements $\underline{Y}_{RS,meas}(t_j)$ can be disturbed by an additive, zero mean noise $\underline{n}(t)$:

$$\underline{Y}_{RS,meas}(t_j) = \underline{Y}_{RS}(t_j) + \underline{n}(t_j). \tag{5.13}$$

The unknown parameters of the parameter vector $\underline{\Theta}$ expressed as

$$\underline{\Theta} = \left[\frac{\underline{a}}{\underline{b}}\atop\underline{c}\right] \in \mathfrak{R}^l, \quad l = \dim \underline{a} + \dim \underline{b} + \dim \underline{c}. \tag{5.14}$$

can be identified for an input $\underline{u}(t)$ with

$$\underline{u}(t_j) = \underline{U}_{RS}(t_j); j = 1,...,k,$$ (5.15)

hence we receive the output function $\underline{Y}_{RS}(t)$ of the model, described by (5.9), (5.10) and (5.11), given as

$$\underline{Y}(t_j) = \underline{Y}_{RS}(t_j).$$ (5.16)

This can be achieved exactly for $\underline{u}(t) = \underline{U}_{RS}(t)$, $t \in [0, T]$ if the structure of the mathematical model has the same structure as the real-world dynamic system, meaning there is no measurement noise added to the model, which means no stochastic part, which is related to the noise. Consider the measurement noise of random type, characterized by its probability density function.

5.3 Identification Task

The identification of the parameters of a p-dimensional vector $\underline{\Theta}_{RS}$ of a real-world system can be characterized by an error criterion, defining the way in which the components of the parameter vector $\underline{\Theta}_{NN}$ of the mathematical (identification) model can be adjusted to coincide with

$$\underline{\Theta}_{RS} = \underline{\Theta}_{MM}.$$ (5.17)

Due to the implementation of the identification method on a computer, a time-discrete description of the time-continuous model will be used subsequently. For linear dynamic systems and of piecewise-constant system inputs a description can easily be deduced from the set of n first-order differential equations as an equivalent set of n first-order difference equations or alternatively, as one difference equation of n-th order, which is a simplification in the computation of the model output for identification purposes. This is not the case for nonlinear time-continuous models.

The identification model of the identification task applied to linear systems can be described by n-th order difference equations

$$y_k + a_{n-1} \cdot y_{k-1} +,..., a_0 \cdot y_{k-n} = b_{n-1} \cdot u_{k-1} +,..., b_0 \cdot u_{k-n},$$ (5.18)

and

$$y_k := y(k \cdot t_a); k = 0,1,....,n \,, \tag{5.19}$$

with t_a as sampling time, y as output variable, and $\underline{\Theta}$ as the p-dimensional parameter vector defined as

$$\underline{\Theta}^T := [a_{n-1}, a_{n-2},..., a_0; b_{n-1},..., b_0] \,, \tag{5.20}$$

with $p = 2n$. Defining the polynomials

$$A(q^{-1}) := 1 + a_{n-1} \cdot q^{-1} +,..., a_0 \cdot q^{-n} \,, \tag{5.21}$$

and

$$B(q^{-1}) := b_{n-1} \cdot q^{-1} +,..., b_0 \cdot q^{-n} \,, \tag{5.22}$$

with q^{-1} as so-called shift operator, given by

$$q^{-1} \cdot Y_k = Y_{k-1} \tag{5.23}$$

the difference equation, (5.18), can be written in the simplified form

$$A(q^{-1}) \cdot Y_k = B(q^{-1}) \cdot U_k \,. \tag{5.24}$$

(5.24) can be rewritten as

$$Y_k = \frac{B(q^{-1})}{A(q^{-1})} U_k \,. \tag{5.25}$$

Fig. 5.3 illustrates the relationships between the real-world system and the mathematical model, representing a model with adjustable parameters q.

Fig. 5.4 shows the relationships with the real-world system, in this case the human cardiovascular system. The mathematical model of Fig. 5.3 is called the identification model, marked by a head sign, and a third block element, called the true model, shown as dashed lines in Fig. 5.4, because its existence is an assumption for theoretical reasons.

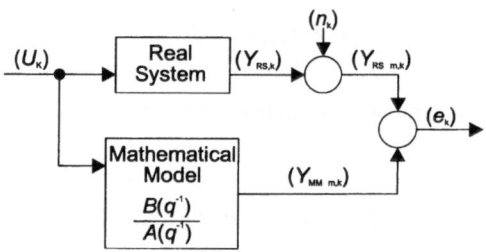

Fig. 5.3. Relationship between the real-world system and the mathematical model

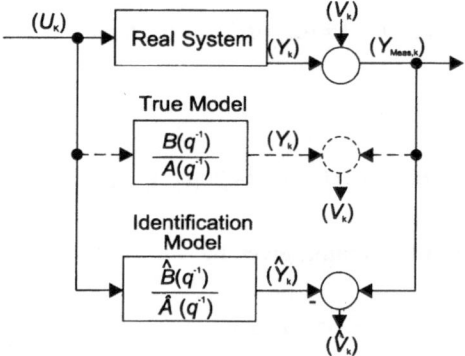

Fig. 5.4 Relationship between the real-world system, the true model, and the identification model

The identification model, shown in Fig. 5.4, is an adaptive model with the adaptive parameter vector

$$\hat{\underline{\Theta}}^{T} := [\hat{a}_{n-1}, \hat{a}_{n-2}, ..., \hat{a}_0; \hat{b}_{n-1}, ..., \hat{b}_0] , \qquad (5.26)$$

and is of the same structure as the true model by definition. All quantities of the identification model are marked by a head sign due to the corresponding quantities of the true model, i.e.

$$\hat{A}(q^{-1}) \cdot \hat{Y}_k = \hat{B}(q^{-1}) \cdot \hat{U}_k , \qquad (5.27)$$

and

$$\hat{Y}_k = \frac{\hat{B}(q^{-1})}{\hat{A}(q^{-1})} \hat{U}_k , \qquad (5.28)$$

corresponding to (5.24) and (5.25). Inputs and outputs are represented by the sequences of their sampled values, e.g. $Y(t)$ is represented by $\{Y_k\}$.

The identification task can now be regarded as the task to adjust the parameter vector $\hat{\underline{\Theta}}$ of the identification model such as that its output sequence $\{\hat{Y}_k(\hat{\underline{\Theta}})\}$ coincide with the output of the true model $\{Y_k\}$. This is possible by definition for

$$\hat{\underline{\Theta}} = \underline{\Theta}_{RS} .\tag{5.29}$$

Consider, this is only the case if

$$\{\hat{Y}_k(\hat{\underline{\Theta}})\} = \{Y_k\},\tag{5.30}$$

the system is called identifiable in its parameters.

The explanation of the identification task given above is admissible if the measured system output $Y_{Meas,k}$ is corrupted by measurement noise v_k, i.e.

$$Y_{Meas,k} = Y_k + v_k ,\tag{5.31}$$

which is shown in Fig. 5.4. If $\{\hat{Y}_k\}$ coincides with $\{Y_k\}$ the difference between the measured system output and the model output, defined by the error functional, given in (5.32),

$$v_k(\hat{\underline{\Theta}}) := Y_{Meas,k} - \hat{Y}_k(\hat{\underline{\Theta}})\tag{5.32}$$

will coincide with the output measurement error v_k, i.e.

$$\hat{v}_k(\underline{\Theta}_{RS}) := v_k.\tag{5.33}$$

Interpreting $\hat{v}_k(\hat{\underline{\Theta}})$ according to (5.33), i.e. as an estimate of the output measurement error, called output error, the identification task can be characterized such that $\hat{\underline{\Theta}}$ has to be adjusted such that the estimated measurement error $\hat{v}_k(\hat{\underline{\Theta}})$ will coincide with the real measurement error v_k.

Remark 5.6

Consider the identification task such that certain known properties of $\{v_k\}$ are impressed on $\{v_k(\hat{\underline{\Theta}})\}$ where $\{v_k\}$ can be a deterministic but generally unknown measurement-error sequence. The estimated error sequence $\{v_k(\hat{\Theta})\}$ can be considered as reconstruction of the error sequence.

Remark 5.7
Consider the error reconstruction has to be determined such that the parameter vector estimate $\hat{\Theta}$ approximates the true parameter vector $\hat{\Theta}_{RS}$.

Remark 5.8
The output-error estimate $\{v_k(\hat{\Theta})\}$ discussed so far is a nonlinear function of $\hat{\Theta}$ even in the case of a linear true model. For linear systems the so-called equation-error is linear in the parameters to be identified if these are defined to be coefficients of the difference equation, given in (5.18). Due to this property in some cases it is possible to find a parameter estimate in closed form.

5.4 Output-Error Least Squares Method*

A method for identifying the system-parameter vector $\hat{\Theta}_{RS}$ is characterized by an error criterion defining the sense in which the components of the parameter vector $\hat{\Theta}$ of the identification model can be adjusted. For the output error least square method the error criterion chosen is

$$J_N(\Theta_{MM}) = \sum_{k=n}^{N}(\hat{v}_k(\hat{\Theta}) - \xi(v_k))^2 \rightarrow Min, \tag{5.34}$$

where N is the number of measurements. Defining the vectors

$$\underline{\hat{v}}^N(\hat{\Theta}) := [\hat{v}_n(\hat{\Theta}), \hat{v}_{n+1}(\hat{\Theta}),...,\hat{v}_N(\hat{\Theta})]^T \tag{5.35}$$

and

$$\underline{v}^N := [v_n, v_{n+1},...,v_N]^T, \tag{5.36}$$

of the estimated and the real-world output errors, respectively, (5.34) can be written as

$$J_N(\hat{\Theta}) = (\underline{\hat{v}}^N(\hat{\Theta}) - \xi\{\underline{v}^N\})^T (\underline{\hat{v}}^N(\hat{\Theta}) - \xi\{\underline{v}^N\}) \rightarrow Min. \tag{5.37}$$

Consider a stationary stochastic process $\{v_k\}$ with $\xi(v_k) = 0$, (5.34) and (5.37) can be simplified as

$$J_N(\hat{\Theta}) = \sum_{k=n}^{N} (\hat{v}_k^2(\hat{\Theta})) \rightarrow Min, \qquad (5.38)$$

and

$$J_N(\hat{\Theta}) = (\hat{\underline{v}}^N(\hat{\Theta}))^T (\hat{\underline{v}}^N(\hat{\Theta})) \rightarrow Min. \qquad (5.39)$$

Substituting $\hat{\underline{v}}^N(\hat{\Theta})$ in (5.38) we obtain the least squares output-error criterion in its well-known notation

$$J_N(\hat{\Theta}) = \sum_{k=n}^{N} (Y_{Meas,k} - \hat{Y}_k(\hat{\Theta}))^2 \rightarrow Min. \qquad (5.40)$$

Defining $\hat{\underline{Y}}^N(\hat{\Theta})$, and $\underline{Y}_{Meas}^N$ in an analogous way, as $\hat{\underline{v}}^N(\hat{\Theta})$, and $\underline{v}^N$, respectively, the corresponding formulation of the criterion in (5.40) in matrix form becomes

$$J_N(\hat{\Theta}) = (\underline{Y}_{Meas}^N - \hat{\underline{Y}}^N(\hat{\Theta}))^T (\underline{Y}_{Meas}^N - \hat{\underline{Y}}^N(\hat{\Theta})) \rightarrow Min. \qquad (5.41)$$

$\hat{\underline{Y}}^N(\hat{\Theta})$ is a nonlinear function of $(\hat{\Theta})$ the parameter vector $(\hat{\Theta})_{Min}^N$, which minimizes $J_N(\hat{\Theta})$, which can be determined by numerical optimization methods.

Remark 5.9
A reasonable interpretation of the error criterion in (5.39) can be given if the output error $\{v_k\}$ is assumed to be Gaussian. Hence the probability of $\underline{v}^N$ has the form

$$p(\underline{v}^N) \approx \exp[-\frac{1}{2}(\underline{v}^N - \xi(\underline{v}^N))^T \sum_{v^N}^{-1} (\underline{v}^N - \xi\{\underline{v}^N\})]. \qquad (5.42)$$

Consider the estimated output-error sequence $\{\hat{v}_k(\hat{\Theta})\}$ as a realization of the output error criterion and determine $(\hat{\Theta})$ such as

$$p(\hat{\underline{v}}^N(\hat{\Theta})) \rightarrow Max, \qquad (5.43)$$

$(\hat{\Theta})$ is the so-called maximum likelihood estimate, since we find from (5.42) that (5.43) is equivalent to

$$(\hat{\underline{v}}^{N}(\hat{\Theta}) - \xi(\underline{v}^{N}))^{T} \sum_{\underline{v}^{N}}^{-1} (\hat{\underline{v}}^{N}(\hat{\Theta}) - \xi\{\hat{\underline{v}}^{N}\}) \rightarrow Min. \qquad (5.44)$$

Assuming a white-noise process $\{v_k\}$ stationary with

$$\mathrm{var}(v_k) = \sigma_v^2, \qquad (5.45)$$

yields

$$\sum_{v} N = \sigma_v^2 \cdot \underline{I} \quad , \qquad (5.46)$$

hence (5.44) is equivalent to (5.39) and the parameter estimate $(\hat{\Theta})_{Min}^{N}$, which minimizes the sum of the squared distances of $\{\hat{v}_k(\hat{\Theta})\}$ from the expected value of v_k becomes the so-called maximum-likelihood estimate.

The output error least squares criterion can be given as follows:

$$\underline{e}(t_j) := \underline{Y}_{MM}(t_j) - \underline{Y}_{RS}(t_j); j = 1,...,k , \qquad (5.47)$$

with $\underline{e}(t_j)$ as the error function to be minimized. The performance criterion can be

$$J_N = \frac{1}{k} \cdot \sum_{i=1}^{m} d_i \sum_{j=1}^{k} |e_i(t_j)|^{q} \quad , \quad d_i > 0. \qquad (5.48)$$

Usually one selects $q = 2$, which results in the output error least square estimation. If $q > 2$ the maximum error is minimized. By means of the weighting coefficients d_i, $i = 1,...,$ m, different error variances of each component of $\underline{Y}_{MM}$ may be taken into account. The model output $\underline{Y}$ is also a function of the model parameters $\underline{p}$, the performance criterion in (5.48) can be rewritten in the form

$$J_N(\underline{p}) = \frac{1}{k} \cdot \sum_{i=1}^{m} d_i \sum_{j=1}^{k} |Y_{MMi}(t_j) - Y_{RSi}(t_j, \underline{p})|^{q}. \qquad (5.49)$$

Hence, the identification problem requires the solution of the mathematical problem

$$\overset{!}{J_N\,(\underline{p})} = Min. \tag{5.50}$$

The structure of the output-error least squares method, based on the assumptions made above, is shown in Fig. 5.5. The following remarks are to be noted:

- The input function $\underline{U}_M(t)$, $t \in [t, T]$ can be either an input signal, available during normal system operation, or a specific generated test signal, like a pseudorandom binary sequence.
- The measurable initial conditions $\underline{x}_M$ will be stored as well as the sampled input and output functions. Unknown initial conditions must be considered as parameters to be identified.
- The nonlinear parameter-optimization problem $\overset{!}{J_N\,(\underline{p})} = Min.$ has to be solved iteratively. Starting with the iteration the initial values $\underline{p}$ must be available. They can be chosen as close as possible to the optimum values in order to allow a faster convergence. Such parameters $\underline{p}$ can usually be obtained from physical considerations.
- Using the stored input measurements and the actual parameters $\underline{p}$, the model output $\underline{Y}(t_j)$, $j = 1,\ldots, k$ has to be simulated during every optimization iteration step. Therefore, a fast and sufficiently accurate numerical integration procedure is necessary.
- The algorithm integrating the differential equations usually requires values $\underline{U}(t)$ where $t \neq t_j$, $j = 1, \ldots, k$. For this reason it is necessary to calculate – by using the stored values $\underline{U}_M(t_j)$ – a function $\underline{u}(t)$ that approximates the input signal $\underline{U}_M(t)$, for $t \in [0, T]$. This approximation may be done exactly for specially generated input signals $\underline{U}_M(t)$. For this purpose piecewise-constant functions have been used.

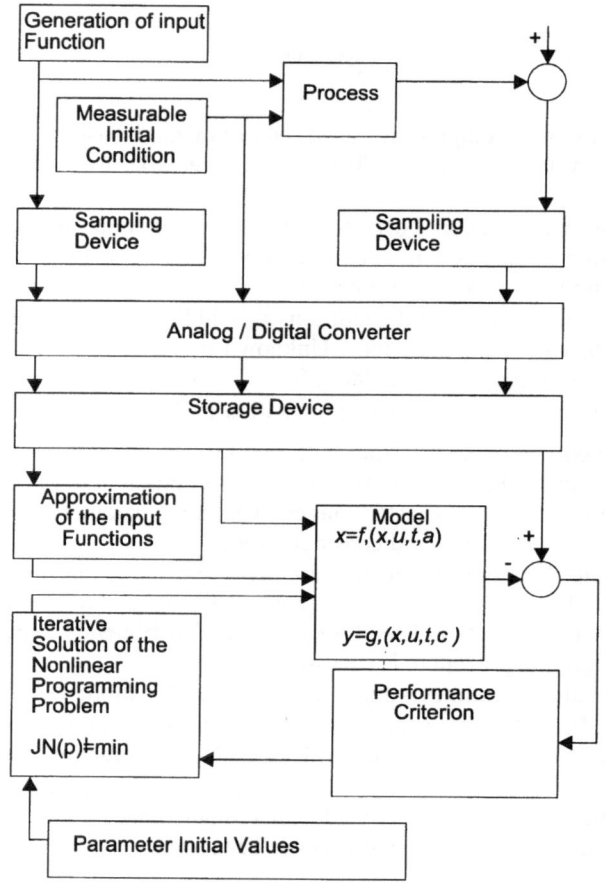

Fig. 5.5 Structure of the output-error least squares method

5.5 Equation-Error Least Squares Method*

Substituting the output variable $\underline{Y}_k$ of the true model in the difference equation given in (5.24) for

$$Y_k = Y_{Meas,k} - v_k, \tag{5.51}$$

as a consequence of the measurement equation

$$Y_{Meas,k} = Y_k + v_k \, ,\qquad(5.52)$$

yields

$$A(q^{-1})Y_{Meas,k} = B(q^{-1})U_k + e_k \, ,\qquad(5.53)$$

where e_k is defined by

$$e_k = A(q^{-1})v_k \, .\qquad(5.54)$$

(5.53) is the mathematical formalism of the true model depending on the measured output variable $Y_{Meas,k}$, described in (5.52), considering an equation-error e_k, given in (5.54). Consider the equation-error process $\{e_k\}$ is white and the measurement process can be deduced from a white-noise process $\{n_k\}$ by

$$A(q^{-1})v_k = n_k \, ,\qquad(5.55)$$

which is of importance due to the consistency properties of $(\hat{\Theta})^N_{Min}$, while the whole class of equation-error methods can be traced back to it. Before discussing this in detail in Sect. 5.6 it can be shown that in the case of a quadratic-error criterion a simply expression for $(\hat{\Theta})^N_{Min}$ can be found. This is due to the linearity of the parameters, i.e. in the components of $(\hat{\Theta})$ of the equation-error. Therefore, a time-consuming iterative search algorithm, necessary while using the output-error methods, can be avoided.

The identification model equation corresponding with the true model equation can be written as

$$\hat{A}(q^{-1})Y_{Meas,k} = \hat{B}(q^{-1})U_k + \hat{e}_k \, ,\qquad(5.56)$$

which is shown in Fig. 5.6

The corresponding true model and the differential equations of the identification model, given in (5.53) and (5.56), can be written as

$$Y_{Meas,k} = \underline{m}^T_{Meas,k-1}\underline{\Theta} + e_k \, ,\qquad(5.57)$$

and

$$Y_{Meas,k} = \underline{m}^T_{Meas,k-1}\underline{\hat{\Theta}} + \hat{e}_k \,,$$

(5.58)

respectively, with

$$\underline{m}^T_{Meas,k-1} := [-Y_{Meas,k-1}; -Y_{Meas,k-2}; \ldots; -Y_{Meas,k-n}; U_{k-1}; \ldots; U_{k-n}] \,.$$

(5.59)

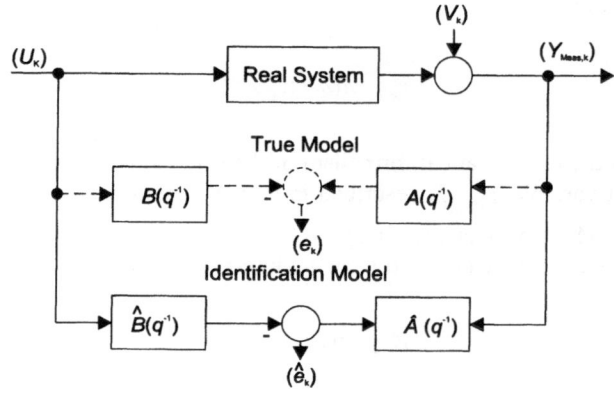

Fig. 5.6. Equation-error criterion realized by the use of the true model and the identification model

Furthermore, defining the measurement matrix as

$$\underline{M}^N_{Meas} := \begin{bmatrix} \underline{m}^T_{Meas,n} \\ \underline{m}^T_{Meas,n+1} \\ . \\ . \\ . \\ \underline{m}^T_{Meas,N} \end{bmatrix},$$

(5.60)

and the corresponding vectors as

$$\underline{Y}^T_{Meas} := [Y_{Meas,n}; Y_{Meas,n+1}; \ldots; Y_{Meas,N}]^T$$

(5.61)

$$\underline{e}^N := [e_n, e_{n+1}, \ldots, e_N]^T \,,$$

(5.62)

and

$$\hat{\underline{e}}^N := [\hat{e}_n, \hat{e}_{n+1}, ..., \hat{e}_N]^T . \tag{5.63}$$

(5.57) and (5.58) can be comprised of

$$k = n, n+1, ..., N , \tag{5.64}$$

in the matrix form

$$\underline{Y}^N_{Meas} = \underline{M}^{N-1}_{Meas} \underline{\Theta} + \underline{e}^N , \tag{5.65}$$

and

$$\underline{Y}^N_{Meas} = \underline{M}^{N-1}_{Meas} \hat{\underline{\Theta}} + \hat{\underline{e}}^N , \tag{5.66}$$

respectively. Analogous to the output error the norm of the difference of

$$\hat{\underline{e}}^N (\hat{\underline{\Theta}}) , \tag{5.67}$$

and $\xi\{\underline{e}^N\}$ can be minimized, i.e.

$$J_N(\hat{\underline{\Theta}}) = (\hat{\underline{e}}^N (\hat{\underline{\Theta}}) - \xi(\underline{e}^N))^T (\hat{\underline{e}}^N (\hat{\underline{\Theta}}) - \xi(\underline{e}^N)) \rightarrow Min. \tag{5.68}$$

Consider $\xi\{\underline{e}^N\}$ to be not known, while being dependent on the unknown true parameter vector $\underline{\Theta}_{RS}$. For a stationary measurement error, with $\xi(v_k) = 0$, we obtain

$$\xi\{\underline{e}^N\} = \underline{0} , \tag{5.69}$$

hence (5.68) can be rewritten as

$$J_N(\hat{\underline{\Theta}}) = [\hat{\underline{e}}^N (\hat{\underline{\Theta}})]^T \hat{\underline{e}}^N (\hat{\underline{\Theta}}) \rightarrow Min. , \tag{5.70}$$

or, alternatively

$$J_N(\hat{\underline{\Theta}}) = \sum_{k=n}^{N} (\hat{e}_k^2(\hat{\underline{\Theta}})) \rightarrow Min. \tag{5.71}$$

If $\hat{\underline{e}}^N(\hat{\underline{\Theta}})$ is replaced in (5.71) by (5.66) we obtain the equation-error least squares criterion in its conventional form

$$J_N(\hat{\underline{\Theta}}) = (\underline{Y}_{Meas}^N - \underline{M}_{MM}^{N-1}(\hat{\underline{\Theta}}))^T (\underline{Y}_{Meas}^N - \underline{M}_{Meas}^{N-1}(\hat{\underline{\Theta}})) \rightarrow Min. \tag{5.72}$$

Remark 5.9

Consider a Gaussian distribution for the equation-error, i.e.

$$p(\underline{e}^N) \approx \exp[-\frac{1}{2}(\underline{e}^N - \xi(\underline{e}^N))^T \sum_{\underline{e}^N}^{-1} (\underline{e}^N - \xi(\underline{e}^N))] , \tag{5.73}$$

we obtain a maximum-likelihood interpretation of (5.68) analogous to the output-error least squares criterion in (5.54).

From (5.73) it follows that the maximum-likelihood criterion

$$p(\hat{\underline{e}}^N(\hat{\underline{\Theta}})) \rightarrow Max \tag{5.74}$$

is equivalent to the weighted equation-error-least squares criterion

$$(\hat{\underline{e}}^N(\hat{\underline{\Theta}}) - \xi(\underline{e}^N))^T \sum_{\underline{e}^N}^{-1} (\hat{\underline{e}}^N(\hat{\underline{\Theta}}) - \xi(\underline{e}^N)) \rightarrow Min, \tag{5.75}$$

which is equivalent to (5.37) if

$$\sum_{\underline{e}^N} = \sigma_e^2 \underline{I} , \tag{5.76}$$

i.e. if $\{e_k\}$ is a white-noise stationary process with variance σ_e^2, which is not the truth in general. On the other hand the criterion in (5.71) can be approximately applied if an estimate $\hat{\sum}_{\underline{e}^N}$ is known, or if $\{e_k\}$ is not given in (5.54) but is a system-noise process with a known statistical behavior.

Remark 5.10

If the identification model, given in (5.58), can be written as

$$Y_{Meas,k} = \hat{\underline{Y}}_{OSP,k}\hat{\underline{\Theta}} + \hat{e}_k, \tag{5.77}$$

with the so-called one-step-ahead prediction (OSP)

$$\hat{Y}_{OSP,k}(\hat{\underline{\Theta}}) := \underline{m}^T_{Meas,k-1}(\hat{\underline{\Theta}}) \tag{5.78}$$

of Y_k, and defining in an analogous way

$$\hat{\underline{Y}}^N_{OSP}(\hat{\underline{\Theta}}) := \underline{M}^{N-1}_{Meas}(\hat{\underline{\Theta}}) . \tag{5.79}$$

(5.72) yields the form

$$J_N(\hat{\underline{\Theta}}) = (\underline{Y}^N_{Meas} - \hat{\underline{Y}}^N_{OSP}(\hat{\underline{\Theta}}))^T (\underline{Y}^N_{Meas} - \hat{\underline{Y}}^N_{OPS}(\hat{\underline{\Theta}})) \rightarrow Min., \tag{5.80}$$

or

$$J_N(\hat{\underline{\Theta}}) = \sum_{k=n}^{N} Y_{Meas,k} - \hat{\underline{Y}}^N_{OSP,k}(\hat{\underline{\Theta}})^2 \rightarrow Min., \tag{5.81}$$

respectively, often found in the literature, which should not be confused with the similar looking output-error criterion in (5.58). The difference between these two criteria becomes clear when writing $\hat{Y}_k(\hat{\underline{\Theta}})$, based on (5.61), in the alternative formula

$$\hat{\underline{Y}}_k(\hat{\underline{\Theta}}) = \hat{\underline{m}}^T_{k-1}(\hat{\underline{\Theta}}), \tag{5.82}$$

where

$$\hat{\underline{m}}^T_{k-1} = [-\hat{\underline{Y}}_{k-1}, -\hat{\underline{Y}}_{n-2}, ..., -\hat{\underline{Y}}_{k-n}; U_{k-1}, ..., U_{k-n}]. \tag{5.83}$$

Therefore, in the equation-error formulation the model outputs $\hat{Y}_k$ in (5.83) appear replaced by the measured-system outputs $Y_{Meas,k}$ in (5.34).

For simplification purposes, the upper index N characterizing the length of the identification interval, can be dropped if misunderstandings are unlikely.

Consider the performance index $J_N(\hat{\underline{\Theta}})$ as a minimum for the parameter vector

$\hat{\underline{\Theta}} = \hat{\underline{\Theta}}^N_{Min}$ satisfying

$$\underline{M}^T_{Meas}\,\underline{M}_{Meas}\,\underline{\hat{\Theta}}^N_{Min} = \underline{M}^T_{Meas}\,\underline{Y}_{Meas}\,,\tag{5.84}$$

as a necessary condition, meaning the assumption of a sufficient performance

$$\underline{M}^T_{Meas}\,\underline{M}_{Meas} > \underline{0}\tag{5.85}$$

is given. From (5.84) and (5.85) minimizing the parameter vector yields

$$\underline{\hat{\Theta}}^N_{Min} = (\underline{M}^T_{Meas}\,\underline{M}_{Meas})^{-1}\,\underline{M}^T_{Meas}\,\underline{Y}_{Meas}\,.\tag{5.86}$$

If the structures of the identification model and the true model coincide and the input sequence persistently excites the real-world dynamic system (see also Sect. 1.1), (5.85) will hold and therefore the inverse in (5.86) will exist.

Remark 5.11
If $\{e_k\}$ is stationary with unknown expectation

$$\xi\{e_k\} = \overline{e}\,,\tag{5.87}$$

$\overline{e}$ can be treated in (5.81) and (5.86) as an additional parameter to be estimated, which can be done by rewriting the problem, given in (5.81), while introducing an extended measurement matrix

$$\underline{\widetilde{M}}^{N-1}_{Meas} := [\,\underline{M}^{N-1}_{Meas},\begin{bmatrix}1\\1\\\cdot\\1\end{bmatrix}]\,,\tag{5.88}$$

and an extended parameter vector

$$\underline{\widetilde{\Theta}} := \begin{bmatrix}\underline{\Theta}\\\overline{e}\end{bmatrix},\tag{5.89}$$

instead of $\underline{M}^N_{Meas}$ and $\underline{\Theta}$, respectively.

Due of the fact that the entity of all measured data along the identification interval are included in (5.86) this solution is called the direct solution. Alternatively, (5.86) can be calculated recursively by incorporating the additional data

$$\{Y_{Meas,k};U_k \mid k = 3n,3n+1,...,N\}\,,\qquad(5.90)$$

step-by-step starting with the direct solution for the identification interval

$$[0,(3n-1)T_a]\,.\qquad(5.91)$$

It can be shown that $\hat{\underline{\Theta}}^k$ can be derived from $\hat{\underline{\Theta}}^{k-1}$ introducing an additive correction term proportional to the estimated equation-error $\hat{\underline{e}}^k$, i.e.

$$\hat{\underline{\Theta}}^k = \hat{\underline{\Theta}}^{k-1} + \underline{K}^k \cdot \hat{\underline{e}}^k\,,\qquad(5.92)$$

with

$$\hat{\underline{e}}^k = \underline{Y}_{Meas}^k - \underline{m}_{Meas,k-1}^T \hat{\underline{\Theta}}^{k-1}\qquad(5.93)$$

$$\underline{K}^k = \sum^{k-1}\underline{m}_{Meas,k-1}[1+\underline{m}_{Meas,k-1}^T \sum^{k-1}\overrightarrow{m}_{Meas,k-1}]^{-1}\,,\qquad(5.94)$$

and

$$\sum^k = \sum^{k-1} - \underline{K}^k \underline{m}_{Meas,k-1}^T\,,\qquad(5.95)$$

where $\sum^k$ is given as

$$\sum^k := (\underline{M}_{Meas}^{k-1})^T \underline{M}_{Meas}^{k-1}\,.\qquad(5.96)$$

For $k = N$ the recursive as well as the direct solution coincide, i.e $\hat{\underline{\Theta}}^N = \hat{\underline{\Theta}}_{Min}^N$. The importance of the recursive formulation, given in (5.92 – 5.96), is due to the possibility of calculating the parameter estimate $\hat{\underline{\Theta}}_{Min}^N$ online.

5.6 Consistency of the Parameter Estimates*

We have shown how to determine the error criteria, given in (5.55) and (5.81). Based on the output error and the equation-error, respectively, the error criteria determine the minimized value of $\hat{\underline{\Theta}}$ such that the identification model output $\hat{Y}_k$ approximate the measured system output $Y_{\text{Meas},k}$ instead of the noise-free output $\underline{Y}_k$, which for some reasons may be not available. This has several consequences due to the statistical properties of the parameter estimates based on the different methods.

Consider the properties of the equation-error estimates we can replace Y_{Meas} in (5.86), obtaining

$$\hat{\underline{\Theta}}_{Min}^N = (\underline{M}_{Meas}^T \underline{M}_{Meas})^{-1} \underline{M}_{Meas}^T \underline{M}_{Meas} \underline{\Theta}_{RS} + (\underline{M}_{Meas}^T \underline{M}_{Meas})^{-1} \underline{M}_{Meas}^T \underline{e} \qquad (5.97)$$
$$= \underline{\Theta}_{RS} + (\underline{M}_{Meas}^T \underline{M}_{Meas})^{-1} \underline{M}_{Meas}^T \underline{e}$$

For convenience, the true model

$$A(q^{-1}) = 1, \qquad (5.98)$$

in (5.48) yields

$$Y_{Meas,k} = B(q^{-1})U_k + e_k, \qquad (5.99)$$

with

$$e_k = v_{k.} \qquad (5.100)$$

Note that for this regression problem the equation-error and the output error are identical because the output components of $\hat{\underline{m}}_k$ in (5.83) and of $\underline{m}_{Meas,k}$ in (5.59) are dropped, meaning $\hat{\underline{\Theta}}_{Min}^N$ is an unbiased estimate of $\underline{\Theta}_{RS}$, i.e.

$$\xi(\hat{\underline{\Theta}}_{Min}^N) = \underline{\Theta}_{RS}, \qquad (5.101)$$

if the input $\{U_k\}$ and the equation-error process $\{e_k\}$ are statistically independent, hence $\xi(e_k) = 0$. Since $\underline{M}_{Meas}$ is independent of $\{Y_{Meas,k}\}$ and therefore of $\{v_k\}$, it is independent of $\{e_k\}$, yields

$$\xi(\underline{M}^T_{Meas}\underline{M}_{Meas})^{-1}\underline{M}^T_{Meas}\underline{e}) = \xi\{(\underline{M}^T_{Meas}\underline{M}_{Meas})^{-1}\underline{M}^T_{Meas}\}\xi(\underline{e}) = \underline{0}. \qquad (5.102)$$

Consider the true model equation

$$A(q^{-1})Y_{Meas,k} = B(q^{-1})U_k + e_k, \qquad (5.103)$$

$\underline{M}_{Meas,k}$ depend on $\{e_k\}$ hence the factorization of the expectation in (5.102) is not allowed.

Let $\{e_k\}$ be a white-noise process independent on $\{U_k\}$, $\underline{\hat{\Theta}}^N_{Min}$ is consistent in probability, given as

$$p\lim_{N\to\infty} \underline{\hat{\Theta}}^N_{Min} = \underline{\Theta}_{RS}. \qquad (5.104)$$

This property is similar to an asymptotic unbiased estimate, which can be unbiased due to limit $N \to \infty$. Using (5.97) we can write

$$p\lim_{N\to\infty} \underline{\hat{\Theta}}^N_{Min} = \underline{\Theta}_{RS} + p\lim_{N\to\infty}[(\underline{M}^T_{Meas}\underline{M}_{Meas})^{-1}\underline{M}^T_{Meas}\underline{e}] \qquad (5.105)$$

$$= \underline{\Theta}_{RS} + p\lim_{N\to\infty}[(\frac{1}{N}\underline{M}^T_{Meas}\underline{M}^T_{Meas})^{-1}]p\lim_{N\to\infty}[\frac{1}{N}\underline{M}^T_{Meas}\underline{e}]$$

Equation (5.104) yields

$$p\lim_{N\to\infty}[\frac{1}{N}\underline{M}^T_{Meas}\underline{e}] = p\lim_{N\to\infty}(\xi[\frac{1}{N}\underline{M}^T_{Meas}\underline{e}]) = \underline{0}. \qquad (5.106)$$

Therefore, the consistency property is a weaker property than the asymptotic unbiased property, hence $\underline{\hat{\Theta}}^N_{Min}$ is consistent but not asymptotically unbiased.

Assuming $\{e_k\}$ is not a white-noise process the equation-error least squares estimate, given in (5.86), will be a nonconsistent estimate.

5.7 Consistency Modifications of the Equation-Error Method*

Suppose that a white-noise process $\{e_k\}$ of the error criteria, given in (5.40) and (5.81) is a restriction that results in the extension of the ordinary equation-error least squares method. Solving this problem means transforming the model, given in (5.54), such that the transformed equation-error will be white. Consequently, we use the assumption that the equation-error can be expressed through a true model

$$D(q^{-1})e_k = w_k , \qquad (5.107)$$

where

$$D(q^{-1}) = 1 + d_1 q^{-1} + \ldots + d_m q^{-m} , \qquad (5.108)$$

is known and $\{w_k\}$ is a white random process. Multiplying the true model (5.54) by $D(q^{-1})$ from the left we receive the transformed model

$$D(q^{-1})A(q^{-1})Y_{Meas,k} = D(q^{-1})B(q^{-1})U_k + \tilde{e}_k, \qquad (5.109)$$

the equation-error of which is defined by

$$\tilde{e}_k := D(q^{-1})e_k = w_k. \qquad (5.110)$$

with $\{\tilde{e}_k\}$ as a white random process. Changing the polynomial factors in (5.109) and defining

$$\tilde{U}_k := D(q^{-1})U_k \qquad (5.111)$$

$$\tilde{Y}_{Meas,k} := D(q^{-1})Y_{Meas,k}, \qquad (5.112)$$

(5.109) can be rewritten as

$$A(q_{-1})\tilde{Y}_{Meas,k} = B(q^{-1})\tilde{U}_k + \tilde{e}_k . \qquad (5.113)$$

Consequently, the transformed input and output sequence can be described by the original model, but with a white equation-error sequence $\{\widetilde{e}_k\}$ as necessary for $\hat{\Theta}^N_{Min}$ to be consistent. This equation is shown in Fig. 5.7, in which the real-world system is assumed to be the cardiovascular system (see Sect. 5.10) .

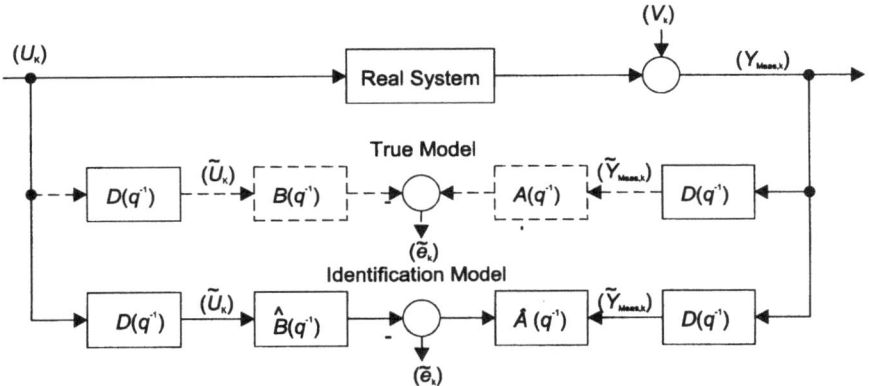

Fig. 5.7. Relationships between the real system, the true model, and the identification model of the modified equation-error method

Remark 5.12
The output-error method, as described in (5.109), can be interpreted as a trans-formed equation-error with

$$D(q^{-1}) = \frac{1}{A}(q^{-1}) . \qquad (5.114)$$

Since $D(q^{-1})$ is unknown, different methods were proposed to estimate $D(q^{-1})$ in addition to $A(q^{-1})$ and $B(q^{-1})$. The most fundamental ones being the generalized least squares method, the extended model method, the extended matrix method and the combined instrumental variable approximate maximum-likelihood method.

Using the noise model of the generalized least square method for estimating $D(q^{-1})$, as shown in (5.56), can only be done if e_k is known. If e_k is not known one has to start with an estimate of e_k and improve it iteratively so that it becomes consistent after a sufficient number of iteration steps. This situation is shown in Fig. 5.8, where the upper indices characterize the number of the actual iteration step.

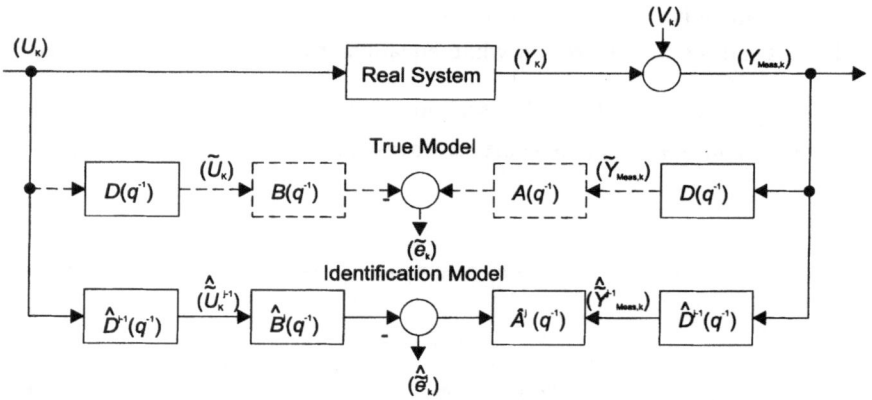

Fig. 5.8. Parameter estimation scheme for the generalized least squares method

Consider an extended model method, the iteration is avoided by using the extended model, shown in (5.54), without transforming the input and output of the system using direct identification of the coefficients of the product polynomials $D \cdot A(q^{-1})$ and $D \cdot B(q^{-1})$, the coefficients of which may be interpreted as the parameters of a true model of the increased order $\tilde{n}$, where

$$\tilde{n} = n + m.\tag{5.115}$$

Neglecting the common zeros of $D\hat{A}(q^{-1})$ and $D\hat{B}(q^{-1})$ the estimates $\hat{A}(q^{-1})$ and $\hat{B}(q^{-1})$ can be determined.

Remark 5.13

An essence of the extended matrix method is that this method is more advantageous relative to the generalized least squares and extended model method in the sense that it can properly be used together with a more general noise model

$$D(q^{-1})e_k = C(q^{-1})w_k,\tag{5.116}$$

with

$$C(q^{-1}) = 1 + c_1 q^{-1} + \ldots + c_m q^{-m}.\tag{5.117}$$

Defining the modified polynomials

$$D'(q^{-1}) = D(q^{-1}) - 1,$$
(5.118)

and

$$C'(q^{-1}) = C(q^{-1}) - 1,$$
(5.119)

the equation-error can be derived from (5.116) as follows

$$e_k = -D'(q^{-1})e_k + C'(q^{-1})w_k + w_k.$$
(5.120)

Substituting this expression for e_k in (5.54) we obtain the modified true model equation

$$A(q^{-1})Y_{Meas,k} = B(q^{-1})U_k - D'(q^{-1})e_k + C'(q^{-1})w_k + w_k.$$
(5.121)

Consider e_k and w_k in the second and third term at the right side of (5.121) are known, (5.121) can be interpreted as a modified true model with an equation-error of

$$\tilde{e}_k = w_k,$$
(5.122)

so that A, B, C' and D' can be consistently estimated using the equation-error least square method. Again using the estimates of e_k and w_k, consistent estimates of the parameters can be approximated in an iterative way.

A further class of consistent estimation methods are the instrumental variable approximate maximum likelihood methods. Consider the random sequence $\{w_k\}$ is Gaussian, the maximum-likelihood estimates of A and B, with C and D fixed, and of C and D, with A and B fixed, can be calculated using a relaxation-type iteration.

The primary problem due to the nonlinearity in the coefficients of the polynomials $\hat{A}, \hat{B}, \hat{C}, \hat{D}$ of the identification model can be separated into two problems linear in the coefficients of $\hat{A}$ and $\hat{B}$ or $\hat{C}$ and $\hat{D}$, which can be solved sequentially. It can be shown by introducing suitable transformed variables, the solutions of these problem will be of the instrumental variable type as well as the approximate maximum-likelihood type, respectively.

The instrumental variable approximate maximum-likelihood method is based on the following system model

$$Y_{Meas,k} = \frac{B(q^{-1})}{A(q^{-1})} U_k + v_k \,, \tag{5.123}$$

and the measurement noise model

$$v_k = \frac{C(q^{-1})}{D(q^{-1})} w_k \,, \tag{5.124}$$

with the Gaussian, white and stationary random sequence $\{w_k\}$ with $\xi\{w_k\} = 0$ and $\mathrm{var}(w) = \sigma_w^2$, i.e.

$$p(w_k) \approx \exp[-\frac{1}{2\sigma_w^2} \sum_k w_k^2] \,. \tag{5.125}$$

Introducing v_k from (5.124) into (5.123) and premultiplying (5.123) by

$$\frac{D(q^{-1})}{C(q^{-1})} \,, \tag{5.126}$$

we obtain, after changing the polynomial factors

$$\frac{D(q^{-1})}{C(q^{-1})} Y_{Meas,k} = \frac{B(q^{-1})}{A(q^{-1})} \cdot \frac{D(q^{-1})}{C(q^{-1})} U_k + w_k \,, \tag{5.127}$$

hence w_k can be interpreted as the output error of the transformed system model

$$\tilde{Y}_{Meas,k} = \frac{B(q^{-1})}{A(q^{-1})} \tilde{U}_k + w_k \,, \tag{5.128}$$

with

$$\tilde{Y}_{Meas,k} = \frac{D(q^{-1})}{C(q^{-1})} Y_{Meas,k} \,, \tag{5.129}$$

and

$$\tilde{U}_k = \frac{D(q^{-1})}{C(q^{-1})} U_k \,. \tag{5.130}$$

Instead of solving the complex maximum-likelihood problem

$$p(\hat{w}_k) \approx \exp[-\frac{1}{2\sigma_w^2}\sum_k \hat{w}^2 \underline{\hat{\Theta}}] \rightarrow Max, \qquad (5.131)$$

where

$$\hat{w}_k = \frac{\hat{D}(q^{-1})}{\hat{C}(q^{-1})}Y_{Meas,k} - \frac{\hat{B}(q^{-1})}{\hat{A}(q^{-1})}\cdot\frac{\hat{D}(q^{-1})}{\hat{C}(q^{-1})}, \qquad (5.132)$$

and

$$\underline{\hat{\Theta}}^T := [\underline{\hat{\Theta}}_{AB}^T; \underline{\hat{\Theta}}_{CD}^T] = [\hat{a}_{n-1},...,\hat{a}_0; \hat{b}_{n-1},...,\hat{b}_0; \hat{c}_{m-1},...,\hat{c}_0; \hat{d}_{m-1},...,\hat{d}_0], \qquad (5.133)$$

the two subproblems

$$p(\hat{w}_k(\underline{\hat{\Theta}}_{AB}^T)) \approx \exp[-\frac{1}{2\sigma_w^2}\sum_k \hat{w}^2(\underline{\hat{\Theta}}_{AB}^T)] \rightarrow Max, \qquad (5.134)$$

with

$$\hat{w}_k(\underline{\hat{\Theta}}_{AB}^T) = \frac{\hat{D}^{j-1}(q^{-1})}{\hat{C}^{j-1}(q^{-1})}Y_{Meas,k} - \frac{\hat{B}^j(q^{-1})}{\hat{A}^j(q^{-1})}\cdot\frac{\hat{D}^j(q^{-1})}{\hat{C}^j(q^{-1})}, \qquad (5.135)$$

and

$$p(\hat{w}_k(\underline{\hat{\Theta}}_{CD}^T)) \approx \exp[-\frac{1}{2\sigma_w^2}\sum_k \hat{w}^2(\underline{\hat{\Theta}}_{CD}^T)] \rightarrow Max, \qquad (5.136)$$

with

$$\hat{w}_k(\underline{\hat{\Theta}}_{CD}^T) = \frac{\hat{D}^j(q^{-1})}{\hat{C}^j(q^{-1})}Y_{Meas,k} \stackrel{\triangle}{-} \frac{\hat{B}^j(q^{-1})}{\hat{A}^j(q^{-1})}\cdot\frac{\hat{D}^j(q^{-1})}{\hat{C}^j(q^{-1})}, \qquad (5.137)$$

can be solved sequentially and iteratively in j, which approximate the complex maximum-likelihood solution. It can be shown that the equations for $\underline{\hat{\Theta}}_{AB}$ have the typical instrumental variable form

$$\underline{\tilde{M}}^T \underline{M}_{Meas}\underline{\hat{\Theta}}_{AB,Min}^N = \underline{\tilde{M}}^T \underline{Y}_{Meas}. \qquad (5.138)$$

On both sides of (5.138) the multiplying matrix factor $\underline{M}^T_{Meas}$ of the equation-error have been replaced by $\underline{\tilde{M}}^T$. $\underline{\tilde{M}}$ can be deduced from $\underline{M}_{Meas}$ by replacing the measurement variables $\underline{Y}_{Meas,k}$ using prefiltered variables $\hat{Y}^*_k$ in $\underline{M}_{Meas}$.

We have shown how to determine the solution for the first subproblem. Solving the second subproblem, defined by (5.135) and (5.136), the necessary conditions for $\hat{\underline{\Theta}}^T_{CD}$ can be found from the equations resulting from the application of the extended-matrix method to the noise model

$$D(q^{-1})v_k = C(q^{-1})w_k , \tag{5.139}$$

using suitably defined prefiltered estimates $\hat{v}^*_k$ and $\hat{w}^*_k$ instead of $\hat{v}_k$ and $\hat{w}_k$.

We may note the existence of a class of consistent methods, known as bias-correcting methods that try to circumvent the iterative character of the methods discussed above to preserve the closed-form character of the equation-error least squares solution. For this purpose the difference of the probability limit of the estimate $\hat{\underline{\Theta}}^N_{Min}$ of the noise-free solution has to be used to correct the equation-error least squares solution, shown in (5.86).

Consider the right side of (5.140)

$$p \lim_{N \to \infty} \hat{\underline{\Theta}}^N_{Min} = [p \lim_{N \to \infty} (\frac{1}{N} \underline{M}^T_{Meas} \underline{M}_{Meas})]^{-1} \cdot p \lim_{N \to \infty} (\frac{1}{N} \underline{M}^T_{Meas} \underline{M}_{Meas}) , \tag{5.140}$$

satisfy the probability limits

$$p \lim_{N \to \infty} (\bullet) = \lim_{N \to \infty} \xi(\bullet), \tag{5.141}$$

we obtain

$$p \lim_{N \to \infty} (\frac{1}{N} \underline{M}^T_{Meas} \underline{M}_{Meas}) = \lim_{N \to \infty} (\frac{1}{N} \underline{M}^T \underline{M}) + \sum_n , \tag{5.142}$$

and

$$p \lim_{N \to \infty} (\frac{1}{N} \underline{M}^T_{Meas} \underline{Y}_{Meas}) = \lim_{N \to \infty} (\frac{1}{N} \underline{M}^T \underline{Y}) + \underline{\gamma} . \tag{5.143}$$

The adapted parameter estimate

$$\hat{\underline{\Theta}}^N_{Min,corr} = [(\underline{M}^T_{Meas} \underline{M}_{Meas}) - N \sum_n]^{-1} (\underline{M}^T_{Meas} \underline{Y}_{Meas} - N\underline{\gamma})$$ (5.144)

will then be a consistent estimate, i.e. it holds that

$$p \lim_{N \to \infty} \hat{\underline{\Theta}}^N_{Min,corr} = \underline{\Theta}_{RS}.$$ (5.145)

It should be noted that (5.142) and (5.143) can also be used in the case of noisy input measurements. Moreover, if the input and output noise processes are independent and white, $\sum_n$ will be diagonal, and $\underline{\gamma} = \underline{0}$. $\hat{\underline{\Theta}}^N_{Min,corr}$ can be calculated recursively in a similar way as $\hat{\underline{\Theta}}^N_{Min}$, given in (5.140).

5.8 Identifiability*

The problem of identifiability has been discussed from a general point of view in Chap. 2 together with the problems of controllability and observability. In this section a concept of identifiability that does not rely on a special parameter-estimation method and its application to time-discrete systems will be discussed. The real-world system to be identified can be described by (5.146)

$$Y_{k+1} = (\Theta_1 + \Theta_3)Y_k + \Theta_2 U_k,$$ (5.146)

where infinitely many parameter vectors

$$\underline{\Theta}^T = [\Theta_1, \Theta_2, \Theta_3]$$ (5.147)

exist that describe the actual state of the real-world system based on the error criterion method. Hence for all parameter vectors that satisfy the condition

$$\tilde{\Theta}_1 := \Theta_1 + \Theta_3 = \Theta_{RS1} + \Theta_{RS3} = \tilde{\Theta}_{RS1} = const.,$$ (5.148)

for all

$$\underline{\Theta} = [\underline{\Theta}_1, \underline{\Theta}_{RS2}, \widetilde{\underline{\Theta}}_{RS1} - \underline{\Theta}_1],$$ (5.149)

the true model outputs $\{Y_k(\underline{\Theta})\}$ are indistinguishable. Thus, the real-world system, the true model of which given in (5.146) with

$$\underline{\Theta}^T = \underline{\Theta}_{RS}^T = [\underline{\Theta}_{RS1}, \underline{\Theta}_{RS2}, \underline{\Theta}_{RS3}],$$ (5.150)

is not identifiable. But a reduced parameter vector, given in (5.151),

$$\underline{\Theta}^T = [\widetilde{\underline{\Theta}}_1, \underline{\Theta}_2]$$ (5.151)

would be uniquely identifiable.

This example shows that real-world systems identifiability can be assumed to be coupled with the problem of model-structure selection, which has to be solved before starting to estimate the real-world system parameters from noisy measurements. It also shows that real-world systems identifiability depends on the systems input sequence $\{U_k\}$. If, for instance, $U_k = 0$, the true model output $\{Y_k\}$ is indistinguishable for all Θ_2.

Remark 5.14
Consider Θ_{RS} does not exist, meaning for no Θ true model output the real system output can be made identical which means it can be excluded due to the assumption that if a true model exists the structure of which is known. This can happen in such cases that too-low model order was chosen

Moreover, identifiability is of importance due to cases when a time-discrete version of the true model is derived, which is for piecewise-constant inputs from a time-continuous model, the parameters of which can be determined from the time-discrete model parameter estimates. In such cases the distinguishability of the time-continuous model outputs is necessary but not sufficient for the distinguishability of the time-discrete model outputs.

Definition 5.2
A real-world system is called – relative to the true model structure at Θ_{RS} –

(i)   Parameter identifiable if there exists an input sequence $\{U_k\}$ such that Θ and Θ_{RS} are distinguishable for all $\Theta \neq \Theta_{RS}$

(ii)   system identifiable if there exist an input sequence $\{U_k\}$ such that $\underline{\Theta}$ and Θ_{RS} are distinguishable for all $\Theta \neq \Theta_{RS}$ but a finite set

(iii)   Unidentifiable in all other cases

A single parameter is said in case (i) to be uniquely identifiable, in case (ii) only identifiable. ∎

Definition 5.2 is independent of the existence of an identification method to be used in estimating the single parameter or the parameter vector consistently, but identifiability in this sense is a necessary condition to identify the real-world system at all.

Example 5.1
As an example of Definition 5.1, part (ii), consider a true model given by

$$Y_{k+1} = \Theta_1^2 Y_k + \Theta_2 U_k \, . \tag{5.152}$$

This system can be at best system identifiable, but not parameter identifiable, at

$$\underline{\Theta} = \underline{\Theta}_{RS} = [\underline{\Theta}_{RS1}, \underline{\Theta}_{RS2}] \, , \tag{5.153}$$

because there is no input sequence $\{U_k\}$ for which the two parameter vectors

$$\underline{\Theta}_1 = [\underline{\Theta}_{RS1}, \underline{\Theta}_{RS2}] \, , \tag{5.154}$$

and

$$\underline{\Theta}_2 = [-\underline{\Theta}_{RS1}, \underline{\Theta}_{RS2}] \, , \tag{5.155}$$

are distinguishable.

The condition for local identifiability of real-world systems, described by linear time-discrete models is considered to be based on the fact that the output sequence of the linear time-discrete model are indistinguishable for two different parameter vectors $\underline{\Theta}_1$ and $\underline{\Theta}_2$ for any input sequence $\{U_k\}$, if the impulse responses for $\underline{\Theta}_1$ and $\underline{\Theta}_2$ are identical. This will be the case if and only if the Markov parameters of the system are identical. The Markov parameters are defined for a single-input, single-output system, as coefficients of the expansion of z-transfer function $H_z(z)$ in z^{-1}. For a true model in the state notation

$$\underline{x}_{k+1} = \underline{\Phi}(\underline{\Theta})\underline{x}_k + \underline{h}(\underline{\Theta})u_k \, , \tag{5.156}$$

$$y_{k+1} = \underline{c}^T \underline{x}_{k+1} \, , \tag{5.157}$$

with the n-dimensional state vector $\underline{x}_k$, the z-transfer function is given by

$$H_z(z) = \underline{c}^T (z\underline{I} - \underline{\Phi})^{-1}\underline{h} = \underline{c}^T \underline{h}z^{-1} + \underline{c}^T \underline{\Phi}\underline{h}z^{-2} + \ldots = g_0 z^{-1} + g_1 z^{-2} + \ldots \quad . \tag{5.158}$$

Taking into account that the Markov parameters for $k \geq 2n$ are identical if they are identical for $k = 0, 1, \ldots, 2n-1$, the local identifiability problem of the system described in (5.156) and (5.157) at $\underline{\Theta} = \underline{\Theta}_{RS}$ can be examined by testing the distinguishability of the Markov parameter matrix

$$\underline{G}(\underline{\Theta}) = [\underline{c}^T \underline{h}, \underline{c}^T \underline{\Phi}\underline{h}, \ldots, \underline{c}^T \underline{\Phi}^{2n-1}\underline{h}]^T = [g_0, g_1, \ldots, g_{2n-1}]^T \tag{5.159}$$

at $\underline{\Theta}_{RS}$. In a sufficiently small vicinity of $\underline{\Theta}_{RS}$ we obtain

$$\Delta\underline{G}(\underline{\Theta}_{RS} + \Delta\underline{\Theta}) = \underline{G}(\underline{\Theta}_{RS} + \Delta\underline{\Theta}) - \underline{G}(\underline{\Theta}_{RS}) = \frac{\partial \underline{G}(\underline{\Theta}_{RS})}{\partial \underline{\Theta}} \Delta\underline{\Theta} \tag{5.160}$$

$$= \begin{bmatrix} (\partial g_0 / \partial \underline{\Theta})^T \\ (\partial g_1 / \partial \underline{\Theta})^T \\ \cdot \\ (\partial g_{2n-1} / \partial \underline{\Theta})^T \end{bmatrix}_{\underline{\Theta} = \underline{\Theta}_{RS}} \cdot \Delta\underline{\Theta} \; .$$

If the $2n \times p$ matrix $\dfrac{\partial \underline{G}(\underline{\Theta}_{RS})}{\partial \underline{\Theta}}$ has full rank p for all $\Delta\underline{\Theta} \neq \underline{0}$, we obtain with (5.160) $\underline{G}(\underline{\Theta}_{RS} + \Delta\underline{\Theta}) \neq \underline{0}$ so that $(\underline{\Theta})$ is distinguishable from $\underline{G}(\underline{\Theta}_{RS})$.

Theorem 5.1
A necessary and sufficient condition for global as well as local parameter identifiability of a linear time-discrete system at $\underline{\Theta}_{RS}$ is

$$\text{rank}\left[\frac{\partial \underline{G}(\underline{\Theta}_{RS})}{\partial \underline{\Theta}}\right] = p. \tag{5.161}$$

Example 5.2
Theorem 5.1 can be applied to the true model examples mentioned before. The state-space notation of the model in (5.160) is

$$x_{k+1} = \Phi x_k + hU_k, \tag{5.162}$$

$$y_{k+1} = c x_{k-1}, \tag{5.163}$$

with $\Phi = (\Theta_1 + \Theta_2)$, $h = \Theta_2$, and $c = 1$, hence the Markov parameter matrix $\underline{G}(\underline{\Theta})$ is given as follows

$$\underline{G}(\underline{\Theta}) = [\underline{\Theta}_2, \underline{\Theta}_2 \cdot (\underline{\Theta}_1 + \underline{\Theta}_3)].$$ (5.164)

Following (5.160) $\dfrac{\partial \underline{G}}{\partial \underline{\Theta}}$ has $p = 3$ columns, but only $2n = 2$ rows, because $n = 1$, hence the

maximum rank of $\dfrac{\partial \underline{G}}{\partial \underline{\Theta}}$ is 2 ($< p$). Theorem 5.1 means that the system described by

(5.162) and (5.163) is locally unidentifiable.

Consider a reduced-parameter vector, given in (5.150), i.e. if $\Phi = \widetilde{\Theta}_1$, $h = \Theta_2$, and $c = 1$, we get

$$\underline{G}(\underline{\Theta}) = [\underline{\Theta}_2, \underline{\Theta}_2 \cdot \widetilde{\Theta}_1],$$ (5.165)

and

$$\frac{\partial \underline{G}}{\partial \underline{\Theta}} = \begin{bmatrix} 0 & 1 \\ \Theta_2 & \widetilde{\Theta}_1 \end{bmatrix},$$ (5.166)

the rank of which is equal to 2 ($= p$) if $\Theta_2 \neq 0$. Thus the system represented by the true model in Equations (5.162) and (5.163), with the reduced parameter vector given in (5.151), is locally parameter identifiable for $\Theta_2 \neq 0$. The exception $\Theta_2 \neq 0$ reflects that for $\Theta_2 \neq 0$ the model cannot be excited by U, so that changes in $\widetilde{\Theta}_1$ do not show any change in the output Y_k.

For the model given in (5.152) we have $\Phi = \Theta_1^2$, $h = \Theta_2$, and $c = 1$, yields

$$\underline{G}(\underline{\Theta}) = [\Theta_2, \Theta_2 \cdot \Theta_1^2]^T,$$ (5.167)

and

$$\frac{\partial \underline{G}}{\partial \underline{\Theta}} = \begin{bmatrix} 0 & 1 \\ 2\Theta \ \Theta_2 & \Theta_1^2 \end{bmatrix},$$ (5.168)

The corresponding system is locally parameter identifiable if $\Theta_1 \cdot \Theta_2 \neq 0$, i.e. $\Theta_1 \neq 0$ and $\Theta_2 \neq 0$.

The additional condition on Θ_1 is due to the fact that the Markov parameters do not change at $\Theta_1 = 0$ for sufficiently small deviations $\Delta\Theta_1$ because

$$\left(\frac{\partial g_1}{\partial \Theta_1}\right)_{\Theta_1=0} = 0. \tag{5.169}$$

It should be noted that for all $\underline{\Theta} = \begin{bmatrix} \Theta_1 \\ \Theta_2 \end{bmatrix}$ under the conditions $\Theta_1 \neq 0$ and $\Theta_2 \neq 0$

the system is locally parameter identifiable although globally there exists always the undis-

tinguishable-parameter vector $\overline{\underline{\Theta}} = \begin{bmatrix} -\Theta_1 \\ \Theta_2 \end{bmatrix}$.

Remark 5.15
If the parameters of a time-continuous model in state-space notation

$$\frac{d\underline{x}}{dt} = \underline{A}(\Theta)\underline{x} + \underline{b}(\Theta)\vec{u}, \tag{5.170}$$

$$y = \underline{c}^T(\Theta)\underline{x}, \tag{5.171}$$

have to be identified, Theorem 5.1 holds for the time-continuous Markov parameter matrix

$$\underline{G}(\Theta) = [\underline{c}^T \underline{b}, \underline{c}^T \underline{Ab}, ..., \underline{c}^T \underline{A}^{2n-1} \underline{b}]^T. \tag{5.172}$$

A sufficient condition for preserving the property of local distinguishability of the system response $y(t, \Theta)$ for the sampled response $\{y_k(\Theta)\}$ is that the sampling frequency

$$\omega = \frac{2\pi}{T_a} \tag{5.173}$$

satisfy the sampling theorem, i.e. that $\omega_a > 2\omega_{max}$, where ω_{max} is the largest imaginary part of the system eigenvalues.

5.9 System Input Properties*

The importance of the system input for identification purposes has become apparent in connection with the various problems discussed in the preceeding sections. At the very first the input sequence $\{U_k\}$ has to be chosen in such a way that the system is identifiable at all, e.g. in the sense of Definition 5.1. Furthermore, it may be found that the equation-error least squares estimate $\hat{\Theta}_{Min}^N$ in (5.86) only exists if the inverse of the matrix $(\underline{M}_{Meas}^N)^T (\underline{M}_{Meas}^N)$ exists, i.e. this matrix must be positive-definite:

$$(\underline{M}_{Meas}^N)^T (\underline{M}_{Meas}^N) > \underline{0} . \tag{5.174}$$

A necessary and sufficient condition for (5.174) to hold is that

$$\underline{M}_{Meas}^{N-1} \tag{5.175}$$

be of full rank. Partitioning of

$$\underline{M}_{Meas}^{N-1} \tag{5.176}$$

by

$$\underline{M}_{Meas}^N = [\underline{Y}_{Meas}^N, \underline{U}^N], \tag{5.177}$$

we find

$$\underline{U}^N := \begin{bmatrix} U_{n-1} & U_{n-2} & . & U_0 \\ U_n & U_{n-1} & . & U. \\ . & . & . & . \\ U_N & U_{N-1} & . & U_{N-n} \end{bmatrix} \tag{5.178}$$

has to be of full rank, i.e. the columns of $\underline{U}^N$ must be linearly independent, so that we found, as a necessary condition,

$$(\underline{U}^N)^T \underline{U}^N > \underline{0} \tag{5.179}$$

for (5.174) to hold.

Proving the consistency of the estimate $\hat{\underline{\Theta}}_{Min}^{N}$ we need a stochastic equivalent of (5.179) which could hold the form

$$p \lim_{N \to \infty} (\frac{1}{N} (\underline{U}^{N})^{T} \underline{U}^{N}) > \underline{0}. \tag{5.180}$$

A measure of the quality of the parameter estimation is its variance, which depends on the input sequence $\{U_k\}$. The variance of a parameter estimate can be reduced significantly if the system is exited by a pulse sequence instead of a step impulse.

5.10 Parameter Identification of the Cardiovascular System*

Consider, that the human cardiovascular system consists of the heart, the lung, the systemic vascular bed, the pulmonary vascular bed, and an adaptation of the system due to physical exercises, based on the respective control loops. Such a type of real-world system can be described through a nonlinear model based on compartments that represent the circulatory fluid dynamic and the central nervous control, shown in the block diagram in Fig. 5.9, which contain nine first-order differential equations and 27 nonlinear algebraic equations.

The model, shown in Fig. 5.10, illustrates the signal flow between the various subsystems of the model of the cardiovascular system and the various interacting control loops. Here, thin lines show the operation flow of the cardiovascular system including the baroreceptor feedback loop; thick lines show the operation flow of oxygen requirement under ergometric (physical) workload, and its influence on the cardiovascular system. Furthermore, dashed lines indicate the nonlinear pressure volume relationship of the compliances of the vascular beds.

From Fig. 5.9 we can derive the nonlinear vector differential equation

$$\frac{dX}{dt} = f(X, U, Z, \Theta_S), \tag{5.181}$$

with X as state vector

$$X := [PAS, PVS, PAP, PVP]^{T}, \tag{5.182}$$

the components of which are the mean systemic pressure (PAS), the mean pulmonary pressure (PAP), the mean venous systemic pressure (PVP) and the mean pulmonary venous pressure (PVP), and U as the control vector

$$U := [\text{HF, RA}]^{\text{T}}, \qquad (5.183)$$

the components of which are HF and RA, as the heart frequency and the peripheral resistance, respectively, Z as system disturbance, and Θ_S as the system-parameter vector.

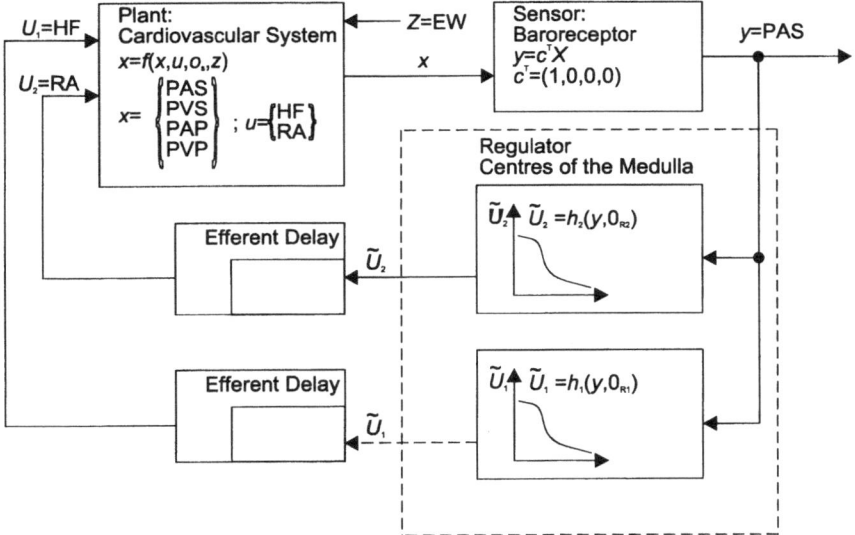

Fig. 5.9. Block diagram of the nonlinear model of the human cardiovascular system

Sensing of PAS by the so-called pressure or baroreceptors at the afferent pathway to the centers of the medulla can be modeled by the linear output equation

$$y = c^{\text{T}} \cdot X \qquad (5.184)$$

Finally, the centers of the medulla can be interpreted as the controller of the system with PAS as control variable, which can be modeled as follows

$$\tilde{U}_1 = h_1 \cdot (y, \Theta_{R1}) \qquad (5.185)$$

$$\tilde{U}_1 = h_1 \cdot (y, \Theta_{R1}), \tag{5.186}$$

with Θ_{R1} and Θ_{R2} as the respective controller-parameter vectors. The control variables U_1 and U_2 are given by the controller outputs $\tilde{U}_1$ and $\tilde{U}_2$ respectively, delayed by the efferent pathway.

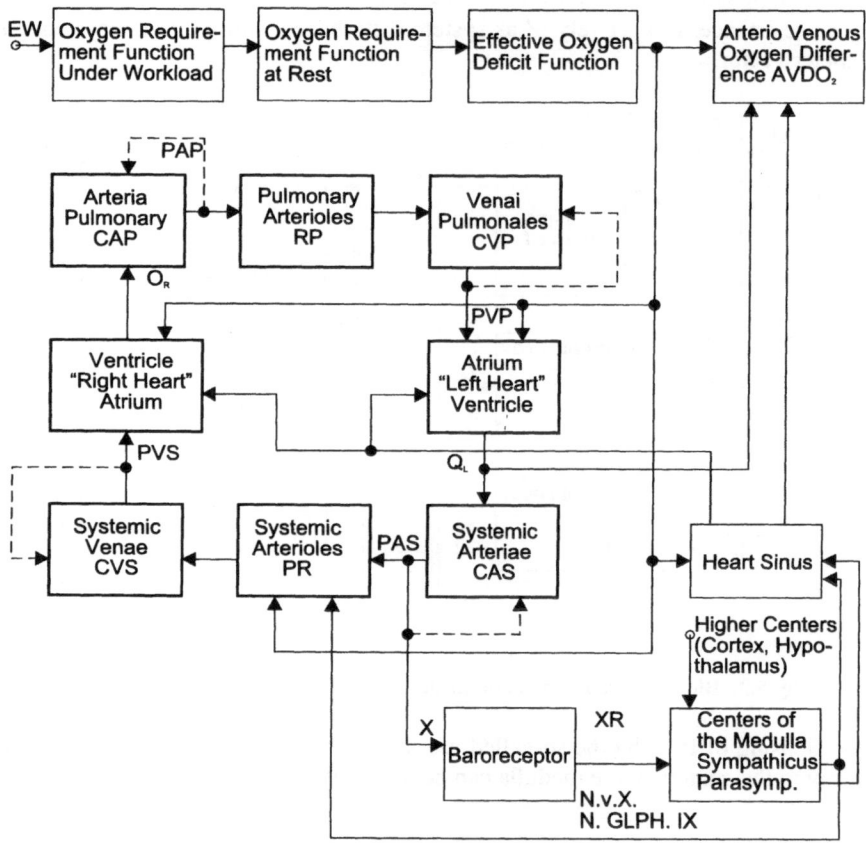

Fig. 5.10. Block diagram of the physiological structure of the cardiovascular system

Using the disturbance input Z, shown in Fig. 5.9, the system can be exited in particular with an ergometric workload (EW).

Consider several cases we first try to identify the contractility of the heart muscle, introduced through the contractility parameters *KKL* and *KKR*, as well as the tonic activity of the sinus node of the heart, expressed as the transfer constant *KHF*, while the model is exited due to an ergometric workload (EW). *KKL* and

KKR are defined as the proportionality constants in the relations of the stroke volumes of the left (*SVL*) and right (*SVR*) chamber of the heart

$$SVL = KKL \cdot \frac{VD}{PAS} ,$$

(5.187)

and

$$SVR = KKR \cdot \frac{VD}{PAP} ,$$

(5.188)

which describe the interdependence between the stroke volume of the heart and the ratio of the end diastolic filling volume (*VD*) and the mean pressure of the corresponding vascular compartment against which the heart has to stroke.

Furthermore, the identifiability of the vascular beds compliance parameters will be examined. For this purpose the other parameters of the system are assumed to be known a priori.

For identification purposes the output-error least squares method was used, with $N = 95$ measurements distributed over the whole identification interval. In order to reduce rounding-error effects in the vicinity of the minimum and to avoid the computation of gradients, needing a significant simulation time, the error functional is minimized using the Rosenbrock direct search method, rather than a gradient method. The computation for the numerical integration of the differential equations

$$\frac{d}{dt} \begin{bmatrix} PAS \\ PVS \\ PAP \\ PVP \end{bmatrix} = \mathbf{A} \begin{bmatrix} PAS \\ PVS \\ PAP \\ PVP \end{bmatrix} + \mathbf{U},$$

(5.189)

over the whole identification interval was done using a 5-th order Runge Kutta method with variable step length.

For the first experiment, the identification of the parameters *KHF*, *KKL*, and *KKR* is attempted relative to a reduced identification interval of 1 minute length and $N = 12$ equally spaced measurements taken from the model step response. It can be seen from Table 5.1 that, by this experiment, only *KHF* can be identified with reasonable precision. *KHF* is the transfer constant of the tonic activity at the sinus node of the heart, influenced by sympathetic changes under an ergometric workload, *KKL* is the contractility constant of the left ventricle, and *KKR* is the contractility constant of the right ventricle.

Table 5.1 Results of simultaneous identification of *KKL*, *KKR*, and *KHF* over an interval of 1 min. length and $N = 12$ equally spaced measurements of the model step response, $\sigma_v = 1\%$. The compliance parameter values are given in parameter set (b) of Table 5.2.

Experiment	Parameter	True value	Initial value	Estimated value
1	*KKL*	50,00	70,00	44,90
1	*KKR*	7,00	10,00	10,40
1	*KHF*	0,60	0,30	0,586

A more detailed examination of the identification results of *KKL* and *KKR* shows that the exactness of corresponding estimates depends significantly on the assumed compliance parameter values. For illustration, the error functional (in logarithmic scale) is shown as a function of *KKR*, parameterized in *KKL* in Figs. 5.11 and 5.12 for the two compliance-parameter sets given in Table 5.2, respectively.

Table 5.2 Compliance-parameter sets assumed in the model-to-model identification

Parameter set	CASN	KCVS	KCAP	KCVP
(a)	2,9087	3575,00	51,542	1087,15
(b)	1,5	371,25	34,361	43,48

For the compliance-parameter set (a) the error functional is of the form of a deep crater over the two-dimensional *KKL* - *KKR* parameter space with a unique minimum for the true parameter values. If, however, the second set of compliance parameters, which is parameter set (b), is assumed, $J_N(\hat{\underline{\Theta}})$ is valley-like with the line of largest depth showing a small ascent from a unique deepest point at the true parameter value. Therefore, Θ_{RS} is identifiable in general, but if the ascent of the bottom of the valley reaches zero, the parameters would no longer be identifiable. In practice it is very difficult to find the search direction along the valley using the Rosenbrock method. Unlike the gradient methods like the Davidon Fletcher Powell method appear to be not applicable at all because of the valley. If, during the minimization procedure, a point of this bottom of the valley is reached, the search hangs up, in general. The graphical representation of the error functional shows that the identification results obtained can be recognized as points of the bottom edge of the valley.

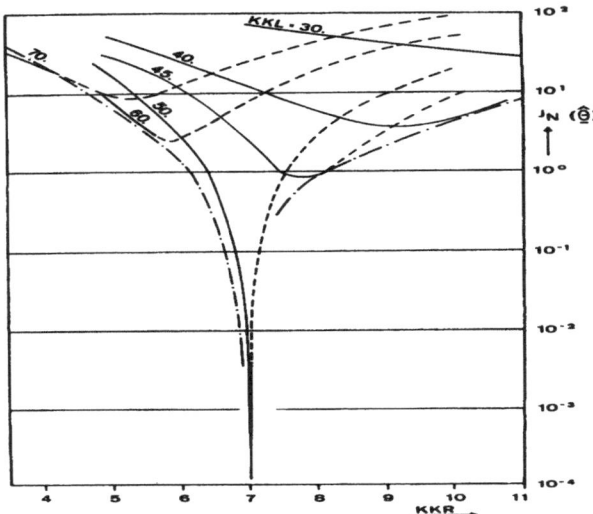

Fig. 5.11. Error functional $J_N(\hat{\Theta})$ as a function of *KKR*, parameterized in *KKL* for $N =$ 12 and compliance parameter set (a) of Table 5.2

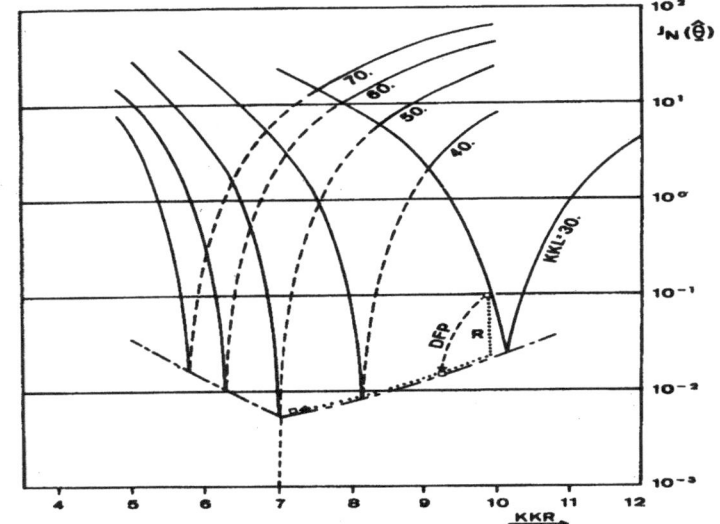

Fig. 5.12. Error functional $J_N(\hat{\Theta})$ as a function of *KKR*, parameterized in *KKL*, for $N =$ 12 and compliance parameter set (b) of Table 5.2. The trace of a typical Davidon Fletcher Powell search (DFP) and corresponding Rosenbrock search (R) are shown as dotted lines

5.11 Error-Functional Minimization by Gradient Methods*

Writing the equation-error functional as follows

$$J_N(\hat{\Theta}) = \hat{\underline{e}}^T(\hat{\Theta})\hat{\underline{e}}(\hat{\Theta}) \tag{5.190}$$

$$= \underline{Y}_{Meas}^T \underline{Y}_{Meas} - 2\underline{Y}_{Meas}^T \underline{M}_{Meas}\,\hat{\underline{\Theta}} + (\hat{\underline{\Theta}})^T\,\underline{M}_{Meas}^T\,\underline{M}_{Meas}\,(\hat{\underline{\Theta}})\;,$$

and if

$$J_N(\hat{\underline{\Theta}} = \underline{0}) = \underline{Y}^T{}_{Meas}\,\underline{Y}_{Meas}\;, \tag{5.191}$$

$$\underline{q}(\hat{\underline{\Theta}}) := \frac{\partial J_N(\hat{\underline{\Theta}})}{\partial(\hat{\underline{\Theta}})} = 2\cdot\underline{M}_{Meas}^T\,\underline{Y}_{Meas} + 2\cdot\underline{M}_{Meas}^T\,\underline{Y}_{Meas}\cdot(\hat{\underline{\Theta}})\;, \tag{5.192}$$

and

$$\underline{H}(\hat{\Theta}) := \frac{\partial J_N(\hat{\underline{\Theta}})}{\partial(\hat{\underline{\Theta}})} = 2\cdot\underline{M}_{Meas}^T\,M_{Meas}\;, \tag{5.193}$$

are introduced in (5.146), $J_N(\hat{\underline{\Theta}})$ can be written in the form of the Taylor series expansion about $\underline{\Theta} = 0$:

$$J_N(\hat{\underline{\Theta}}) = J_N(\underline{0}) + \underline{q}^T(\underline{0})(\hat{\underline{\Theta}}) + \frac{1}{2}(\hat{\underline{\Theta}})^T\,\underline{H}(\underline{0})(\hat{\underline{\Theta}})\;. \tag{5.194}$$

The necessary condition for an extremum of $J_N(\hat{\underline{\Theta}})$ yields

$$\frac{\partial J_N(\hat{\underline{\Theta}})}{\partial(\hat{\underline{\Theta}})} = \underline{q}(\underline{0}) + \underline{H}(\underline{0})(\hat{\underline{\Theta}}) = \underline{0}\;, \tag{5.195}$$

hence the expression for $\hat{\underline{\Theta}}{}^N_{Min}$ can be written as

$$\hat{\underline{\Theta}}{}^N_{Min} = -\underline{H}^{-1}\underline{q}(\underline{0})\;, \tag{5.196}$$

where the argument of $\underline{H}(\hat{\Theta})$ has been dropped, due to (5.80), while $\underline{H}$ is independent of $\hat{\underline{\Theta}}$.

Using the Taylor expansion of $J_N(\hat{\underline{\Theta}})$ up to the quadratic term in the nonquadratic case, in order to receive an approximation of $\hat{\underline{\Theta}}{}^N_{Min}$, i.e. solving the problem

$$J_N(\hat{\underline{\Theta}}) = J_N(\hat{\underline{\Theta}}_i) + \underline{q}^T(\hat{\underline{\Theta}}_i)(\hat{\underline{\Theta}} - \hat{\underline{\Theta}}_i) + \frac{1}{2}(\hat{\underline{\Theta}} - \hat{\underline{\Theta}}_i)^T \underline{H}(\hat{\underline{\Theta}}_i)(\hat{\underline{\Theta}} - \hat{\underline{\Theta}}_i) \rightarrow Min. \qquad (5.197)$$

The closer $\hat{\underline{\Theta}}_i$ is to $\hat{\underline{\Theta}}_{Min}^N$ the better the approximation of $J_N(\hat{\underline{\Theta}})$ through $J_N(\hat{\underline{\Theta}})$ at $\hat{\underline{\Theta}}_{Min}^N$ and the closer the minimum argument $\hat{\underline{\Theta}} = \hat{\underline{\Theta}}_{i+1}$, resulting from the necessary condition

$$\frac{\partial J_N(\hat{\underline{\Theta}}_{i+1})}{\partial(\hat{\underline{\Theta}})} = \underline{q}(\hat{\underline{\Theta}}_i) + \underline{H}(\hat{\underline{\Theta}}_i)(\hat{\underline{\Theta}}_{i-1} - \hat{\underline{\Theta}}_i) = \underline{0} \qquad (5.198)$$

to the minimum argument of $J_N(\hat{\underline{\Theta}})$. From (5.198) we obtain

$$\hat{\underline{\Theta}}_{i+1} - \hat{\underline{\Theta}}_i = -\underline{H}^{-1}(\hat{\underline{\Theta}}_i) \cdot \underline{q}(\hat{\underline{\Theta}}_i), \qquad (5.199)$$

or

$$\hat{\underline{\Theta}}_{i+1} = \hat{\underline{\Theta}}_i - \underline{H}^{-1}(\hat{\underline{\Theta}}_i) \cdot \underline{q}(\hat{\underline{\Theta}}_i), \qquad (5.200)$$

if the inverse of $\underline{H}(\hat{\underline{\Theta}}_i)$ exists. If $\underline{H}(\hat{\underline{\Theta}}_i)$ is positive-definite, i.e. $\underline{H}(\hat{\underline{\Theta}}_i)\underline{0}$ we obtain

$$J_N(\hat{\underline{\Theta}}_{i+1}) < J_N(\hat{\underline{\Theta}}_i). \qquad (5.201)$$

However,

$$J_N(\hat{\underline{\Theta}}_{i+1}) < J_N(\hat{\underline{\Theta}}_i) \qquad (5.202)$$

is a necessary condition for the convergence of the method based on (5.200) to be reached.

A sufficient condition for $\underline{0}_{i+1}$, generated by the iteration step

$$\hat{\underline{\Theta}}_{i+1} = \hat{\underline{\Theta}}_i - p_i \underline{\Pi}_i \underline{q}(\hat{\underline{\Theta}}_i), \qquad (5.203)$$

may be successful in the sense of (5.202) is

$$\underline{\Pi}_i > \underline{0} \qquad (5.204)$$

for a sufficiently small step of length p_i in the search direction

$$\underline{v}_i := \underline{\Pi}_i \underline{q}(\hat{\underline{\Theta}}_i) \, .$$

(5.205)

This means that the modified iteration step in (5.200)

$$\hat{\underline{\Theta}}_{i+1} = \hat{\underline{\Theta}}_i - p_i \underline{H}^{-1}(\hat{\underline{\Theta}}_i) \cdot \underline{q}(\hat{\underline{\Theta}}_i)$$

(5.206)

will be successful if $\underline{H}(\hat{\underline{\Theta}}_i)$ is positive-definite, and if the step length p_i is sufficiently small. When using the Newton Raphson method, $p_i = 1$ is assumed fixed so that even for $\underline{H}(\hat{\underline{\Theta}}_i) > \underline{0}$ a successful step is not guaranteed by the sufficient condition mentioned above, but if the functional is quadratic in $\underline{0}$, this method may give the optimum solution in one step. If $\underline{\Pi}_i = \underline{I}$ is chosen, (5.203) becomes the steepest descent iteration which is known to be of very slow convergence in the vicinity of the minimum, since, in general, only a very small step length will be successful. Optimization methods relying on the iteration of (5.203) are called gradient methods.

5.12 Error-Functional Minimization by Direct Search Methods*

In comparison to the gradient methods, the success of a parameter change relative to a chosen search direction in the sense of (5.202) is not known in advance. Hence the difference of the functional values has to be calculated only in order to decide whether an iterative step was successful or not. This decision may be done correctly even if only the sign of the calculated result is correct.

Due to the fact that the direct search methods do not need the gradient values themselves, no restrictions relative to the differentiability properties of $J_N(\hat{\underline{\Theta}})$ have to be taken into account. This is particularly important if the methods are to be applied in the case of nonlinear systems. In spite of this nonlinear system, there are possible situations where the direct search methods are unable to improve a parameter estimate $\hat{\underline{\Theta}}_i$, although it is still significantly different from the true parameter vector $\hat{\underline{\Theta}}_{RS}$, and although $\hat{\underline{\Theta}}_{RS}$ is locally parameter identifiable in the sense of definition 5.8-1. It may be seen that such problems are different from

gradient methods, which may be of a principal nature, although parameter insensitivities of the error functional may cause smaller effects, too.

The direct search methods are comprehensively described in the literature. Hence we may restrict ourselves to some principal problems when applying the Rosenbrock method. The most basic search method is the so-called relaxation method where an initial parameter vector $\hat{\underline{\Theta}}_0$ is changed in linearly independent directions of the parameter space, mostly the coordinate directions, in a successive way. If the change in a direction

$$\hat{\underline{\Theta}}_1 := \hat{\underline{\Theta}}_0 + \Delta(\hat{\underline{\Theta}}) \qquad (5.207)$$

is successful the search will be continued with the new initial parameter vector defined by

$$\hat{\underline{\Theta}}_0 := \hat{\underline{\Theta}}_1 . \qquad (5.208)$$

The convergence of this method is slow. Therefore, the more sophisticated search methods, e.g. the Hooke Jeeves and the Rosenbrock methods tried to accelerate the convergence by continuing the search in a new direction, taking into account the direction of the overall success of the last iteration step. The successive parameter changes along all basis vectors of the parameter space, relative to a predefined criterion, is said to be an exploration step. The result of an exploration, which may be a new point in the parameter space, is called a base point.

The relaxation method may be interpreted as a sequence of simple exploration steps to the coordinate basis. When applying the Hooke Jeeves method, a successful exploration step is followed by a search in the direction of the difference of the last two base points, followed by a new exploration step in the coordinate basis. Using the Rosenbrock method, in comparison, the first exploration step is followed immediately by the next exploration step. These strategies are illustrated in the two dimensional parameter space in Fig. 5.13,

From Fig. 5.13 it can be seen, on applying the relaxation method, that the step length will be doubled from base point to base point in the case of success, otherwise multiplied by –0.5, while in the Rosenbrock method, case (c) in Fig. 5.13, this strategy is followed in the search of a new base point too. The new point is defined by the property that in every search direction a success was followed by a failure. The points p_i in Fig. 5.13 characterize the base points as results of exploration steps relative to the criteria defined. Noninterconnected points indicate unsuccessful trials.

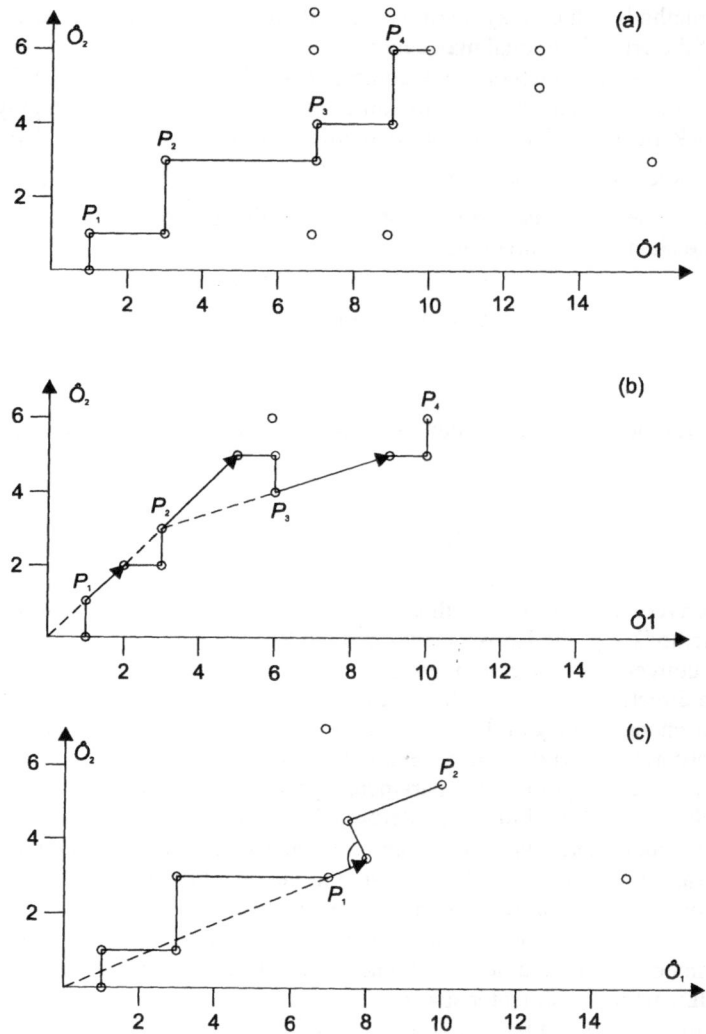

Fig. 5.13: Illustration of the relaxation method (a), the Hooke Jeeves method (b) and the Rosenbrock method (c)

Figure 5.14 illustrates a situation in which a direct search method may fail. The point $\underline{\hat{\Theta}}_i$ is assumed to lie on an edge of the error functional $J_N(\underline{\hat{\Theta}})$ which is characterized by its lines of constant height forming acute angles along the edge, This angle is shown in particular in $\underline{\hat{\Theta}}_i$ by crossing tangents of $J_N(\underline{\hat{\Theta}}_i)$. If the search directions lie in the shaded area between these tangents, which is the truth, i.e. for the coordinate directions in this case an exploratory step can not be successful even by using a step length as small as possible. The Rosenbrock method may be only successful in $\hat{\Theta}_i$ if the direction of overall success from the last step, or the basis vector orthogonal to this direction, points into the unshaded area between the two tangents. Otherwise, it will fail and hang up in $\hat{\Theta}_i$.

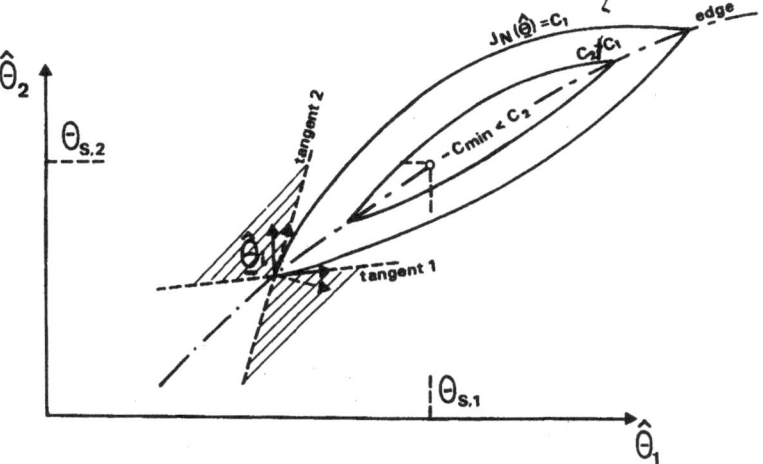

Fig. 5.14: Illustration of a situation where a direct search method in $\hat{\Theta}_i$ fails if the search directions lie inside the shaded area.

Remark 5.12-1
The search directions in the Rosenbrock method change only if at least in two of them the last exploration was successful.

In the example, shown in Fig. 5.14 it was assumed that $J_N(\hat{\Theta})$ was not differentiable along the edge. In this case the gradient methods would not be applicable at all. If, in spite of this, the method of the steepest descent were used, the method would hang up in $\hat{\Theta}_i$ because the two possible calculated search directions lie in the shaded area.

Similar problems may appear in the case of a differentiable functional that is flat around the minimum point so that the sensitivity of $J_N(\hat{\Theta})$, relative to one parameter, is very small. Such a case is illustrated in Fig. 5.15. In $\hat{\Theta}_i$ there is no improvement possible in one coordinate direction $\hat{\Theta}_2$ at all because it is tangent to $J_N(\hat{\Theta}) = c_1$. In the $\hat{\Theta}_i$ direction an improvement is relative to the numerical precision used only realizable if the changes of $\Delta\hat{\Theta}_1$ are sufficiently large. If the initial step length of the exploration is too small, $J_N(\hat{\underline{\Theta}}_i)$ appears as a minimum although $\hat{\underline{\Theta}}_1$ is significantly different from $\hat{\underline{\Theta}}_{RS}$.

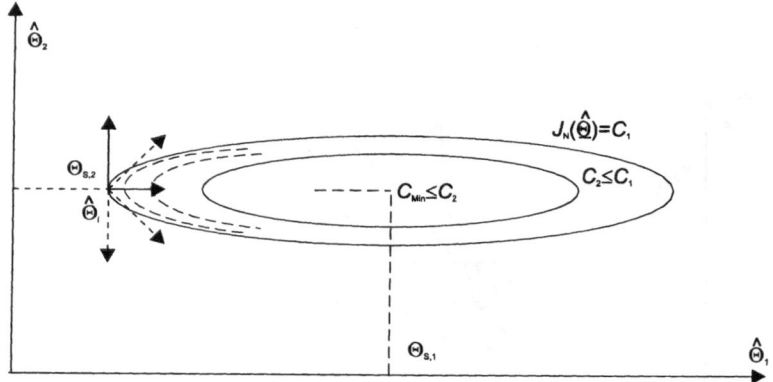

Fig. 5.15: Illustration of a situation where a search method may fail because of the insensitivity of the error functional $J_N(\hat{\underline{\Theta}})$ relative to a parameter component

Remark 5.12-2

This effect becomes even more apparent if the search directions are rotated against the coordinate directions, because the possible changes in $\hat{\underline{\Theta}}_1$ are even smaller. This effectively results in a search along a line $\hat{\underline{\Theta}}_{1i} \approx$ constant, without any improvement. A wrongly calculated gradient in $\hat{\underline{\Theta}}_1$ or too small starting step length may cause similar effects when using the gradient methods in such a situation.

5.13 Identifiability and the Output-Error Least Squares Method*

The definition of local identifiability, based on the distinguishability of the system output, is equivalent to the so-called least squares identifiability. A system is called locally least squares parameter identifiable in $\hat{\underline{\Theta}}_{RS}$ if $J_N(\hat{\underline{\Theta}})$ has a unique minimum at $\hat{\underline{\Theta}} = \hat{\underline{\Theta}}_{RS}$.

Remark 5.13-1
In the stochastic case, the distinguishability of the system outputs may be defined by the distinguishability of the probability densities of the output.

If $J_N(\hat{\underline{\Theta}})$ is calculated for a sufficiently large N relative to one coordinate $\hat{\underline{\Theta}}^j$, it is possible to test the property of identifiability due to these two parameters by searching for a unique minimum, without the necessity of determining the exact value of $\hat{\underline{\Theta}}^N_{Min}$. If it is, in principle, possible to estimate $\hat{\underline{\Theta}}_{RS}$ consistently with a sufficient small variance for the chosen N, i.e. if N is large enough, this may only be the case if $J_N(\hat{\underline{\Theta}})$ shows the unique local minimum, stated by the equivalence of output distinguishability and least squares identifiability.

Remark 5.13-2
The local identifiability of a system, relative to a linearized nonlinear model, implies identifiability of the original nonlinear system due to the linearization. Since the linearization of a complex model may be involved, the computer-aided determination of identifiability is of great importance.

A graphical representation of $J_N(\hat{\underline{\Theta}})$ in the vicinity of the local minimum also allows a pseudominimum to be detected, which is a minimum pretended by the minimization algorithm.

Due to the possibility of a systematic or random variation of the search directions for successfully continuing the search, hanging up in a pseudominimum, it may be possible to vary the initial parameter values in order to find a possibly closer approximation of the true minimum, i.e. a smaller value of the error functional. Hence the detection of pseudominima will become of importance since the numerical results may have similar characteristics in the case of a really unidentifiable system. This may be demonstrated based on the model given in equation (5.146) with the assumption of noise-free measurements $Y_{Meas,\, k} = Y_k$. In this case, the identification model is represented by

$$Y_{k+1} = (\hat{\Theta}_1 + \hat{\Theta}_3)Y_k + \hat{\Theta}_2 U_k , \qquad (5.209)$$

for which no unique equation-error identification solution exists as a consequence of the fact that the matrix $\underline{M}^{N-1}$ in

$$\underline{Y}^N = \underline{M}^{N-1}\hat{\underline{\Theta}} , \qquad (5.210)$$

i.e. in

$$
\begin{bmatrix} Y_1 \\ Y_2 \\ \cdot \\ Y_N \end{bmatrix} = \begin{bmatrix} Y_1 & U_0 & Y_0 \\ Y_2 & U_1 & Y_1 \\ \cdot & \cdot & \cdot \\ Y_{N-1} & U_{N-1} & Y_{N-1} \end{bmatrix} \begin{bmatrix} \hat{\Theta}_1 \\ \hat{\Theta}_2 \\ \hat{\Theta}_3 \end{bmatrix} \tag{5.211}
$$

has no full rank. Disregarding the numerical minimization method, applied to the equation-error least squares problem, as a best case an element of the infinite solution set

$$
\{ \ \underline{\hat{\Theta}}_{Min}^{N} \big| \hat{\Theta}_1 ; \hat{\Theta}_2 = \Theta_{S2} ; \hat{\Theta}_3 = \Theta_{S1} + \Theta_{S3} - \hat{\Theta}_1 \ \} \tag{5.212}
$$

may be found. On the other hand, the particular solution, as an element of this set, depends on the initial parameter vector for the iteration and the special strategy of the method chosen for minimization purposes. In the more general case of a non linear model, e.g. of

$$
Y_{k+1} = f(\hat{\Theta}_1, \hat{\Theta}_3) Y_k + \hat{\Theta}_2 U_k , \tag{5.213}
$$

the solution set would be characterized by the condition

$$
f(\hat{\Theta}_1, \hat{\Theta}_3) = const. = f(\Theta_{S1}, \Theta_{S2}). \tag{5.214}
$$

If the corresponding parameter set is infinite, the system may be unidentifiable relative to this model, and the error functional $J_N(\hat{\underline{\Theta}})$ forms a valley along this set.

Remark 5.13-3
A system, identifiable in principle, may appear practically unidentifiable if $J_N(\hat{\underline{\Theta}})$ is valley-based over the parameter space, due to the insensitivity of the error functional along this valley.

5.14 References and Further Reading

Aoki M, (1971), Introduction to Optimization Techniques, McMillan Company, New York
Bard Y, (1974), Nonlinear Parameter Estimation, Academic Press, New York

Hsia TC, (1977), System Identification, Lexington Books
Iserman R, (1974), Process-Identification (in German), Springer, Berlin, Heidelberg, New York
Jacoby GLS, Kowolik Jg, Pizzo JT, (1972), Iterative Methods for Nonlinear Optimization Problems, Proentice Hall Publ., New Jersey
Möller DPF, (1992), Modeling, Simulation and Identification of Dynamic Systems (in German), Springer, Berlin, Heidelberg, New York
Möller DPF, Popovic D, Thiele G, (1983), Modeling, Simulation and Parameter-Estimation of the Human Cardiovascular System, Vieweg Publ., Braunschweig, Wiesbaden
Richter O, Söndgerath D, (1990), Parameter Estimation in Ecology, VCH Publishers, Weinheim
Rufer DF, (1977), General Purpose Nonlinear Programming Package, In: Lecture Notes in Control and Information Sciences, Springer, Berlin, Heidelberg, New York
Wong KY, Polah E, (1967), Identification of Linear Discrete-Time Systems Using the Instrumental Variable Method, In: IEEE Trans. Automat. Control, Vol 12, pp.707–718

5.15 Exercises

5.1 What is meant by the term parameter identification?
5.2 Give the models for the direct and the adaptive parameter identification schemes.
5.3 What is meant bysaying a real-world system is called identifiable in its parameters?
5.4 The identification model of the identification task can be described by a set of linear differetial equations. Give the mathematical description for a p-dimensional parameter vector and define the polynomials $A(q^{-1})$ and $B(q^{-1})$.
5.5 What is meant by the term true model?
5.6 A method for identifying the system parameter vector is characterized by the error criterion. For the output-error least square method what does the error criterion look like?
5.7 What is meant by equation-error least squares method?
5.8 Give a model for the equation-error criterion realized by the use of the true model and an identification model.
5.9 What is meant by the term consistency of the parameter estimates?
5.10 Give an example of the generealized least squares method.

5.11 Give an example of the maximum-liklehood method.

5.12 What is meant by the term identifiability?

5.13 Describe the error functional for monimization by using the gradient method.

5.14 What is meant by the term direct search method?

5.15 The most basic search method is the relaxation method. How does this method works?

5.16 Explain the Hooke Jeeves search method.

5.17 Explain the Rosenbrook search method.

5.18 Explain the Davidon Fletcher Powel search method.

5.19 Compare the search methods mentioned in Exercises 5.14 to 5.18 as fit, fitter, and fittest for use in nonlinear parameter identification.

5.20 The identification model $y_{meas,k} = \underline{m}_{meas,k-1}^T \underline{\Theta} + \widehat{e}_k$ can be reformulated for the case of the one-step-ahead-prediction (OSP). Give the formula for y_k.

6 Soft-Computing Methods

6.1 Introduction

The formalizations of modeling are only useful if they succeed in seizing the essential features of the dynamic system under test. They permit extrapolations that allows one to generalize, often correctly, from past experience to future events from which we can learn how the dynamic system can be manipulated for ones purposes, which is a kind of uncertainty. In our world, which is more or less precisely understandable or predictable, we are more conscious of uncertainty. This uncertainty appears in the form of imprecision, vagueness, and ill-defined, ill-separable, and doubtful data. Using nonprecise information, called soft-information, needs a specific form of computation, called soft-computing. There are four main classes of methods that form soft-computing:

- Neural networks
- Fuzzy logic
- Genetic algorithms
- Probabilistic reasoning

Although each of these classes of methods can be used to resolve certain types of applications, they are in fact complementary to each other, and in many cases it can be better to employ them in combination rather than exclusively.

Usually, the approaches in uncertainty, combining soft information with classical mathematical methods in an ad hoc manner, can be investigated by using simulation to show the validity of the approaches to the specific problem under test. Therefore, during past years, processing of uncertainty, or soft-information processing, had been applied due to the different disciplines for a large variety in formal representations of models in the different scientific domains. But this task has become its own view, impressed by the respective domain. Introducing for those formalizations, soft-computing techniques, one can impart an understanding that the formalization itself can not provide. Because soft-computation is more of a simulation science discipline, it is a collection of methods that can be expressed in terms of algorithms, belonging to the respective disciplines, that has proved to be of vital importance to progress in all fields of endeavor.

In practice, the formalization of models itself is an iterative process, as introduced in Chap. 1, consisting of measurements, if possible, and computing strategies, by changing the structure of the formal description in an effort to closely match the complex system behavior. The computing strategy behind this is based on the category in the nearest-neighbor sense, if the adapted representation is close enough to the previous one. In fact, the formalization has served its purpose when an optimal match is obtained between the computed results and the data obtained from the investigated real-world system.

Soft-information processing in modeling and simulation generate the basic insight that categories are not absolutely clear cut; they belong to lesser or greater degree to that category. Hence, soft-computing systems break with the tradition, that the real world can be precisely and unambiguously characterized, which means divided into categories, and then manipulating these formalizations according to precise and formal rules. From the mathematical point of view soft-computing means multivaluedness or multivalence. Logical paradoxes and the Heisenberg uncertainty principle led to the development of multivalence, and in the 1930s quantum theorists allowed for indeterminacy by including a third or middle truth value in the bivalent logical framework. In 1965, the American systems scientist Zadeh, born 1921 in Baku in Azerbaijan, introduced the term fuzzy into the literature, and inaugurated a second wave of interest in multivalued mathematical structures – from systems to topologies.

6.2 Fuzzy Logic

The classical two-valued logic represents the meaning of a proposition as true or false. It is possible to combine simple propositions through the use of operators such as and, or, and not, into more complex ones. Whether this new resulting proposition is true or false depends not only on the truth of each simple proposition, but also on the connectives used. Several propositions may be used to perform reasoning, but the kind of logic we need is not classical logic, it is fuzzy logic. This is because classical logic can not represent a proposition with imprecise meaning. However, in fuzzy logic, which can be viewed as an extension of multivalued mathematical structures, a proposition can be true or false and have an intermediate value such as very true or less false. In general, fuzzy logic is concerned with formal principles of approximately reasoning, while classical two-valued logic is concerned with formal principles of reasoning, as systems scientist Zadeh mentioned in 1965, introduced the term fuzzy into the literature, and inaugurated a second wave of interest in multivalued mathematical structures – from systems to topologies – extending a bivalent indicator function iA of nonfuzzy subset A of X to a multivalued indicator or membership function $mA:X{\rightarrow}[0,1]$. This allows one to combine multivalued or fuzzy sets with the point wise operators of indicators functions for the large variety of fuzzy-logic systems.

Let X be a collection of objects denoted generically by x; that is, $X = \{x\}$. A fuzzy set A in X is a set of ordered pairs:

$$A = \left\{\left(x, f_A(x)\right)\middle| x \in X\right\},$$

(6.1)

where $f_A(x)$ is the generalized form of a membership function that associates with each $x \in X$ a real number in the interval $[0, 1]$. The value $f_A(x)$ indicates the grade of membership of x in A. When $f_A(x) = 1$, it means that x strongly belongs to A. As the value of $f_A(x)$ gets close to zero, the grade of membership of x in A becomes lower, while a value $f_A(x) = 0$ indicates x does not belong to A.

Example 6.1
A classed set A of real numbers greater than 8 can be expressed as

$$A = \{x | x > 8\},\qquad(6.2)$$

where there is a clear, unambiguous boundary 8 such that if x is greater than this number, then x belongs to the set A; otherwise x does not belong to the set.

If X is a collection of objects denoted generically by x, then a fuzzy set A in X is defined as a set of ordered pairs, as follows:

$$A = \{(x, \mu_A(x)) | x \in X\},\qquad(6.3)$$

where $\mu_A(x)$ is called the membership function for the fuzzy set A. The membership function maps each element of X to a membership grade (or membership value) between 0 and 1.

Example 6.2
Let $X = \{$Berlin, Hamburg, Stuttgart, Munich$\}$ be the set of cities one may chose to live in. The fuzzy set $C =$ desirable city to live in can be described as follows:

$$C = \{(Berlin, 0.75)(Hamburg, 0.5)(Stuttgart, 0.5)(Munich, 0.75)\}.\qquad(6.4)$$

Apparently the universe of discourse X is discrete and it contains nonordered objects, four big cities in Germany.

The membership functions should be designed in such a way that they model precisely observed values in the real world. However, often, it is difficult to derive membership functions with such characteristics. In practice, membership functions are often defined based on the data collected from measures or from experiments and a set of well-shaped functions. The most commonly used membership functions are:

- Linear membership function, which is specified by two parameters $\{a, b\}$ as follows:

$$\mu_L(x) = \begin{cases} 0; x \le a \\ \dfrac{x-a}{b-a}; a \le x \le b \\ 1; x \ge b \end{cases} \qquad (6.5)$$

- Piecewise linear or triangle membership function, which is specified by three parameters $\{a, b, c\}$ as follows:

$$\mu_P(x) = \begin{cases} 0; x \le a \\ \dfrac{x-a}{b-a}; a \le x \le b \\ \dfrac{c-x}{c-b}; b \le x \le c \\ 0; x \ge c \end{cases} \qquad (6.6)$$

- Trapezoidal membership function, which is specified by four parameters $\{a, b, c, d\}$ as follows:

$$\mu_T(x) = \begin{cases} 0; x \le a \\ \dfrac{x-a}{b-a}; a \le x \le b \\ 1; b \le x \le c \\ \dfrac{d-x}{d-c}; c \le x \le d \\ 0; x \ge d \end{cases} \qquad (6.7)$$

An alternative concise expression of the trapezoidal membership function using min and max is

$$\text{Trapezoidal}(a,b,c,d) = \max\left(\min\left(\frac{x-a}{b-a}, 1, \frac{d-x}{d-c}\right), 0\right). \qquad (6.8)$$

The parameters $\{a, b, c, d\}$ with $a < b \le c < d$ determine the x coordinates of the four corners of the underlying trapezoidal membership function.

- S-membership function, which is specified by three parameters $\{a, b, c\}$ as follows:

$$\mu_S(x) = \begin{cases} 0; x \le a \\ 2(\dfrac{x-a}{c-a})^2; a \le x \le b \\ 1 - 2(\dfrac{x-c}{c-a})^2; b \le x \le c \\ 1; x \ge c \end{cases} \qquad (6.9)$$

In the preceding functions a, b, c, and d are some constants, where

$$a \le b \le c \le d. \qquad (6.10)$$

In real-world systems the union and the intersection of two fuzzy sets A and B are often used, describing the model behavior.

The union of two fuzzy sets A and B is a fuzzy set C, written as $C = A \cup B$, whose membership function is defined by

$$\mu_U(x) = \max\{\mu_A(x), \mu_B(x)\}; x \in X, \qquad (6.11)$$

called the maximum of the fuzzy sets or T-conorm maximum operator, the model of which will use fuzzy sets as follows:

$$A \ or \ fuzzy \ set \ B. \qquad (6.12)$$

The intersection of two fuzzy sets A and B is a fuzzy set C, written as $C = A \cap B$, whose membership function is defined by

$$\mu_I(x) = \min\{\mu_A(x), \mu_B(x)\}; x \in X, \qquad (6.13)$$

called the minimum of the fuzzy sets or the T-norm minimum operator, the model of which will use fuzzy set as follows:

$$A \ and \ fuzzy \ set \ B. \qquad (6.14)$$

In practice, fuzzy relations are of importance. A fuzzy relation is a fuzzy set defined on the Cartesian product of crisp sets X_1, X_2, ..., X_n, where tuples $(x_1, x_2, ..., x_n)$ have varying degrees of membership within the relation. Two binary relations can be combined to produce a new binary relation, called composition. Given the binary relations $P(X,Y)$ and $Q(Y,Z)$, their composition $R(X,Z)$ is represented as

$$R(X,Z) = P(X,Y) \circ Q(Y,Z). \qquad (6.15)$$

The relation $R(X,Z)$ is a subset of the Cartesian product of X and Z, where $(x,z) \in R$ if and only if there exists at least one $y \in Y$ such that $(x,y) \in P$ and $(y,z) \in Q$.

There are different ways for calculation of the composition. The most well-known method is the max-min composition. Given $R(X,Z) = P(X,Y) \circ Q(Y,Z)$, the max-min composition can be thought of as the strength of the relational tie between elements X and Y. For this type of composition, the membership degree for each tuple $(x, y) \in R$ is defined as follows:

$$\mu_R(x,z) = \max_{y \in Y} \left\{ \min\left[\mu_P(x,y), \mu_Q(y,z) \right] \right\}; x \in X, z \in Z. \tag{6.16}$$

Two of the main concepts that play an important role in many applications of fuzzy logic are the concepts of linguistic variables and fuzzy if-then rules. A linguistic variable is a variable whose values are words or sentences in a language. The set of the linguistic values of a linguistic variable is called a term set. In general, fuzzy if-then rules can be represented as

$$if \; x_1 \; is \; A_1 \; and \; x_2 \; is \; A_2 \; and \; ... \; x_n \; is \; A_n \tag{6.17}$$

$$then \; y_1 \; is \; B_1 \; and \; y_2 \; is \; B_2 \; and \; ... \; y_n \; is \; B_m$$

where $x_1, x_2, \ldots, x_n, y_1, y_2, \ldots, y_m$ are linguistic variables, and $A_1, A_2, \ldots, A_n, B_1, B_2, \ldots, B_m$ are their respective linguistic values.

The importance of the calculus of fuzzy if-then rules results from the fact that much of the human knowledge lends itself to representation in the form of a hierarchy or fuzzy if-then rules. Furthermore, the inference mechanisms in the calculus of fuzzy if-then rules are relatively simple and in harmony with the modes of human reasoning, which are approximate rather than exact. As a consequence the calculus of fuzzy if-then rules is easy to master and apply.

6.2.1 Pure Fuzzy-Logic Systems

The basic configuration of a pure fuzzy-logic systems is based on the fuzzy rule base, which consists of a collection of fuzzy if-then rules, and the fuzzy-inference mechanisms that uses these fuzzy if-then rules in order to determine a mapping output universe of discourse $U \subset R^n$ to fuzzy sets in the output universe of discourse $V \subset R$ based on fuzzy-logic principles. Fuzzy if-then rules are of the following form:

$$R(k): \; IF \; x_1 \; is \; A_1^{(k)} \; AND...AND \; x_n \; is \; A_n^{(k)} \; THEN \; y \; is \; B^k , \tag{6.18}$$

where $A_i^{(k)}$ and $B_i^{(k)}$ are the respective fuzzy sets, $x = (x_1,...,x_n)^T \in U$ and $y \in V$ are input and output linguistic variables, respectively, and $k = 1,2,...,l$.

Each fuzzy if-then rule defines fuzzy set $A_1^{(k)} x_1, ..., x_n A_n^{(k)} \rightarrow B^{(k)}$ in the product space $U \times V$. Let A be an arbitrary fuzzy set in U, then the output determined by each fuzzy if-then rule of (6.1) is a fuzzy set $A \circ R^{(k)}$ in V whose membership function is

$$\mu_A(x) \circ R^{(k)}(\mu) = \sup x \mu U \left[\mu_A(x) \cdot \mu_A^{(k)} x_1, ..., x_n A_n^{(k)} \rightarrow B^{(k)}(x, y) \right], \qquad (6.19)$$

with $\cdot$ as operator that can be MIN, MAX, PRODUCT, or others. μ_A is used to represent the membership function of a fuzzy set A.

The final output of a pure fuzzy-logic system is a fuzzy set $A \circ (R_1, ..., R_n^{(k)})$ in V that is a combination of the respective fuzzy set. Hence a pure fuzzy-logic system constitutes the essential part of fuzzy-logic systems as a general framework in which linguistic information is quantified and fuzzy-logic principles are used to realize systematic use of linguistic information.

6.2.2 Takagi and Sugeno fuzzy logic systems

Instead of considering fuzzy if-then rules in the form of (6.1), Takagi and Sugeno proposed the Sugeno fuzzy model in an effort to develop a systematic approach to generating fuzzy rules from a given input-output data set, using fuzzy if-then rules in the following form:

$$L(k): \quad IF \; x_1 \; is \; A_1^{(k)} AND...AND \; x_n \; is \; A_n^{(k)} \qquad (6.20)$$

$$THEN \; y^{(k)} = c_0^{(k)} + c_1^{(k)} + ... c_n^{(k)} \; x_n$$

where $A_i^{(k)}$ are fuzzy sets, c_i are real-valued parameters, called crisp values, $y^{(k)}$ is the Takagi-Sugeno fuzzy system output due to the rule $L^{(k)}$, and $k=1,2,...l$. That is, they considered rules whose if-part is fuzzy but whose then-part is crisp. For a real-valued input vector $x = (x_1, ... , x_n)^T$, the output $y(x)$ of a Takagi Sugeno fuzzy systems is a weighted average of $y^{(k)}$:

$$y(x) = \frac{\displaystyle\sum_{k=1}^{l} w^{(k)} y^{(k)}}{\displaystyle\sum_{l} w^{(k)}}, \qquad (6.21)$$

where weight $w^{(k)}$ implies the overall truth value of the premise of rule $L^{(k)}$ for the input and is calculated as

$$w^{(k)} = \prod_{i=1}^{n} \mu A_i^{(k)}(x_i), \qquad (6.22)$$

which is shown in the rule base $L^{(k)}$ for example

$$L^{(1)} : if \ x_1 \ is \ A_1^{(1)} \ and \ ,..., \ x_n \ is \ A_n^{(1)} \ then \ y^{(1)} = c_0^{(1)} + c_1^{(1)} x_1 + ,..., c_n^{(1)} x_x \qquad (6.23)$$

$$w^{(1)} y^{(1)}$$

$$X \in U \qquad\qquad\qquad \rightarrow \ y(x) \in V$$

$$w^{(k)} y^{(k)}$$

$$L^{(k)} : if \ x_1 \ is \ A_1^{(k)} \ and \ ,..., \ x_n \ is \ A_n^{(k)} \ then \ y^{(k)} = c_0^{(k)} + c_1^{(k)} x_1 + ,..., c_n^{(k)} x_x$$

6.2.3 Fuzzy-Logic Systems with Fuzzification and Defuzzification

Compared to the pure fuzzy-logic system of Sect. 6.2.1 we now add a fuzzifier to the input and a defuzzifier to the output of the pure fuzzy-logic system. The fuzzifier maps crisp points in U to fuzzy sets in U, and the defuzzifier maps fuzzy sets in V to crisp points in V. The fuzzy rule base as well as the fuzzy-inference mechanisms are called the fuzzy-inference engine, and are the same as those shown in Fig. 6.1 for the pure fuzzy-logic system.

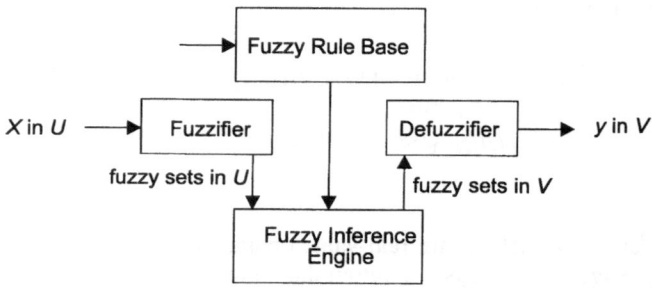

Fig. 6.1. Fuzzy system with fuzzifier and defuzzifier

6.2.4 Fuzzy Modeling of a Soccer Playing Mobile Robot

Soccer playing robots are being used in education to foster understanding and interest in artificial intelligence, multimodal systems, engineering, and science in order to solve complex problems through active learning comprehensive knowledge undergoing real-world applications. RoboCup, previously known as the Robot Cup World Initiative, has initiated a broad international program of research and education, which has the aim to promote artificial intelligence and intelligent robot research by providing a simple problem to be solved by the integration of different technologies and the collaboration of various resources. The idea to have soccer-playing mobile robots dates from 1993, with the first official conference and tournament held in Japan four years later in 1997. There are some mobile robots used for soccer, which are remotely controlled. These robots cannot be categorized as autonomous robots, as these systems do not behave autonomously. Mobile autonomous robots are systems that do not need any guidance. They are programmed to work in a specific environment and they work independently. As a mobile autonomous robot did not had any clear solutions to offer for industries they have been mostly developed in research laboratories for very specific task execution, like pathfinder. The very first mobile autonomous systems was developed in 1968 by the Stanford Research Institute. Today the Lego® Mindstorms™ are very convenient to use for mobile autonomous robots.

To represent how fuzzy logic can be used in a control system, an example for moving a soccer-playing mobile robot, is discussed. Fig. 6.2 shows the playground with several mobile robots.

From a more general point of view the position of the mobile robot movement is determined by two linguistic variables, the direction angle, denoted as α, and the distance from the object, which is for a soccer-playing robot the ball, denoted as x. The direction of the mobile robot movement, denoted as β, is determined by the angle of the wheels steering position. For a given initial mobile robot position within the specific area, the soccer playground, the goal for the mobile robot is to move toward the center of the ball. The desired final position is to let the mobile robot moving the ball on a track toward the goal. α, β, and x are the respective linguistic variables for this purpose. To each of these linguistic variables, a set of linguistic values can be assigned as follows:

Table 6.1. Distance variables

Distance x	Input variable
L	left side of the track toward the goal
C	center of the track toward the goal
R	right side of the track toward the goal

Table 6.2. Direction angle variables

Direction angle α	Input variable
N	north direction toward the goal
W	west direction toward the goal
S	south direction toward the goal
E	East direction toward the goal

Table 6.3. Wheels angle variables

Wheels angle β	Output variable
TR	turn right toward the goal
SF	straight forward toward the goal
TL	turn left toward the goal

As shown in Fig. 6.3, a range of numerical values can be assigned to each linguistic value of a linguistic variable. In Fig. 6.3, each graph, called a membership function, indicates the degree to which an input value belongs to a particular linguistic value. Such a degree of membership ranges from 0 to 1. The value 0 indicates no membership, and the value 1 represents full membership. Hence a value between 0 and 1 represents a partial membership.

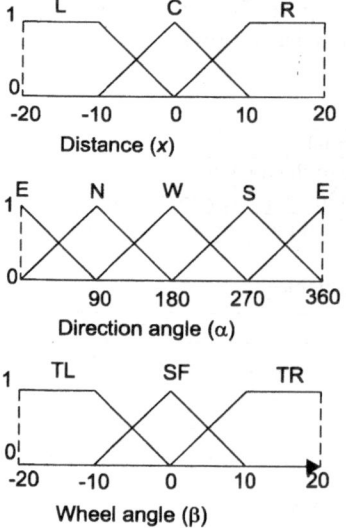

Fig. 6.2. Soccer playing mobile robots

Fig. 6.3. Membership functions for the distance x, direction angle α, and wheels angle β

Based on the steering and control concepts of the different chassis construction such as mobile robots with small maneuver space, or such as a 3-wheel-driving system, or such as a mobile robot system using 2 or 4 wheels, the driving system of the soccer playing robot uses 2 wheels as shown in Fig. 6.4.

Fig. 6.4. Steering-strategy and control of a specific chassis construction for a soccer playing mobile robot

The rules have to be defined, describing the dynamic behavior of the soccer playing robot. In general, each rule produces some output linguistic values based on some input linguistic values. In the case of the mobile robot, some of the rules can be defined as

$$if \ \alpha \ = \ N \ and \ x \ = \ L \ then \ \beta \ = \ TR \qquad (6.24)$$

$$if \ \alpha \ = \ N \ and \ x \ = \ C \ then \ \beta \ = \ SF$$

$$if \ \alpha \ = \ N \ and \ x \ = \ R \ then \ \beta \ = \ TL$$

These rules can be extended to consider all the possible values for α; thus there will be 12 rules in all, which can be represented in the fuzzy associative memory (FAM), shown in Fig. 6.5.

Table 6.4. FAM set of rules for determining the mobile robot movement

		Distance x		
		L	C	R
	N	TR	SF	TL
Direction α	W	TR	TR	SF
	S	TL	TL	TR
	E	SF	TL	TL

For given input values for x and α, the fuzzy-logic controller can determine an output value for β, the angle of the steering wheel(s). For this purpose, for each input value the fuzzy controller determines the membership degree of its corresponding linguistic values. As a next step, for each rule, as shown for example in Table 6.1, the minimum of the membership degrees of its antecedents is chosen as a membership degree for the rules consequent, which is considered as a weight for the rules consequent. When there is more than one membership degree for a consequent, the MAXIMUM operator is chosen for that consequent. Hence the membership degree is assigned to each linguistic value. If a crisp output is required from the fuzzy rule base rather than the fuzzy output set, a process called defuzzification is used to compress this information. The crisp output is generally obtained using a mean of maxima or a center of gravity defuzzification strategy. The most widely adopted method for defuzzifying a fuzzy set A of a universe of discourse Z, is the centroid defuzzification or center of gravity method, which is based on the centroid of area z_{COA}

$$z_{COA} = \frac{\int_Z \mu_A(z)z\,dz}{\int_Z \mu_A(z)\,dz} , \qquad (6.25)$$

where $\mu_A(z)$ is the aggregated output of the membership function and z_{COA} is the control output, which equals the fuzzy centroid of A, where the limits of integration correspond to the entire universe of discourse Z of angular values of the steering wheel(s) velocity values.

The center of gravity method, COG, provides a weighted average of all linguistic output values. A simplified calculation is as follows:

$$COG = \frac{\sum_{i=1}^{n} c_i \cdot L_i}{\sum_{i=1}^{n} L} , \qquad (6.26)$$

where the L_i are the weights of linguistic output values and the c_i are the weighting factors.

As an illustration of the information process between the fuzzification and defuzzification, Fig. 6.6 shows the signal flow through a continuous fuzzy-logic system using the center of gravity defuzzification method. There exist p multivariate fuzzy input sets and q univariate fuzzy output sets.

Example 6.3
Let the starting point of the mobile robot be at direction $x = -10.0$, and the direction angle $\alpha = 90°$. For these initial values the membership degree of the linguistic input values are for the distance $x = -10$: $\mu_L = 1.0$; $\mu_C = 0$; $\mu_R = 0$, and for the direction angle $\alpha = 90°$: $\mu_N = 1.0$; $\mu_W = 0$; $\mu_S = 0$; $\mu_E = 0$. Combining distance and direction, as shown in the fuzzy associative memory in Fig. 6.5, with the respective membership degree μ_i for each rule consequently results in the membership matrix, shown in Fig. 6.7.

Table 6.5. Membership degree for each FAM rule based on the initial conditions

	L	C	R
N	1	0	1
S	0	0	0
W	0	0	0

E	0	0	0

The system output-value calculation can be evaluated by using the center of gravity method, and the MAX operator, while there is more than one membership degree, as follows:

$$COG= \frac{((20.0 \cdot MAX(\mu_{TR}(\cdot)) + (-20.0 \cdot MAX(\mu_{SF}(\cdot)) + (0.0 \cdot MAX(\mu_{TL}(\cdot)))}{MAX(\mu_{TR}(\cdot)) + MAX(\mu_{SF}(\cdot)) + MAX(\mu_{TL}(\cdot))} \quad (6.27)$$

In this formula the maximum degree of each of the four membership degrees is chosen, 1.0 for TR, and 0.0 and 0.0 for ST and TL, respectively. Based on these degrees we finally receive the mobile robot system output value as follows:

$$COG = \frac{(20.0 \cdot 1.0) + (-20.0 \cdot 0.0) + (0.0 \cdot 0.0)}{1.0} = 20.0 . \quad (6.28)$$

That is, that the wheels of the mobile robot will turn to the right with an angle of 20.0°. The robot moves for a short distance and then the process repeats for the new position.

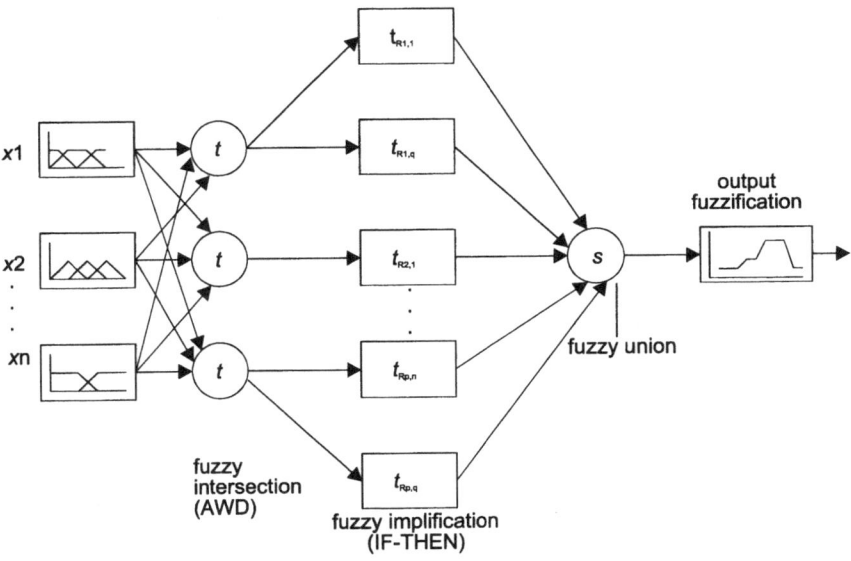

Fig. 6.6. Information flow through a continuous fuzzy system with p multivariate fuzzy input sets and q univariate fuzzy output sets

Example 6.4
Let the slope of a terrain range between –45 and +45 degrees, which can be divided into several memberships in between large negative and large positive. We will further assume that the terrain can vary between very rough, rough, moderate, and smooth, and the output speed of the fuzzy-logic system may range between 0 and 20 mph, divided into very slow,

slow, medium, fast, and very fast. The rules of the fuzzy-logic mobile robot system are as follows:

> **if** slope is large-positive **and** terrain is very-rough
>
> **then** speed of the robot is very-slow
>
> **if** slope is large-positive **and** terrain is rough
>
> **then** speed of the robot is slow
>
> **if** slope is large-positive **and** terrain is moderate
>
> **then** speed of the robot is medium
>
> **if** slope is large-positive **and** terrain is smooth
>
> **then** speed of the robot is high
>
> **if** slope is positive **and** terrain is very-rough
>
> **then** speed of the robot is slow
>
> **if** speed is positive **and** terrain is rough
>
> **then** speed of the robot is medium
>
> **if** slope is positive **and** terrain is moderate
>
> **then** speed of the robot is slow
>
> **if** ...

Using the fuzzy logic toolbox of MATLAB SIMULINK, one can customize the fuzzy system adapted to the needs of the specific application. Customizing the membership function is based on using an M-file. To define a membership function one has to follow the guideline, given in "Fuzzy Logic Toolbox, Users Guide Version 2, The Mathworks". Constructing rules can be done using the graphical Rule Editor interface, as described in the "Fuzzy Logic Toolbox, Users Guide Version 2". For example to insert rules in the Rule Editor proceed as follows:

> **large-positive** under the variable **slope**
> **very-rough** under the variable **terrain**
> the ratio button **and** in the **connection block**
> **very-slow** under the output variable **speed**

The resulting rule is:

> **if** slope is large-positive **and** terrain is very-rough
> **then** speed of the robot is very-slow.

6.2.5 Fuzzy Modeling of a Wastewater Treatment Plant*

Wastewater treatment plants are constructed on data, obtained from a priori knowledge, on-line, and laboratory side measures, as well as from tank models. In 1983 the IAWPRC (International Association on Water Pollution Research and Control) formed a task group to facilitate the application of mathematical models for the design and operation of biological wastewater treatment plants. The models, which are developed for single-activated sludge systems performing carbon oxidation, nitrification, and denitrification, are representing complex biological processes, as shown in Fig. 6.8 (see Example 4.15 in Sect. 4.3.5).

The model includes several fundamental processes such as aerobic growth of heterotrophic biomass – meaning different kinetic states – anoxic growth of heterotrophic biomass, aerobic growth of autothophic biomass. The increase in water quality demands results in the construction of different reactor types for waste water treatment plants, the most common ones are the three-stage reactor types, as shown in Fig. 6.9.

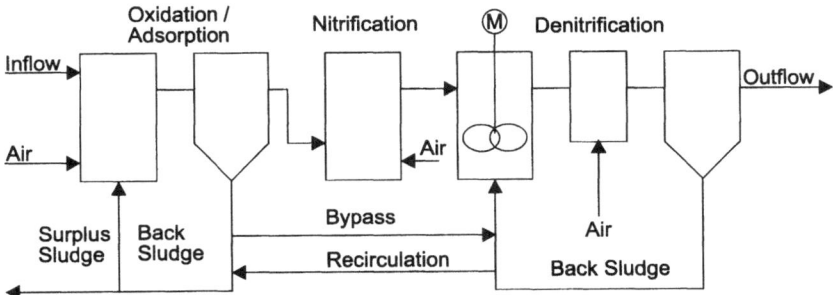

Fig. 6.8. Model of a wastewater treatment plant with oxidation, nitrification, and denitrification processes

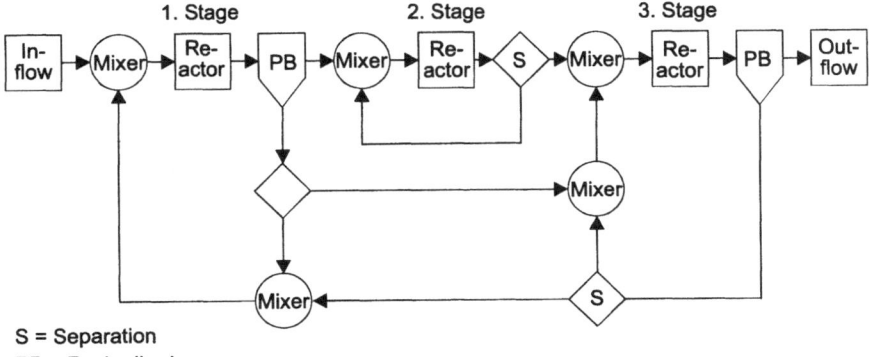

S = Separation
PB = Packedbed

Fig. 6.9. Model of a wastewater treatment plant with three-stage reactor type

While short-term, mid-term and long-term behavior of wastewater treatment plants influence the measurement accuracy as well as the measures themselves, modeling and simulation are of importance when designing optimized wastewater treatment plants. Qualitative modeling, which describes the relations of the several processes of a wastewater treatment plant in the context of a fuzzy-logic system approach, is a way to handle vagueness and ill-defined systems. The fuzzy-logic approach covers fuzzification and defuzzification. Inferences can be deduced from the analysis of the real-world process. The rule-based reasoning system represents an adequate state transition description of the complex nonlinear system behavior, which can not be solved using common algebraic equations. Using fuzzy-logic control systems, the design and operation of wastewater treatment plans can become more efficient. The goal of using fuzzy-knowledge-based controllers in single-sludge wastewater treatment plants is based on the purpose of how to achieve the following objectives

- Removing significant errors in the systems output by appropriate adjustment of the systems control output
- Preventing the systems output from exceeding constraints
- Producing control in a close vicinity of the set-point-area

The rule base of a fuzzy-knowledge-based controller can be divided into two groups of rules:

- Active rule groups, i.e., incremental change in control output, determined by the rules applied for every time slot
- Constraint rules, which become active only when the systems output is within the constraint bounds

Fuzzy if-then rules can be represented in the form shown in Table 6.1. Based on real-world measurements, a fuzzy-logic model of a gas furnance can be derived, the input is the methane feed rate, and the output is the CO_2 concentration in the outlet gases which can consist of rule sets in the following semantic notation:

if the current CO_2 concentration is medium
and the previous methame feed rate was low
then the next CO_2 concentration will be high

In contrast for a fuzzy-knowledge-based controller the if-then rules for this process result in:

if the previous CO_2 concentration was medium
and the current CO_2 concentration is high
then change the methane feed rate with a small increase

Next, similar to an expert system, a set of rules has to be defined. In general each rule produces some output linguistic values based on some input linguistic values. For example, this rule can be transformed into a general semantic fuzzy-controller syntax. A simplified representation of which is as follows:

if e is low and edot is zero then system output is low

*if e medium and edot is zero **then** system output is medium*

*if e is high and edot is positive **then** system output is medium*

where e denotes the error, and *edot* the change of error, for the chosen PI-fuzzy-knowledge based controller.

Based on these rules, a fuzzy-knowledge-based controller for optimal control of a single-sludged wastewater treatment plants can be derived and tested based on modeling and simulation in order to keep the system optimally controlled due to its different boundary conditions. The single-sludge wastewater-treatment plant models, expanded by the fuzzy-logic-knowledge based controller, can be integrated in an object-oriented modeling and simulation environment, meaning that it will be possible to create new models using inheritance mechanisms, applied to the object-oriented models data base, suited to a specific problem by means of inheritance and refinement. Based on such a data base, an object-oriented modeling and simulation toolbox has to be created that supports engineers in the design process of adaptive control of single-sludge wastewater-treatment plants, with embedded fuzzy-logic controller systems, i.e. to solve problems dealing with wastewater-treatment plants like aerobic bulking.

6.2.6 Fuzzy-Logic Control System*

Fuzzy control is the industrial application domain of fuzzy-logic. The basic architecture of a fuzzy controller consists of an inference engine that operates on proportional and derivative input signals, and produces an output that is either the control action or the calculated change in control. Thus it works as either a P (proportional), a PD (proportional and derivative), a PI (proportional and integral), or a PID (proportional, integral and derivative) controller. For a learning fuzzy controller, the learning layer, which is built around the fuzzy controller, acts as a critic, assessing the current state of the plant and recommending changes to the control signal via the performance index and the model.

The pure fuzzy controller operates by measuring the current deviation of the plant from the desired set point, producing an error, and the change in error, the derivative, is given by the difference between two successive errors. Other inputs, such as an integral term, can be included as inputs to the rule base, although it is sufficient just to consider the two variables. Both inputs are multiplied by the scaling or gain factors GE and GCE, which can be used to alter the gain of the system, where GE represents the gain of the proportional part of the controller, and GCE represent its integral part, while E and C represent the input signal the capacitor. respectively. The input membership functions for the pure fuzzy controllers are defined on a discrete input space, hence a quantization layer is necessary, which maps each scaled input to an integer, and fuzzy input membership functions are then defined on this space. A control action is calculated, e.g. based on a fuzzy- reasoning algorithm, after being scaled by the factor GU, the control is applied to the plant, by means of look-up tables, and any old rules are modified when the plant not operating is required. This process can be repeated for a range of different step inputs until the plant is operating satisfactorily. The pure fuzzy controller with an add-on adaptive learning mechanism is shown in Fig. 6.10.

The fuzzy controller is able to control systems that previously could only be controlled by skilled operators such as in power plant systems, process engineering control, chemical engineering control, etc.

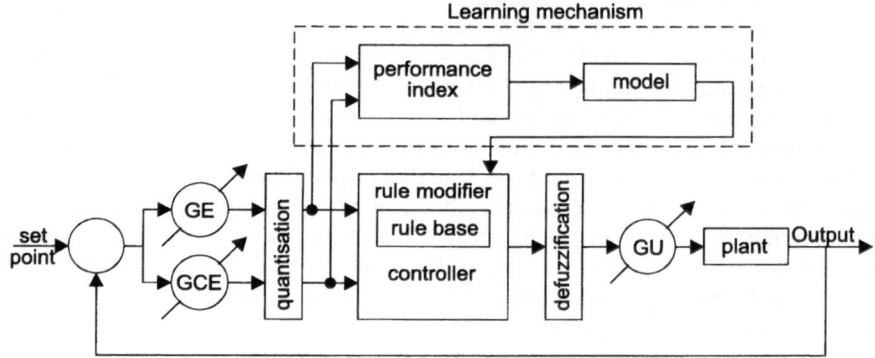

Fig. 6.10. Pure fuzzy-control architecture with add-on learning mechanism

Example 6.5
A popular nonlinear fuzzy control problem is the inverted pendulum. Let the pendulum be positioned by a joint on top of a mobile system, as shown in Fig. 6.11, where l is the length of the pendulum shaft, m_s is the mass of the pendulum shaft, m is the mass of the mobile system, $F(t)$ is the pulling force of the mobile system, and Θ is the angle. The task is to adjust the cost function of the mobile system to balance the inverted pendulum in two dimensions.

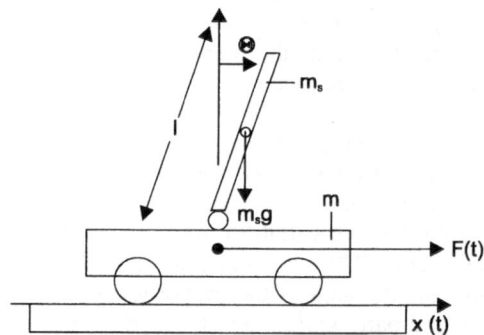

Fig. 6.11. Fuzzy control applied to the inverted pendulum problem

The classical system model for control of the direction of the movement can be described using differential equations of second order, as follows:

$$l \cdot \ddot{\Theta} = F_V \cdot \sin \dot{\Theta} - F_H \cdot l \cdot \cos \dot{\Theta} \qquad (6.29)$$

$$F(t) - F_H = m \cdot \ddot{x}$$

where F_v is the vertical joint force, F_H is the horizontal joint force, $F(t)$ is the pulling force of the mobile system, and m_S is the mass of the pendulum and x is the position.

There are two state fuzzy variables and one control fuzzy variable. The first state fuzzy variable is the angle Θ the pendulum shaft holds with the vertical. Hence zero angle corresponds to the vertical position, and positive angles are to the right of the vertical and negative angles are to the left. The second fuzzy variable forms the angular velocity $\Delta \Theta$. In practice we approximate the instantaneous angular velocity $\Delta \Theta$ as the difference between the present angle measurement Θ_t and the previous angle measurement Θ_{t-1}, yields

$$\Delta \Theta_t = \Theta_t - \Theta_{t-1}. \qquad (6.30)$$

The control fuzzy variable is the driving system current or angular velocity v_t, which can be positive or negative. Let the pendulum fall to the right, the driving-system velocity may be negative to compensate, and when the pendulum falls to the left, the driving system velocity may be positive. If the pendulum successfully balances at the vertical, the driving-system velocity will be zero.

Let the real line R be the universe of discourse of the fuzzy variables. In practice we restrict each universe of discourse to a comparatively small interval, such as [−90, 90] for the pendulum angle, centered about zero, means quantize each universe of discourse into several overlapping fuzzy-set values. Consider that the fuzzy variables can be positive, zero, or negative, we can quantize the magnitudes of the fuzzy variables finely or coarsely. Suppose we quantize the magnitudes as small, medium, and large yielding in seven fuzzy-set values:

> *NL:* Negative Large
> *MM:* Negative Medium
> *NS:* Negative Small
> *ZE:* Zero
> *PL:* Positive Large
> *PM:* Positive Medium
> *PS:* Positive Small

The fuzzy rules of the inverted pendulum are triples, such as *NM, ZE, PM*. They describe how to modify the control variable for observed values of the pendulum state variables. Hence we can interpret the triple *NM, ZE, PM* as the set-level implication

> ***if*** *the pendulum angle Θ is negative but medium*
>
> ***and*** *the angular velocity $\Delta \Theta$ is about zero,*
>
> ***then*** *the driving velocity has to be positive but medium*

which can be expressed in the short linguistic form

> ***if*** $\Theta = NM$ ***and*** $\Delta \Theta = ZE$ ***then*** $v = PM,$

and the steady-state rule can be described by the triple *ZE, ZE, ZE*.

Control problems often require nulling a scalar error measure, which means that control can be done by nulling the norms of the system error vector and the error-velocity vectors.

The error measure can be the angle and the angular velocity, hence $\Theta \approx e(t)$, or, $\Delta\Theta \approx \Delta e(t)$. Adaptive error nulling extends the classical fuzzy-logic methodology to nonlinear estimation and control, introduced as a learning mechanism, as shown in Fig. 6.11.

The inverted pendulum can be described by a 7 by 7 matrix with linguistic fuzzy-set entries, where the columns are indexed by the seven fuzzy sets that quantize the angle Θ, and the rows are indexed by the seven fuzzy sets that quantize the angular velocity $\Delta\Theta$. The 49 entries of the 7 by 7 matrix represent a subset of 7^3 possible two antecedent fuzzy rules, but in practice most of the entries are blank.

Let the angle Θ be zero but the pendulum moves. If the angular velocity $\Delta\Theta$ is negative, the pendulum will overshoot to the left, which means that the driving system velocity should be positive to compensate. If the angular velocity $\Delta\Theta$ is positive, the driving system velocity should be negative. The greater the angular velocity is in magnitude, the greater the driving system velocity in its magnitude. Positive Θ values with negative $\Delta\Theta$ values may result in negative driving system current values, since the pendulum heads toward the vertical, which results in the triple *PS, NS, NS*. Symmetrically, negative Θ values with positive $\Delta\Theta$ values may result in positive driving-system current values, which results in the triple *NS, PS, PS*. Finally we obtain 15 fuzzy-rules altogether, which in practice, successfully balance the inverted pendulum. We can represent the 15 fuzzy-rules as the 7 by 7 linguistic matrix:

Table 6.6. Rule base of the inverted pendulum example

Θ		*NL*	*NM*	*NS*	*ZE*	*PS*	*PM*	*PL*
$\Delta\Theta$	*NL*				*PL*			
	NM				*PM*			
	NS				*PS*	*NS*		
	ZE	*PL*	*PM*	*PS*	*ZE*	*NS*		
	PS			*PS*	*NS*			
	PM				*NM*			
	PL				*NL*			

Suppose the current pendulum angle $\Theta = 15$ degrees and the angular velocity $\Delta\Theta = -10$, the corresponding driving system current value $v = F(\Theta, \Delta\Theta) = (15, -10)$. The fuzzy-rule notation *ZE, ZE, ZE* implicitly assumes that we combine antecedent fuzzy sets conjunctively with *and*. Hence this data satisfies as the antecedent of the fuzzy rule *ZE, ZE, ZE*

$$\min(m_{ZE}^{\Theta}(\Theta), \ m_{ZE}^{\Delta\Theta}(\Delta\Theta)) = \min(m_{ZE}^{\Theta}(15), \ m_{ZE}^{\Delta\Theta}(-10) = \qquad (6.31)$$

$$\min(zero\ angle\ value, \ zero\ angular-velocity)$$

where the respective values of the angle and angular velocity for the angular value and the angular-velocity datum can be found from the respective triangular fuzzy-set membership functions.

6.3 Neural Nets*

Neural networks address the issue of effective information organization and processing. Since biological brains are being interpreted, from a more mechanistically
point of view, as working examples of massively parallel, densely interconnected,
self-organizing computational networks, they represent an ideal prototype with
which special-purpose simulation models can be run.

The basic brain processing unit is the nerve cell, called a neuron, as shown in
Fig. 6.12a. An artificial neural network consists, compared to the neuron, of input
variables and weighting factors, activation layers, and output variables, as shown
in Fig. 6.12b. The neurological pendant of the inputs are called dendrites as part of
the anterior motor-neuron, extend for one-half to one millimeter in all directions
from the neuronal soma. The dendrites can receive signals from a fairly large spatial area around the motor-neuron. This provides vast opportunity for summation
of signals from many separate so-called presynaptic neurons. The neurological
pendant to the weighting functions are called synapses. The synapse could be interpreted as the juncture between one neuron and the next, based on three major
parts, the soma, the main body of the neuron; a single axon, which extends from
the soma into the peripheral nerve; and the dendrites, which are thin projections of
the soma that extend up to one millimeter, into the surrounding areas of the cord.
The output has its neurological pendant in the axon, which is the central core of a
nerve fiber.

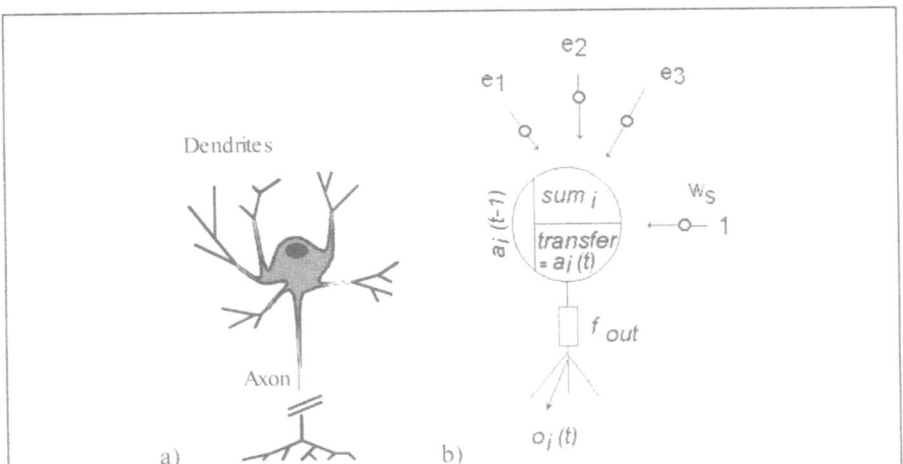

Fig. 6.12. Biological neuron (a) and artificial neuron (b)

Fig. 6.12b shows a neuron with three inputs, e_1, e_2, and e_3 which form the synapses of an artificial neuron that receive an activation x_i with a specific strength w_i
from another artificial neuron, which will be part of the summaries process of the

output, the axon. Each input can take only one of the two binary values, +1 or −1. In addition, the neuron also has a constant input w_S that has the constant value +1 all the time. The weight w_i corresponding to the input e_1 is w_1, and so on. The weight corresponding to the input w_S is w_B and is called the bias weight, which is used for threshold-level control. All inputs are multiplied by their weights and then summed to determine the activation level of the neuron, the real value a. For the output of this neuron, o variables, to be connected to the input of other neurons, the real value a needs to be converted into a binary value of +1 or −1. A hard-limiting activation function can be used to do this conversion. The output of their neuron is computed with a value of +1 if a is greater then zero and a value of −1, otherwise.

The basic concept results in an input vector $e = (e_1, ..., e_n)^T$, a weighting vector $w = (w_1, .., w_n)^T$ and the resulting activity as the sum of the weighted input, which could be assigned as activity function o:

$$o(w,e) = \sum_j w_j e_j = w^T e .$$ (6.32)

Modeling the threshold relation can be formalized as follows:

$$o(w,e) = w^T e - T ,$$ (6.33)

where T indicates the threshold. With

$$e = (e_1,...,e_n,1)^T ,$$ (6.34)

and

$$w = (w_1,...,w_n,-T)^T ,$$ (6.35)

this can be written as

$$o(w,e) = \sum_j w_j e_j - T = (w,...,w,-T)(e,...,e,T) = w^T e ,$$ (6.36)

which can be rearranged as follows

$$o = w^{(0)} + \sum_i w_1^{(1)} e_i ,$$ (6.37)

where the power $^{(i)}$ notation indicates the correlation of the e components. Modeling high-order synapses then can directly be derived from the equations above as follows:

$$o = w^{(0)} + \sum_i w_i^{(1)} e_i + \sum_{i,j,k} w_{i,j,k}^{(2)} e_j e_k + \sum_{i,j,k,l} w_{i,j,k,l}^{(3)} e_j e_k e_l + \dots \ . \tag{6.38}$$

This type of artificial neuron are called sigma-pi units.

Let us consider the use of a two input artificial neurons representing some elementary logic functions such as AND and OR. This requires finding the weights w_S, w_1, and w_2, so that the neuron can represent the desired mapping. By choosing $w_S = -1.5$, $w_1 = +1$, and $w_2 = +1$, we find the neuron represents an AND function, and choosing $w_S = 1.5$, $w_1 = +1$, and $w_2 = +1$, we find the neuron represents an OR function, which can be seen from the truth table in Table 6.7.

Table 6.7. Truth tables for the a) AND and b) OR Functions

e_1	e_2	o_{AND}	e_1	e_2	o_{OR}
-1	-1	-1	-1	-1	-1
-1	$+1$	-1	-1	$+1$	$+1$
$+1$	-1	-1	$+1$	-1	$+1$
$+1$	$+1$	$+1$	$+1$	$+1$	$+1$
	(a)			(b)	

A single divides the input patterns into two classes, one for which the output is $+1$ and the other for which the output is -1. The distinction between outputs $+1$ and -1 occurs when the weighted sum a equals 0. For the OR function we have

$$e_1 w_1 + e_2 w_2 + w_B w_S = 0 , \tag{6.38}$$

$$e_1 w_1 + e_2 w_2 + w_S = 0 , \tag{6.39}$$

for the AND function.

$$e_1 + e_2 - 1.5 = 0 , \tag{6.40}$$

which is the equation of a straight line.

Example: 6.6

A two-layer perceptron model to implement the XOR function with three neurons, the input layer N_1 and N_2, and the output layer N_3, has nine variable weights, as follows:

Table 6.8. Weight variables of the two layer perceptron model

w_{11}	w_{12}	w_{13}
w_{21}	w_{22}	w_{23}
w_{SN1}	w_{SN2}	w_{SN3}

the output of neuron N_i is

$$O_1 = \sigma\left[w_{11} \cdot e_1 + w_{21} \cdot e_2 + (-1.0) \cdot w_{SN1}\right], \qquad (6.41)$$

with σ as the sigmoid function, where -1.0 indicates that N_1 is less than zero. Let $e_1 = e_2 = 0.1$, and the initial weights are

Table 6.9. Valued weight variables of the two layer perceptron model

$w_{11} = -0.1$	$w_{12} = 0.2$	$w_{13} = -0.05$
$w_{21} = -0.2$	$w_{22} = 0.1$	$w_{23} = 0.2$
$w_{SN1} = 0.2$	$w_{SN2} = -0.2$	$w_{SN3} = 0.1$

then we obtain:

$$O_1 = \sigma\left[(-0.1) \cdot 0.1 + (-0.2) \cdot 0.1 + (-1.0) \cdot 0.2\right], \qquad (6.42)$$

$$O_1 = \sigma\left[(-0.01) + (-0.02) + (-0.2)\right] = \sigma[-0,23] = \frac{1}{(1+e^{0.23})} . \qquad (6.43)$$

Similarly, the output of neuron N_2 is:

$$O_2 = \sigma\left[w_{12} \cdot e_1 + w_{22} \cdot e_2 + (-1.0) \cdot w_{SN2}\right], \qquad (6.44)$$

$$O_2 = \sigma\left[(0.2) \cdot 0.1 + (0.1) \cdot 0.1 + (-1.0) \cdot (-0.2)\right], \qquad (6.45)$$

$$O_2 = \sigma\left[(0.02) + (0.01) + (0.2)\right] = \sigma[0,23] = \frac{1}{(1+e^{-0.23})} , \qquad (6.46)$$

Finally, the output of the network is

$$O_3 = \sigma\left[w_{13} \cdot e_1 + w_{23} \cdot e_2 + (-1.0) \cdot w_{SN3}\right], \qquad (6.47)$$

$$O_3 = \sigma\left[(-0.05) \cdot 0.1 + (0.2) \cdot 0.1 + (-1.0) \cdot (0.1)\right], \qquad (6.48)$$

$$O_1 = \sigma\left[(-0.005) + (0.02) + (-0.1)\right] = \sigma[-0,085] = \frac{1}{(1+e^{0.085})} . \qquad (6.49)$$

The preceding functions represent a simple mechanism to calculate the output of the neural network. The error signals can now be calculated starting from the outermost layer. In this case neuron N_3 is an output neuron. The error signal for this neuron is as follows:

$$\varepsilon_3 = o_3(1-o_3)(t-o_3),$$ (6.50)

where o_3 is the actual output and t is the target output. Therefore,

$$\varepsilon_3 = \frac{1}{(1+e^{0.085})} \cdot (1-\frac{1}{(1+e^{0.085})}) \cdot (0.1-\frac{1}{(1+e^{0.085})}).$$ (6.51)

We can now update the weights of w_{13}, w_{23}, and $w_{SN\,3}$.

$$w_{13} = w_{13} + \rho o_1 \varepsilon_3$$ (6.52)

$$w_{13} - 0.05 + [0.5 \cdot \frac{1}{1+e^{0.25}} \cdot \frac{1}{1+e^{0.085}} \cdot (1-\frac{1}{1+e^{0.085}}) \cdot (0.1-\frac{1}{1+e^{0.085}})],$$ (6.53)

$$w_{23} = w_{23} + \rho o_2 \varepsilon_2,$$ (6.54)

$$w_{33} = w_{33} + \rho o_3 \varepsilon_3.$$ (6.55)

Calculating the weight changes for the hidden layer, the error must be propagated back toward the input. As such, the error signal for neuron N_1 becomes

$$\varepsilon_1 = o_1(1-o_1)(\varepsilon_3 w_{13}).$$ (6.56)

The error ε_1 can be used to update the weights coming from the inputs to the neuron N_1, which results in the new weights as follows:

$$w_{11} = w_{11} + \rho \varepsilon_1 e_1,$$ (6.57)

$$w_{21} = w_{21} + \rho \varepsilon_1 e_2,$$ (6.58)

$$w_{SN1} = w_{SN1} + \rho \varepsilon_1(-1.0).$$ (6.59)

The error signal for neuron N_2 becomes:

$$\varepsilon_2 = o_2(1-o_2)(\varepsilon_3 w_{23}).$$ (6.60)

Finally, the weights w_{12}, w_{22}, and w_{SN2} become

$$w_{12} = w_{12} + \rho \varepsilon_2 e_1 , \tag{6.61}$$

$$w_{22} = w_{22} + \rho \varepsilon_2 e_2 , \tag{6.62}$$

$$w_{SN2} = w_{SN2} + \rho \varepsilon_2 (-1.0) . \tag{6.63}$$

Using the preceding weights, a new output value for the neural network can be calculated. This type of network calculation is called a back-propagation algorithm. Multilayer perceptrons and back-propagation networks are currently the most widely used neural networks. Consider an n-input, single-output network with p nodes in the hidden layer, as shown in Fig. 6.13. The $(p+1)$-dimensional weight vector associated with the output node is denoted by w_0, and the $(n+1)$-dimensional weight vector associated with the i-th node in the hidden layer is given by w_i for i = 1, 2, ..., p. Each node has connections with all the nodes in the previous layer plus a bias term. The $(n+1)$-dimensional network input vector at time t is composed of the bias term plus the n-dimensional input vector and is denoted by $x(t)$ and the network output is $y(t)$. The output of the hidden layer nodes is denoted by the $(p+1)$-dimensional vector $a(t)$ and the output of the i-th node in the hidden layer is

$$a_i(t) = \begin{cases} 1 & if \ i \ = \ 0 \\ f(u_i(t)) & otherwise \end{cases} \tag{6.64}$$

$$u_i(t) = \sum_{j=0}^{n} x_j(t) \cdot w_{i,j} = x^T(t) \cdot w_i \quad for \ i \ = \ 1,2,...,p. \tag{6.65}$$

The dependence of the weights on time is neglected for simplification and when the network has an output node with a linear transfer function the output will be

$$y(t) = \sum_{j=0}^{p} a_j(t) \cdot w_{0,j} = a^T(t) w_0 . \tag{6.66}$$

Example 6.7
We now introduce the adaptation of fuzzy systems by using sensitized neuronal nets. The idea of sensitization of neuronal nets means condition of a neuronal net by well-defined distinguishable data sets in order to deepen and to enlarge the stored information in the context of chunking. Chunking means adaptation onto a new fact or an unknown situation based on knowledge in the form of facts or models. Transforming these facts to neuronal nets means that at the very first a net has to learn the respective basic concept. To prevent this the neuronal net includes typical output ranges in its classification behavior, it is neces-

sary to normalize the input data set, by using preprocessing algorithms. For this reason we add the wheel angle β of Sect. 6.2.4 to a sensitized neural-net-trained neural mobile system controller as an input. The controller network contained 24 hidden neurons. The controller network was controlled with 52 training samples from the fuzzy controller; 26 samples for the left half of the plane and 26 samples for the right half of the plane. The training requires more than 100000 iterations. The sensitized neuronal nets classifier can separate all trained states, representing a powerful concept of weaker evolutionary states to be trained, as shown in Fig. 6.14. In Fig. 6.14 time differences between the early warning of trajectory errors of a common neural classifier and a sensitized neural network is shown. Both nets have had the same warning criteria, setting an alarm when the probability for a trajectory error is higher then 85%. It can be seen that a sensitized neural classifier shortens the alarm time by a factor 5, as the net classifies the evolutionary state of the beginning of a trajectory error rather early.

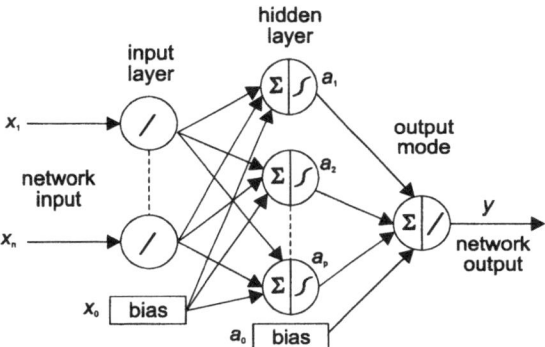

Fig. 6.13. Three-layer network with linear input nodes and a linear output node

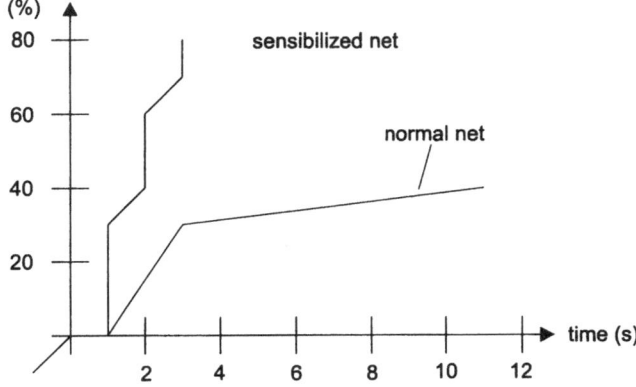

Fig. 6.14. Difference between early warning of a trajectory error of a sensitized and a normal neural net.

6.4 References and Further Reading

Brown M, Harris C, (1994), Neurofuzzy Adaptive Modelling and Control, Prentice Hall, New York, London, Toronto, Sydney, Tokyo, Singapore

Demuth HB, Beale M, (2000), Neural Network Toolbox, The Math Works Inc.

Hagan MT, Demuth HB, Beale M, (1996), Neural Network Design, PWS Publishing Company, Boston

Jang JSR, Sun CT, Mizutani E, (1997), Neuro-Fuzzy and Soft Computing, Prentice Hall

Jungblut J, (1998), Modeling and simulation of biological wastewater-treatment plants based on newest development of wastewater-treatment methods, (in German), PhD Thesis, TU Clausthal

Kaufmann A, Gupta MM, (1991), Introduction to Fuzzy Arithmetic, Van Nostrand Reinhold, New York

Kosko B, (1992), Neural Networks and Fuzzy Systems, Prentice Hall, Englewood Cliffs

Zadeh LA, (1992), The Calculus of Fuzzy If/Then Rules, In: Fuzzy Logic, pp. 84–94, Ed. B. Reusch, Springer, Berlin, Heidelberg, New York

Technical Manuals
Fuzzy Logic Toolbox, Users Guide, Version 2, The Math Works Inc., 2001

6.5 Exercises

6.1   What is meant by the term fuzzy set?
6.2   What is meant by the term membership function?
6.3   Describe the maximum operator by using a simple example.
6.4   Describe the minimum operator by using a simple example.
6.5   Describe the Takagi and Sugeno fuzzy system.
6.6   Describe fuzzyfication by using a simple example.
6.7   Describe defuzzyfication by using a simple example.
6.8   Describe the fuzzy inference engine by using a simple example.
6.9   What is meant by the term fuzzy control?
6.10 Describe the fuzzy control system by using a simple example.
6.11 What is meant by the term fuzzy rule base?
6.12 Describe the fuzzy rule base system by using a simple example.
6.13 What is meant by the term center of gravity?
6.14 Give an example of the center of gravity method.
6.15 What is meant by the term neural network?
6.16 What is meant by the term activity function?
6.17 What is meant by the term perceptron?
6.18 What is meant by the term back-propagation?
6.19 What is meant by the term hidden layer?

7 Distributed Simulation

7.1 Introduction

Conservative simulation algorithms are based on the restriction that an event cannot be executed unless it is proved that no other event shows up during execution. This restriction is fulfilled by using strict synchronization between logical processes. Different methods for the analysis of event-based systems have been introduced for revealing the parallel properties of the several applications, algorithms, as well as environments. To handle parallel-event traces from a more methodological point of view, the distributed-simulation method was been introduced. Distributed simulation is used for real-world system analysis, where events have to be processed in a concurrent way. This results in a speed up of the simulation task. The basic ideas behind distributed simulation are mapping and scheduling, which means that an event can only be executed if it is proved that this event is independent of other prospective executable events. For this reason guarantee messages are exchanged. This concept finally results in

- Classical simulation methods
- Optimistic simulation methods
- Hybrid methods
- Speculative simulation methods
- Deterministic tie-breaking methods
- Distributed shared memory methods
- Fuzzy-based methods

One of the most important points in event-oriented simulation is that every event will be processed in a deterministic way. Consider a tie-breaking method that does not sequentialize simultaneous events. Hence one can assume that sequencing pays attention to the transitive generation sequence. Consider that the simultaneous events e and e' of the same logical process can be processed in such a way that e can be executed before e', if e' can be directly or indirectly generated from e. This constraint of sequentialized simulation is a simple boundary, meaning e' does not exist before e has been executed and the event execution is not repeated, which is not possible at any time for an optimistical method. The constraint which has to be fulfilled is called deterministic tie-breaking.

Definition
A tie-breaking method is called deterministic, if

(1) In several simulation runs, one by one, the respective logical processes (LP) have at each time the same sequence of event handling
(2) The several simulation runs are based on a transitive generation sequence
■

Another important method used for performance prediction and analysis of distributed simulation is the so-called critical-path analysis, based on event traces that present the events and their interactions.

The critical-path analysis is an analytical approach based on an event trace from the simulation of a real-world problem. The obtained trace can be transformed into a directed-event graph whose edges are weighted with real-time values representing computational and communicational delays. Consider that the weights are selected such that they represent the timing characteristics of the execution hardware and the communication network. When constructing the directed-event graph a critical path algorithm can be applied by searching the longest weighted path in the graph. The logical path between the first event and the last event represents the so-called lower bound for the simulation.

The generation of a directed-event graph is limited by two constraints, i.e. predecessor and creator relationships, which restrict the parallel execution of events, since the constraints were fulfilled before an event can be scheduled for execution. Let the creator relationship represent a situation where an event e causes an event e'. Hence the event e has to be executed before the event e'. Let the predecessor relationship represent a situation where events e and e' are scheduled to the same logical process and the time stamp of event e is less than the timestamp of e'. Hence event e is executed before e', and the earliest possible completion time for event e can be calculated using the critical times of its creator and predecessor as follows:

$$\tau_i = \max\left\{\tau_{creat(i)}, \tau_{pred(i)}\right\} + \Delta\tau_i \tag{7.1}$$

where $\Delta\tau_i$ is the required execution time of event e_i and $\tau_{creat(i)}$ and $\tau_{pred(i)}$ are the critical times of the creator and the predecessor, respectively. The critical time of the last event in the critical path represents the lower bound for the execution time of the simulation. Obtaining critical-time results for a given problem results in a procedure that uses the measured time of simulation traces for event execution, send functions, and receive times. Based on these time traces a model configuration can be selected for which the event times in the traces can be updated. With these results a new run of the critical path analysis can be started that yields new results.

Due to the heterogeneous concepts used in distributed simulation Sects. 7.2 and 7.3 will focus on specific aspects of building distributed-simulation environments for traffic-simulation systems.

7.2 Distributed Simulation of Traffic and Transportation

7.2.1 Introduction

The huge increase in motorized individual traffic during the past two decades results in the well-known traffic-jam situations as well as in traffic collapse during the rush hours in the cities. Solutions adapting to the different demands, depending oo the real-world traffic situations, are not available either for local or for global traffic planning. But there are world-wide needs for solutions. Due to the complexity and heterogeneous concepts of traffic systems, local as well as global solutions seem to be difficult to realize. For practical reasons we have to restrict ourselves from the macroscopic to the microscopic aspects of system analysis of traffic flows, based on positions and movements of transport vehicles, traffic lanes and crossings, traffic lights and traffic signs, municipal and individual traffic flows, people as parallel execution of events etc. Hence microscopic traffic structures can be properly analyzed by modeling and simulation for the respective scenario analysis.

Computer modeling of complex and heterogeneous traffic systems is an iterative process, consisting of model building and computer simulation by changing the real structure of the model and its parameters in an effort to match the complex traffic system well. In fact, the derived model has served its purpose when an optimal match is obtained between the simulation results and the data obtained from the real-world traffic system. In general, building a model of a traffic system entails the utilization of three types of information sources:

- Goals and purposes of modeling, i.e. boundaries, components of relevance, level of details
- A priori knowledge of the traffic system being modeled
- Experimental data consisting of measurements on the system inputs and outputs

Due to the possibilities of modeling, a variety of levels of conceptual and mathematical representations of traffic systems are evident, depending on the goals and purposes for which the models are intended, the extent of the a priori knowledge available, data gathered through experimentation, and measurements of the real-world traffic systems. For a real-world traffic simulation a distributed interactive simulator is necessary, which fits to the inherent boundaries, reflecting:

- Simulation of single transport vehicles
- Simulation of transport vehicle bundles
- Simulation of individual lanes
- Simulation of any kind of road network

- Realization as distributed interactive simulator in order to fit large road network situations
- Simple but flexible interface to adapt the control strategies i.e. traffic lights

In order to build a realistic distributed traffic simulator, a library of simulation models is necessary, including

- Different types of traffic situations
- Different types of road conditions
- Different types of car-following behavior
- Different types of velocity profiles
- Different interaction profiles of traffic participants
- Different types of lane changing
- Different types of dangerous traffic situations
- Different types of interactions with pedestrians
- etc.

7.2.2 Traffic-Simulation Model

Model building of traffic situations is based on assumptions as follows:

- Traffic flow can be described as a sequence of situations
- Within each situation the driver has scope for action that could interact with different subsequent situations
- Each driver decides from his point of view on the appropriate action

With these situation-action-model assumptions, for each time segment the specific traffic situation can be analyzed for decision support of the so-called driver-vehicle-element. With this in mind, the following interactions of the distributed traffic simulator can be realized:

- Driver-vehicle-element holds the actual lane, adapting to the actual traffic situation
- Driver-vehicle-element decides for lane changing in case of a slow vehicle in front or fall into line in case of branch off
- Driver-vehicle-element has to adapt to a right-of-way situation

Each situation described above, is based on models for driver-vehicle-elements. For a right-of-way situation the driver-vehicle-element has to adapt, which means decision for in-time reaction. The on-time t depends on the actual velocity v and the distance d from the actual conflict zone, yielding

$$t = \frac{d}{v}, \tag{7.2}$$

where t is the time needed with the actual velocity to arrive at the conflict zone; t can be calculated for the respective driver-vehicle-elements of a right-of-way situation, which yields

$$\Delta t = t_a - t_{na}, \tag{7.3}$$

where t_a means right-of-way authorized direction and t_{na} means right-off-way non authorized direction. Hence Δt can be interpreted as the decision of driving into the actual conflict zone or not. The decision to drive depends on the mathematical notation of a stochastic process, based on a probabilistic distribution. Driving dependents from time yields

$$P(A|Z = z) = p(z), \tag{7.4}$$

where $p(z)$ is a continuous function, $z = -\infty \Rightarrow p = 0$ and $z = +\infty \Rightarrow p = 1$. Investigations show that $p(z)$ holds $p(z) = 0$ at $z = -10$ and $p(z) = 1$ at $z = +10$.

If we use another probabilistic function such as a symmetrical logarithmic function,

$$p(z) = \frac{1}{1 + a \cdot e^{-b \cdot z}}, \tag{7.5}$$

the parameters a and b represent the respective regression model parameters.

Calculating the on-time of a decision, we have to define a vehicle-dependent variable x

$$x = \frac{rpv + ndv}{2}, \tag{7.6}$$

where rpv characterizes a randomized probabilistic variable and ndv a normal distributed variable, reflecting the driver's safety needs. For $x > p(z)$ within the considered in-time segment, the decision means drive. If $x < p(z)$ the driver decides to wait for a gap.

7.2.3 Distributed Traffic-Simulation System

The distributed traffic-simulation system contains several programs running as single modules in a distributed computer network. Distribution of the computational load onto different computers allows an effective simulation of large road net-

works. Because a single computer could not process the burden while the distributed simulator for traffic simulation contains several simulation modules, several
control modules as well as several graphic modules and a communication server,
as shown in Fig. 7.1.

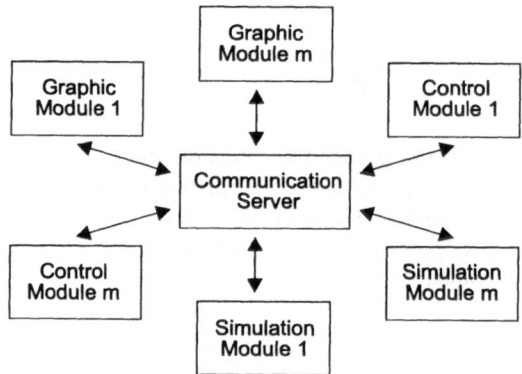

Fig. 7.1. Distributed interactive traffic simulator

The realization as a distributed simulation system is a state-of-the-art approach,
while decision making within the different simulation models depends on load
boundaries for each driver-vehicle-element. This means only driver-vehicle-elements in the near surroundings have to be taken into account. Therefore, parts of
the simulated road network can be simulated independently from each other. Only
in the neighborhood is a transfer of data necessary. The modules tasks are as follows:

- Communication server: data handling and synchronization between
  the different components tasks via a server

The server contains the data of the different modules:

- Simulation modules: computation of the different types of traffic
  flows under the supervision of the communication server
- Control modules: activation and deactivation of sensors and actors of
  traffic-light control systems under the supervision of the communication server
- Graphic modules: display of the actual intrinsic traffic dynamics of
  the involved driver-vehicle-elements within the road network

7.2.4 Description and Implementation of Road Networks

By using the traffic simulator it is possible to optimize the traffic flows at different traffic nodes. Simulating a specific traffic node, the affiliated road network has to be built up. For the respective road networks a specific discrete road network description language had been developed, written in the object-oriented language C++. The description language deals with

- Clear description of road networks
- Separation of a global road network into local road networks as parts of the simulation modules

A road-network description offers the **simmodule** and the **end simmodule** while in between the simulation modules specific road network is described. Based on the town map of the city of Portland, OR, we receive:

```
//road network description
//definition of the first module
simmodule McCall Waterfront Park, Portland
      .
      .
      .

end simmodule

//definition of the second module
simmodule Lownsdale Square, Portland
      .
      .
      .

end simmodule
```

With the definition of a simulation module the road network can be described, based on lane segments that have to be connected. The definition of lane segments is done by **section network**. Connection of lane segments is done by **section connection**.

```
//definition of the first module
simmodule McCall Waterfront Park, Portland
section network
  input SW Front Ave.(x,y,ali,len)
  {
  }
  lane SW Main St.(x,y,ali,len)
  {
  }
  lane SW 4th Ave.(x,y,ali,len)
  {
  }
```

```
section connection
  connect SW Front Ave.[1]->SW Main St.[1];
  connect SW Main St.[1]->SW 4th Ave.[1];
end simmodule
```

Based on the example given above, connections between different simulation modules are necessary, which is realized within the developed distributed simulation system for traffic simulation with the elements **netoutput** and **netinput**.

```
simmodule Chinatown, Portland
    .
    .
    .

//road network output
netoutput Burnside St.(x,y,ali,len)
{
}
    .
    .
    .

end simmodule

simmodule South Park Blocks, Portland
    .
    .
    .

//road network South Park Blocks, Portland
netinput SW 9th Ave.(x,y,ali,len)
{
}
    .
    .
    .

end simmodule
netconnect  Chinatown.Burnside  Street->South  Park  Blocks.SW  9th
Ave.;
```

where the abbreviations used are as follows: x means x-direction of the middle of the road at the very beginning of the element, y means y-direction of the middle of the road at the very beginning of the element, *ali* means align of the lane element, *len* means length of the lane element.

The different simulation modules require additional attributes, like number of lanes, speed limits, demand-dependent sensor distances, crossings, etc.

For example, the simulation module Berlin, Germany, could look like

```
simmodule Berlin, Germany
section network
  lane Unter den Linden(x,y,ali,len)
  {
```

```
numlanes(2);
maxspeed(50.0);
demand sensor(1,50);
}
        .
        .
        .
end simmodule
```

where *numlanes(x)* means numbers of lanes, *maxspeed(s)* means speed limit, demand *name(l,p)* means demand sensor with lane number and position measure from the very beginning of the lane element.

7.2.5 Implementation and Simulation

The implementation and the simulation will be discussed based on two case study examples. The first example deploys on classical traffic-light-control concepts, the second one uses modern fuzzy-set control (see Chap. 6). Simulation is based on the simulation modules described above, control modules, traffic-light modules, sensor-instrumented traffic measures, and animation modules. Hence the application-specific road network can easily be adapted to actual objects of investigation.

Example 7.1
Classical traffic-light control is based on sensor instrumentation with fixed green light duration. Depending on the queue and the actual green light direction, the new traffic dependent green light direction will be determined, with fixed green light duration and fixed duration time. For practical reasons we use a formula that fits the different traffic situations well

$$gld = 5 + spv \cdot len ,$$
(7.7)

where *gld* stands for green light duration, *spv* stands for second per vehicle, and *len* means length.

Example 7.2
The fuzzy-set traffic-light control is based on sensor instrumentation with adaptive green light durations, and is dependent on the actual traffic queues. Fuzzy-set system is a name for systems with direct relationship to fuzzy concepts, as described in Chap. 6 in detail. Based on this assumption the fuzzy-set traffic-control system can be described by rule bases as follows:

*if priority.high = queue.long **or** queueingtime.long*
*if priority.medium = queue.medium **and not** queueingtime.long*
        .
        .

*if priority.low =queue.short **and not** queueingtime.long* which results, together with the defuzzification function GetPriority:P in,

```
//Identification of traffic direction with max.priority
//case study 2Clausthal-Zellerfeld, Germany
intCFuzControl:ClausthalMaxPriority(int dir)
{
int i, j;
//Direction with max. priority
int maxdir;
//Max. Priority
double maxpriority;
double newpriority;
// Fuzzy membership functions
double lowpri, midpri, highpri;
//Priorities for each direction
maxpriority = -1;
maxdir = -1;
for (j = dir; J  dir+4; j++)
{
i = j % 4;
lowpri=FuzzyAND(LengthShort(Clausthallength[i]),
NotWaitLong(Clausthalwaitingtime[i]);
midpri=FuzzyAND(LengthMid(Clausthallength [i]),
NotWaitLong(Clausthalwaitingtime[i]);
highpri = FuzzyOR(LengthLong(Clausthal[i]),
WaitLong(Clausthalwaitingtime[i]);
newpriority = GetPriority(Lowpri, midpri, highpri);
if (newpriority > maxpriority)
{
//New direction has higher priority then former direction
maxpriority = newpriority;
maxdir = 1;
}
}
return maxdir % 4;
}
```

The fuzzy-set traffic simulator can simply be regarded as an input-output transfer system, the transfer operator of which is based on a fuzzy kernel as shown in Fig. 7.2.

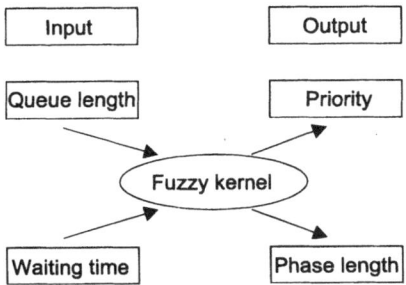

Fig. 7.2 Fuzzy kernel of the fuzzy-set traffic-control system

The control program of the fuzzy-set traffic-control system can be described as follows:

```
//Fuzzy Traffic Control Program
//Fuzzyset for Queuelength
fuzzyset Queuelength
membershipfunction short
start255
(10,255)
(60,0)
end0
.

.
memberfunction long
start0
(70,0)
(120,255)
(170,0)
end0
.

.
//Rulebase
start rulebase
.

.
Priority.Low(OutPriRG)=QueueLength.Short(InQL
RG);
Priority.High(OutPriRG)=QueueLength.Long(InQLRG) OR
WaitingTime.ExtraLong(InWTRG)
.

.
end
```

Comparing the simulation results from the classical traffic-control system with the fuzzy traffic-control system it can be stated that the fuzzy system is more flexible. Moreover, it allows a direct understanding and operating of the traffic-control system in a familiar way using a description language and rules.

7.2.6 Distributed Transportation

The distributed-transportation problem can be described and solved as shown for the case study approach of the traffic simulator, but in this approach we will use mathematical programming as a modeling technique as a powerful tool in decision support. Assume the transportation problem can be described as a product that will be shipped in the amounts $o_1, \ldots, o_m$ from each of the m shipping origins, and received in the amounts $d_1, \ldots, d_n$ by each of the shipping destinations. Hence the distributed transportation problem consists of determining the amounts x_{od} to be shipped from origins o to destinations d, to minimize the cost of transportation. The transportation problem described deals with a decision problem and can be solved using the method of linear programming, which deals with linear objective functions and linear constraints. Hence, at the beginning, one has to identify the possible decisions to be made that leads to identifying the problem variables. Thereafter, one has to determine which decisions are admissible, which results in a set of constraints according to the nature of the described problem. As a final step the cost functional has to be calculated. For the transportation-problem description the data are as follows:

- m: number of origins
- n: number of destinations
- o_i: amount to be shipped from origin i
- d_j: amount to be received in destination j
- c_{ij}: cost of sending a unit of product from origin i to destination j

The variables of the distributed transportation problem are

- x_{ij}: amount to be shipped from origin i to destination j

assuming that these variables are not negative:

$$x_{ij} \geq 0; \quad i = 1, \ldots, m; \quad j = 1, \ldots, n , \tag{7.8}$$

which implies that the direction of the product flow is prefixed by the origin and the destination. However, other assumptions can be made, such as using unre-stricted real variables, i.e. $x_{ij} \in \Re$, when we do not want to prefix origins and destinations.

The constraints of the distributed-transportation problem can be written as fol-lows:

$$\sum_{j=1}^{n} x_{ij} = o_i; \quad i = 1, ..., m \tag{7.9}$$

$$\sum_{i=1}^{m} x_{ij} = d_j; \quad j = 1, ..., n$$

The first set of constraints states that the total amount that is shipped from origin i has to be equal to the sum of the amounts that are going from origin i to all destinations j, with $j = 1, ..., n$.

The second set of constraints states that the total amount that is received at destination j has to be equal to the sum of the amounts that are shipped to that destination from all origins i, with $i = 1, ..., m$.

Finally, the function to be minimized for the distributed-transportation problem has to be described. For this approach we are interested in minimizing the total cost of transportation, which can be described as the sum of the unit costs times the amounts being shipped. Hence we have to minimize the function

$$F = \sum_{i=1}^{m} \sum_{j=1}^{n} c_{ij} \cdot x_{ij}. \tag{7.10}$$

When we have identified the respective formulas, we are able to solve the linear programming problem.

Example 7.3
Assuming that the transportation problem can be described by $m = 3$ shipping origins and $n = 3$ destinations, and $o_1 = 2, o_2 = 3, o_3 = 4, d_1 = 5, d_2 = 6, d_3 = 7$, we obtain for the constraints the matrix

$$c \cdot x = \begin{pmatrix} 1 & 1 & 1 & 0 & 0 & 0 & 0 & 0 & 0 \\ 0 & 0 & 0 & 1 & 1 & 1 & 0 & 0 & 0 \\ 0 & 0 & 0 & 0 & 0 & 0 & 1 & 1 & 1 \\ 1 & 0 & 0 & 0 & 1 & 0 & 1 & 0 & 0 \\ 0 & 1 & 0 & 0 & 0 & 1 & 0 & 1 & 0 \\ 0 & 0 & 1 & 0 & 0 & 0 & 0 & 0 & 1 \end{pmatrix} \cdot \begin{pmatrix} x_{11} \\ x_{12} \\ x_{13} \\ x_{21} \\ x_{22} \\ x_{23} \\ x_{31} \\ x_{32} \\ x_{33} \end{pmatrix} = \begin{pmatrix} 2 \\ 3 \\ 4 \\ 5 \\ 6 \\ 7 \end{pmatrix}, \tag{7.11}$$

with $x_{ij} \geq 0$; $i, j = 1, 2, 3$.

From this matrix one can conclude that the first three equations correspond to the product balance at the three origins and the last three equations are due to the balance at the three destinations. Consider the particular values

$$C = \begin{pmatrix} 1 & 2 & 3 \\ 4 & 5 & 6 \\ 7 & 8 & 9 \end{pmatrix}, \tag{7.12}$$

we obtain for the minimization problems of the transportation costs

$$F = x_{11} + 2x_{12} + 3x_{13} + 4x_{21} + 5x_{22} + 6x_{23} + 7x_{31} + 8x_{32} + 9x_{33} . \tag{7.13}$$

Using specific software packages, such as the GAMS package, the minimization problem can easily be solved, which means obtaining a minimum value for the objective function F which implies a minimum cost.

7.3 Introduction into HLA*

HLA is the abbreviation for high-level architecture for modeling and simulation, which has been developed by the US Department of Defense (DoD). Initially HLA was created with a special focus on military simulation applications and their special needs for interoperability and reusability of the components (called federates in HLA). Since the general problems in modeling and simulating complex large-scale systems in the military and in the nonmilitary community are more and more comparable, the question of whether HLA can be used in the military as well as in the civilian domains had been answered such that HLA is now available for both parties, after it had been accepted as an IEEE standard for distributed simulation.

Using HLA for distributed simulation, several users, so called federates, of the distributed-simulation environment, a so-called federation, cooperate, and use a common runtime infrastructure, so called RTI, a software, which can be assumed to be a specific part of a distributed operating system. HLA itself defines a bi-directional interface between federates and the RTI.

7.3.1 HLA at the Very First*

HLA is defined by three major elements, which are very similar to VHDL (VHDL = very high speed integrated circuit hardware description language):

* HLA rules or federation rules (FR), which ensure proper runtime interaction of simulations (or federates) in a federation, describing the simulation and federation responsibilities

- Interface specification (IS), which defines the interfaces between federates and the run-time-infrastructure (RTI) services and provides the means for federates to exchange data
- Object model template (OMT) , which describes the data federates exchange providing a common method for recording information, establishing the format of key models:
    - °Federation object model (FOM)
    - °Simulation object model (SOM)
    - °Management object model (MOM)

7.3.2 Federation Rules*

At the highest level, HLA consists of a set of HLA rules that must be obeyed if a federate or federation is to be regarded as HLA compliant.

HLA Rules for Federations

- Federations shall have a FOM, documented in accordance with the OMT.
- All representations of objects in the FOM shall be in the federates, not in the RTI.
- During a federation execution, all exchange of FOM data among federates shall take place via the RTI.
- During a federation execution, federates shall interact with the RTI in accordance with the HLA interface specification.
- During a federation execution, an attribute of an instance of an object shall be owned by only one federate at any given time.

HLA Rules for Federates

- Federates shall have a SOM, documented in accordance with the OMT.
- Federates shall be able to update and/or reflect any attributes of objects in their SOM, and send and/or receive SOM interactions externally, as specified in their SOM.
- Federates shall be able to transfer and/or accept ownership of attributes dynamically during a federation execution, as specified in their SOM.
- Federates shall be able to vary the conditions under which they provide updates of attributes of objects, as specified in their SOM.
- Federates shall be able to manage local time in a way that will allow them to coordinate data exchange with other members of a federation.

7.3.3 Interface Specification*

The interface specification identifies how federates interact with federation and, ultimately, with one another.

Run-Time Infrastructure (RTI)

- Software that provides common services to simulation systems
- Implementation of HLA interface specification
- Architectural foundation encouraging portability and interoperability

RTI Services

- Separate simulation and communication
- Improve on older standards
- Facilitates construction and destruction of federations
- Supports object declaration and management between federates
- Assists with federation time management
- Provides efficient communications to logical groups of federates

Interface Specification Management Areas

- Federation management
- Declaration management
- Object management
- Ownership management
- Data distribution management
- Time management

7.3.4. Object Model Template (OMT)*

Reusability and interoperability require that all objects and interactions managed by a federate, and visible outside the federate, are specified in detail and with a common format. OMT provides a standard for documenting HLA object-model information.

OMT

- Provides a common framework for HLA object-model documentation
- Posters interoperability and reuse of simulations and their components

Required Information

- Object class structure table
- Object interaction table
- Attribute/parameter table
- FOM/SOM lexicon

Optional Information (OMT Extensions)

- Component structure table
- Associations table
- Object model metadata

The OMT defines the FOM, the SOM, and the MOM.

Federation Object Model (FOM)

- One per federation
- Introduces all shared information, e.g. objects, interactions
- Contemplates interfederate issues, e.g. data encoding schemes

Simulation Object Model (SOM)

- One per federate
- Describes salient characteristics of a federate
- Presents objects and interactions that can be used externally
- Focus on the federate's internal operation

Management Object Model (MOM)

- Universal definition
- Identifies objects and interactions used to manage federations

7.3.5 Suggested Steps at the Very First*

Federates shall have a HLA SOM, documented in accordance with the HLA OMT. The suggested steps are as follows:

1. Identify essential objects.
2. Identify attributes used in describing the above objects.
3. Build class hierarchy based on common attribute groupings.
4. Classify each object and prepare object class structure table.
5. Repeat steps 1–4 for interactions. Identify interactions and associated parameters, build hierarchy, classify interactions, and prepare the interaction class table.

6. Prepare initial attribute and parameter tables.
7. While constructing data types, lexicons, and routing space tables, iterate with earlier tables, especially the attribute and parameter tables. Verify that the potential attributes or parameters have not been overlooked and/or modify existing ones as necessary.

7.3.6 Land-based Transportation*

From a practical point of view, as shown in Sect. 7.2, one has to restrict oneself from the macroscopic aspects to a microscopic systems analysis concept when modeling land-based transportation, based on positions and movements of transport vehicles, traffic lanes and crossings, traffic lights and traffic signs, municipal and individual traffic flows, people as discrete events, etc. Microscopic land-based transportation could then be properly analyzed by modeling and simulation.

With respect to the spectrum of modeling, a variety of levels of conceptual and mathematical representation of land-based transportation are evident, depending on the goals and purposes for which the model was intended, the extent of the a priori knowledge available, data gathered through experimentation, and measurements on real land-based transportation. For the HLA land-based transportation simulator we need a similar structure as discussed in Sect. 7.2.

7.3.7 HLA Land-based Transportation Simulator*

The distributed land-based transportation simulation system contains a federation and several federates that could run on a distributed computer network. Distribution of the computational load on different computers allows an effective simulation of large road networks because one single computer could not process the burden. The components of the system, shown in Fig. 7.3, are

- RTIexec, a global process that manages the creation and destruction of federation execution
- FedExec, one running process per executing federation that manages the federation, allows federates to join and resign from the federation and facilitates data exchange between federates
- Federates (simulation modules) that perform the computation of the different types of traffic flows
- A graphic module that displays the actual intrinsic traffic dynamics of the involved driver-vehicle-elements within the road network
-

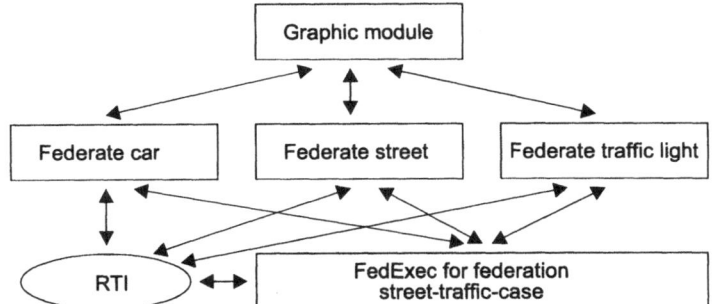

Fig. 7.3. Schematic diagram of the HLA traffic-simulation-system
Example 7.4
The initialization of the RunTime Infrastructure contains the following steps:

- Instantiating of the objects for the RTI ambassador and federate ambassador
- Creating federation execution with specified name
- Joining the federate to the federation
- Setting the initial time management parameter

which can be implemented as follows:

```
//---------------------------------
// Create RTI objects
//
// The federate communicates to the // RTI through the RTIambassador
// object and the RTI communicates
// back to the federate through the

// FederateAmbassador object.
//---------------------------------
RTI::RTIambassador rtiAmb;
// libRTI provided
TrFederateAmbassador fedAmb;
// User-defined
// Named value placeholder for the federates handle
RTI::FederateHandle federateId;


//---------------------------------
// Create federation execution.
// The RTI_CONFIG environment
// variable must be set in the
// shell's environment to the
// directory that contains the
// RTI.rid file and the Traffic.fed
//---------------------------------
try
    {
//---------------------------------
// A successful createFederation
```

```
    // Execution will cause the fedex
    // process to be executed on this // machine.
    // A "Traffic.fed" file must exist
    // in the current directory. This
    // file specifies the FOM object,
    // interaction class structures,
    // default/initial transport and or
    // dering information for object at
    // tributes and interaction classes
    //--------------------------------
        cout << "FED_TR: CREATING FEDERATION EXECUTION" <<
        endl;
        rtiAmb.createFederationExecution( fedExecName, "Traffic.fed" );
        cout << "FED_TR: SUCCESSFUL CREATE FEDERATION
        EXECUTION" << endl;
        }
    catch (RTI:: FederationExecutionAlreadyExists& e )
        {
        cerr << "FED_TR: Note: Federation execution already exists." <<
            &e << endl;
        }
    catch ( RTI::Exception& e )
        {
        cerr<<"FED_TR:ERROR:"<<&e<<endl;
        return -1;
        }
    RTI::Boolean Joined=RTI::RTI_FALSE;
    int numTries  = 0;

    //--------------------------------
    // Join federation execution
    // Here we loop around the
    // joinFederationExecution call
    // until we try too many times or
    // the Join is successful.
    //--------------------------------
    while(!Joined &&(numTries++ < 20))
        {
    //--------------------------------
    try
    {
    cout<<"FED_TR: JOINING FEDERATION EXECUTION: " << ex
    eName << endl;
    federateId =
    rtiAmb.joinFederationExecution (myStreet->GetName(),fedExecName,
    &fedAmb);
            Joined = RTI::RTI_TRUE;
    }
    catch (RTI:: FederateAlreadyExecutionMember& e)
    {
    cerr<<"FED_TR:ERROR:"<<
```

```
myStreet->GetName()<<"already exists in the Federation Exe-
cution"<<fedExecName<<"."<<endl;
cerr << &e << endl;
return -1;
}
catch (RTI::FederationExecutionDoesNotExist&)
{
cerr<<"FED_TR:ERROR:"<<fedExecName <<"Federation Execu
tion"<<"does not exist"<<endl;
rtiAmb.tick(2.0, 2.0);
}
catch ( RTI::Exception& e )
{
cerr<<"FED_TR:ERROR:"<<&e<<endl;
return -1;
}
} // end of while

cout<<"FED_TR:JOINED SUCCESSFULLY:" <<exeName<<": Fed
erate Handle=" << federateId << endl;
```

Each federate has to

- Define what data needs to be published for each update or event
- Declare which updates and interaction (event) it is interested in receiving by subscribing to those attributes/messages
- Specify if the federate is interested in controlling unnecessary message traffic

The simulation consists of

- Calculate state and update to RTI
- Ask for time advance
- Tick the RTI waiting for grant

When a federate has completed its simulation, it deletes the objects it created (streets, lane, intersection, cars, etc.), resigns from the federation execution and tries to destroy the federation.

7.3.8 HLA Description of Road Networks*

By using the land-based transportation simulator it is possible to optimize the traffic flows at different traffic nodes. To simulate a specific traffic node, the affiliated road network has to built up. For easy modeling of road networks a HLA-specific notation was introduced, which deals with:

- Clear description of road networks

- Splitting of a global road network into local road networks as parts of the simulation modules

Example 7.5

The HLA-RTI land-based transportation simulator results on cars, streets, and possible interrupt requests for traffic lights or traffic signs. In accordance with the explanation in Sect. 7.2 streets are assumed, consisting of lanes for each direction, like *west* to *east*, *east* to *west*, *north* to *south*, and *south* to *north*, realized with blocks of equal length. Here the streets are assumed to have one lane per direction. The cars are driving – as a simplification of the simulation system realized in case study 1 – with the same speed. Hence traffic flow can be described as a sequence of blocks, as shown in Fig. 7.4.

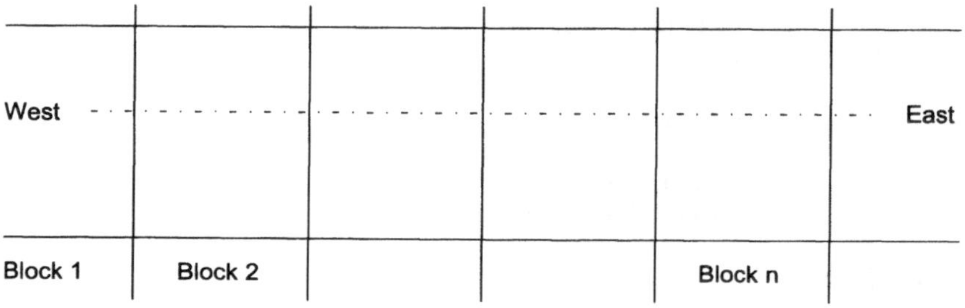

Fig. 7.4. Schematic diagram of federation street traffic *west* to *east* and vice versa

With the HLA land-based transportation simulation environment the federation consists of more federates responsible for traffic flow. Due to that type of federates, the following object classes can be used: **Street, Lane, Block, Car, Traffic Light, Intersection, EntryInIntersection.**

With the HLA land-based transportation simulation environment the object classes can be described as follows:

> *class Street* contains:
> name of the street
> number of lane per direction
>   (1 lane in this case of study)
> number of blocks per lane
> direction ·char[2];
>   (for example: direction (0) = "west"
>   direction (1) = "east")
> lane for direction[0]
> lane for direction[1]
>
> *class Lane* contains:
> direction
> array of Block
>
> *class Block* contains:
> array of TrafficLight
> array of TrafficSign

array of Car

class Car contains:
pointer to Street
pointer to Lane
pointer to Block
max velocity
current velocity

class TrafficLight characterized by:
ID
state(color)
for entry in street
for entry in block

class Intersection contains:
array of EntryInIntersection

class EntryInIntersection characterized by:
street
lane
numberOfBlock

The developed HLA land-based transportation simulator version can be enhanced by more features like:

- Streets with more then one lane for each direction
- Cars do not need to have the same velocity profile; (when one has to taken into account the fact that a driver-vehicle-element can decide to overtake another driver-vehicle-element driving with a slower velocity)
- More complex intersections, traffic sign, etc.

In the case of several streets, the **class Street** must contain as a supplement an array with pointers to each street (a static variable), in order to access them and perform the communication:

static StreetPtr ms_StreetExtent[MAX_STREETS + 1];

In addition, instead of the two attributes of type **Lane**, (lane for direction[0] and lane for direction[1]), the class has two arrays of lanes, each of length "number of lanes per direction".

7.4 References and Further Reading

Castillo E, Conejo AJ, Pedregal P, Garcia R, Alguacil A, (2002), Building and Solving Mathematical Programming Models in Engineering and Science, John Wiley & Sons, New York

Jari P, Ikonen J, Harju J, (1999), Predicting the Performance of Distributed Simulation through Event Traces, In: SCSC-Proceedings, SCS, San Diego

Straßburger S, (2001), Distributed Simulation Based on HLA in Civilian Application Domains; SCS-Europe

Technical Documents
Draft Standard for Modeling and Simulation High Level Architecture: Framework and Rules, Technical report P1516/D1, IEEE, 1998

High Level Architecture Interface Specification, Technical report P1516,1, IEEE, 1998

High Level Architecture Run-Time Infrastructure Programmers Guide, DoD, 1998

Links
http://hla.dmso.mil
http://hla.dmso.mil/docslib/mspolicy/msmp/
http://hla.dmso.mil/RTISUP/hla_soft/rti/rti-3r4/prog.doc
http://www.ecst.csuchico.edu/~hla/courses.html

7.5 Exercises

7.1   What is meant by the term distributed simulation?
7.2   What is meant by the term tie-braecking method?
7.3   What is meant by the term driver-vehicle-element?
7.4   Describe the traffic simulation by using a simple example.
7.5   Describe the components of a traffic simulator.
7.6   What is meant by the term HLA?
7.7   What is meant by the term federation?
7.8   What is meant by the term federate?
7.9   What is meant by the term object model template?
7.10 What is meant by the term federation object model?
7.11 What is meant by the term simulation object model?
7.12 What is meant by the term management object model?
7.13 What is meant by the term RTI?

8 Virtual Reality

8.1 Introduction

Virtual reality can be described as a synthetic 3D computer-generated universe that is a perceived as the real universe. The key technologies behind virtual reality systems (VRS) and virtual-environment systems (VES) are

- Real-time computer graphics
- Color displays
- Advanced software

Computer graphics techniques have been successfully applied for creating synthetic images, necessary for virtual reality and virtual-environmental systems. Creating an image, using computer graphics techniques, can be done by storing 3D objects as geometric descriptions, which can then be converted into an image by specific information of the object, such as color, position, and orientation in space, and from what location it is to be viewed. Real-time computer graphics techniques allow the user to react within the time frame of the application domain, which finally results in a more advanced man machine interface, which is the whole rationale for virtual reality systems and virtual-environments.

Color displays are used for displaying the views of the virtual reality as well as the virtual-environmental universe to provide a visual sensation of the objects from the physical application domain into the virtual domain. The color displays are of great variety, such as monitors fixed to the windows of the simulator cockpit for visual sensation of flying in a flight simulator, or head mounted displays (HMD) which visually isolates the user from the real world. A head-mounted display can provide the left and right eye with two separate images that include parallax differences, which supplies the user's eyes with a stereoscopic view of the computer generated world, which is a realistic stereoscopic sensation.

Advanced software tools are used to support the real-time interactive manipulation of large graphic databases, which can be used to create a virtual environment, which can be anything from 3D objects to abstract data bases. Moreover, 3D modeling and simulation tools are part of the advanced software tools. Hence, a 3D model can be rendered with an illumination level simulating any time of the day, or using OpenGL, a quasistandard for 3D modeling and visualization, one can create geometrical bodies of every shape and size for simulating the different views of the geological and geophysical parameters of a tunnel scenario, as shown in Figs. 8.1 and 8.2, which can be moved in size in real time, using the advanced simulation software tools. The image realism can be improved by incorporating real-world textures, shadows, and complex surfaces, etc. For example, Fig. 8.1 shows the sequence of a "flight through a tunnel". Top left part of Fig. 8.1 shows the scenic view of the landscape, top middle part of Fig. 8.1 shows the top-frontal view of the scenic landscape, top right part of Fig. 8.1 shows the front view of the scenic landscape, and bottom left part of Fig. 8.1 shows the different geological structures and the tunnel inlet and outlet as front view of the scenic landscape, bottom middle part of Fig. 8.1 showing the front view of the tunnel inlet with parts of the geological structure, and bottom right part of Fig. 8.1 shows a scenario inside the tunnel with the end of the tube in front of the view.

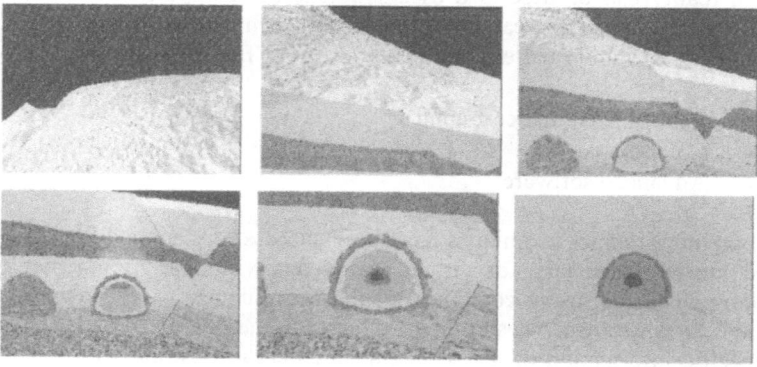

Fig. 8.1. Virtual reality tunnel simulation scenario

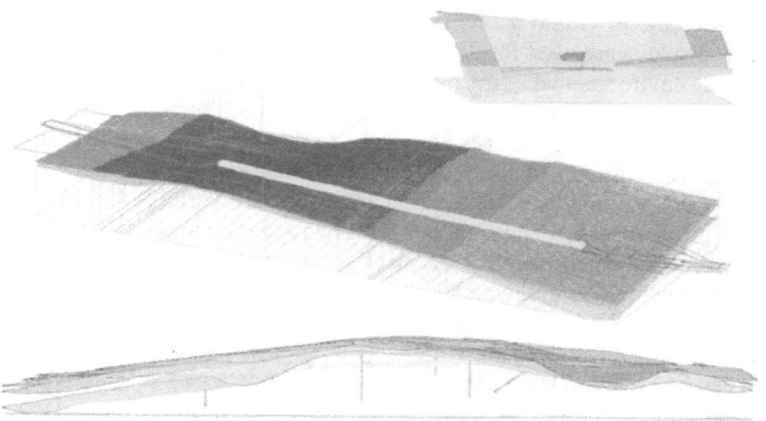

Fig. 8.2. 3D model of the virtual reality tunnel simulation scenario of Fig. 8.1

Due to intuitive interaction with the virtual reality techniques new scenic presentations are possible, as shown in Figs. 8.1 and 8.2, which offers concepts for modeling and simulation of complex real-world systems with parameterized or nonparameterized topologies within a unique framework. This results in rapid prototyping based on flexible modeling tools with concepts for geometry, motion, control, as well as virtual reality components like images, textures, shadowing, rendering, animation, multimedia, etc.

The technical complexity associated with developments in the virtual universe requires the use of metric values, which can then be converted into several important factors that relate to the metric values themselves, especially metric dimensionality, metric attributes, metric types, etc.

An easy and straightforward approach for the design of metric-valuated dimensions could be found using unidimensional scaling. However, methods of unidimensional scaling are generally applied only in cases with good reason to believe that one dimension is sufficient. But metric-valuated accuracy and presentation fidelity lead to a multidimensional scaling. A multidimensional scale is necessary for the adequate description of images, if additional information would probably be required. Therefore, a multidimensional scale has to be developed, where metric-valuated attributes are the actual quality parameters measured along each quality dimension, which are realism, interpretability, and accuracy.

There are a number of possible metric-valuated types that could be used for the dimensions of a quality-assessment metric, such as:

- Criteria-based ones, which are based on a textual scale, prefixing the levels of the scale
- Image-based ones, which are based on a synthetic scene where a rating is assigned by identifying the standard image having a subjective quality that is closest to that being rated
- Physical-parametric-based ones, which are based on measured values such as integrated power spectrum, mensuration error, etc.

The big challenge of virtual reality techniques is that it takes us one step closer to virtual objects by making us part of the virtual domain. Computer graphics techniques applied in the virtual reality systems of today providing visual images of the virtual universe, but the systems of tomorrow will also create acoustic images of the virtual universe, which can be introduced as the 5-th dimension of the virtual reality technique – while time is the 4-th dimension – which can stimulate the sounds in the virtual environments. One could imagine that other more advanced modes of user interaction, such as to touch and feel virtual objects, can complete the sensation of illusion in virtual worlds, which can be introduced as the 6-th dimension of virtual reality techniques. Moreover, smelling and tasting may also become imaginable in virtual environments, enhancing the order of dimension. The benefits of the technique of virtual reality are manifold, which is why this technique is so vital to many different application domains, ranging from automotive and avionics applications in the industry, as well as molecular and medical topics, military applications, catastrophic management, education and training, etc., to the different academic research domains. Based on features offered through computer graphics techniques, meaning visualization of highly realistic models, and through the integration of real-time computing, virtual reality enables the user to move around in virtual environments, such as walking through a tunnel as shown in Fig. 8.1, or to acquire flying skills without involving real airplanes or airports, as realized in virtual training environments for pilots, etc.

Based on the spatial and temporal geometric description, which can then be converted into an image by specifying the respective information behind, virtual reality techniques can be used as the basic concept for virtual-world simulation, as well as for analysis and prognosis of complex processes in virtual worlds.

Furthermore, underlying databases in virtual-environments offer the ability to store and retrieve heterogeneous and huge amounts of data for modeling virtual worlds. Hence, virtual reality can be seen as a specific type of a real-time embedded system combining different technological approaches that are integrated within one environmental solution.

In the case of a flight simulator, as shown in Fig. 8.3, the computer graphics techniques are used to create a perspective view of a 3D virtual world, and the view of this world is determined by the orientation of the simulated aircraft. Simulating the complex behavior of the aircraft requires a sophisticated modeling technique and embedding of several real-time systems, such as engines, hydraulics, flight dynamics, instruments, navigation, etc., as well as weather conditions, and so on, which are components and modes of the flight simulator's virtual-environment. The information necessary to feed the flight simulator with real-world data are available from the databases of the aircraft manufacturer and the manufacturer of the aero engines. They describe the dynamic behavior of the aircraft when taxiing on the ground, or flying in the air, or engine temperature and fuel burn rates, etc. The flight models used in the flight simulator are based on the data obtained from the manufacturer as well as the data describing the flight controls to simulate the behavior of the airplane under regular as well as under non-regular flight conditions.

During flight simulation, the pilot – as well as the co-pilot – sit inside a replica cockpit and gaze through the forward-facing and side-facing windows, which are 200° panoramic displays reflecting the computer-generated graphical virtual universe. The flight simulator creates a realistic sensation of being in a real-world plane flying over some 3D landscape, as shown in Fig. 8.3. But today, the flight-simulator panoramic displays do not contain stereoscopic information, the fact that the images are collimated to appear as though they are located at infinity creates a strong sense of being immersed in a 3D world.

Furthermore, immersion can be enhanced by allowing the users head movements to control the gaze direction of the synthetic images that provides the user's brain with motion-parallax information to complement other cortical pathways of the visual cues in the brain. This requires tracking the user's head in real time, and if the user's head movements are not synchronized with the images, the result will be disturbing.

Fig. 8.3. Flight simulator (bottom) and cockpit view inside the flight simulator (top right) and view from the waiting position for take off (top left)

When visually immersed within a virtual environment there is a natural inquisitive temptation to reach out and touch virtual objects as part of interaction possibilities in the virtual universe, which is impossible, as there is nothing to touch and to feel, when dealing with virtual objects. But, the user's sense of immersion can be greatly enhanced by embedding tactile feedback mechanisms in the virtual environment. Embedding tactile feedback needs some specific hardware components, such as data gloves, which enable the user to grasp or to sense real-time hand gestures. Hence data gloves will provide a simple form of touch-and-feel stimulus where small pads along the fingers stimulate a touching and feeling sensation. Thus, if a collision is detected between the users virtual hand – the data glove – and a virtual object, the data glove is activated to stimulate the touch and feel condition. However, the user may not be suddenly aware of the objects mass, as there is no mechanism for engaging the user's arm muscles. Therefore, it is necessary to transmit forces from the virtual domain to the user interface, meaning there is a need for embedding articulated manipulators in the virtual environment that could create such forces.

There are many advantages of working in the virtual domain, such as:

- Accuracy due to subject specification, which means that the real-world models can be built with great accuracy as they are based upon CAD data of the real-world objects
- Flexibility, which means building virtual representations of anything as well as interacting with this representation via the virtual reality front ends
- Animated features, which means animation of sequences, objects, etc., in space and time

Example 8.1
Combining these three aspects for real-time simulation in virtual environments should be based on the integration of the overall information, but only a few approaches maintain this problem and have been developed like the cave automatic virtual-environment, or the digital mock up (DMU) in the avionic industry, shown in Figs. 8.4 and 8.5, allowing the user a real-time interaction that is not only restricted to the 3D model itself, it also is parameterized, which could lead to a better framework for real-world system analysis, such as

- Statistic and cinematic interference tests
- Development of new methods for DMU application
- Investigation of applicability of new technologies within the virtual product design process

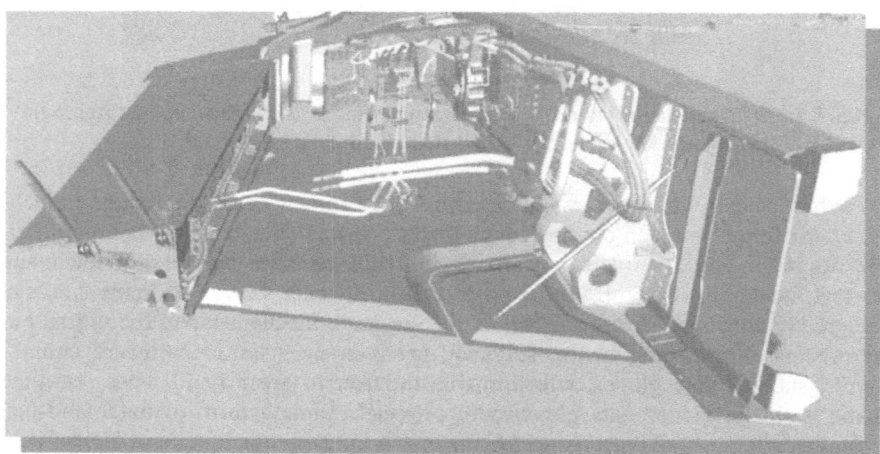

Fig. 8.4. Digital mock up (DMU) of a planes wing[1]

Fig. 8.5. Digital mock up (DMU) of a plane wing showing the application of virtual reality to simulate the possibility of a maintenance procedure within the plane's wing[1]

8.2 Virtual Reality applied to Medicine

8.2.1 Introduction

Applying the virtual reality technique to the medical domain could be stated as combining distributed virtual environments to support collaboration among team members working with space distance, developing plans and procedures, doing measurements and data processing of surgical procedures, medical research projects, clinical-oriented support systems development and evaluation, etc. One of the most interesting new paradigms in virtual reality techniques in this domain is that 3D representations are not the only possibility of a setting.

Many virtual applications in medicine, if not already now, will in the future make use of specific graphics. The virtual space will be visualized in space and time. Users in charge of virtual reality in the medical domain should be able to interact in space and time, which can then be converted into an image such as walking through the vascular bed for inspection of collagen settings at the vessels walls, or interacting with other medical disciplines for consultancy through a graphical user interface in the context of a computer-supported cooperative work, as well as designing the vivid view of cosmetic surgery, etc.

[1] I would like to thank Dr. Roland E. Haas, DaimlerChrysler RTI Bangalore, India, for his support.

The interweaving of functionality, distribution, efficiency, and openness aspects are very noticeable in the computer graphics techniques. The virtual space is graphically visualized flamboyance and for the most part the users in charge of the medical virtual domain should see the same image.

Therefore, for virtual reality applications in medicine, a multiuser virtual environment has been developed, consisting of the following components:

- Space ball and cyber gloves for tactile interaction in the virtual environment
- Head mounted devices for visual interaction in the virtual environment
- 3D geometrical body creation and motion technique for virtual space feeling
- 3D visual interactive interface for definition, manipulation, animation, and performance analysis of medical geometrical bodies
- Object-oriented data base system for efficient data management in virtual reality applications
- Hardware for the necessary computational power in space and time
- Objects organization into inheritance hierarchies for the transparency of the virtual environment
- Simulation software

Created medical object's inherit the properties and verbs of their ancestors. Additional verbs and properties, as well as specializations of inherited components, can be defined to address the new object's unique behavior and appearance. Based on these assumptions a simulator for a virtual environment for medical application can be designed.

8.2.2 Morphing

The time-dependent presentations of processes are of importance, bringing together real-world scenarios and virtual scenarios of real-world objects, to find optimal geometries, which can be calculated using non uniform rational B-splines (NURBS) . This B-spline representation is based on a grid of defined points $P_{i,j}$, which are approximated through bicubic parameterized analytical functions, as given in (8.1) and (8.2).

$$P_{i,j} = \left\{ \begin{matrix} p_{1,1} & p_{1,2} & \cdots & p_{1,n} \\ p_{2,1} & p_{2,2} & \cdots & p_{2,n} \\ \vdots & \vdots & \ddots & \vdots \\ p_{m,1} & p_{m,2} & \cdots & p_{m,n} \end{matrix} \right\}, p_{i,j} = (x, y, z), \tag{8.1}$$

$$S(u,v) = \frac{\sum_{i=0}^{n}\sum_{j=0}^{m} N_{i,p}(u)N_{j,q}(v)w_{i,j}P_{i,j}}{\sum_{i=0}^{n}\sum_{j=0}^{m} N_{i,p}(u)N_{j,q}(v)w_{i,j}}, \qquad 0 \le u,v \le 1. \tag{8.2}$$

The NURBS method allows calculation of the resulting surface or curve points by varying one curve or two surface parameter values u and v of the interval $[0,1]$, respectively, and evaluating the corresponding B-spline basis function $N_{i,p}$ as given in (8.3).

$$N_{i,0}(u) = \begin{cases} 1 & if \ u_i \le u \le u_{i+1} \\ 0 & otherwise \end{cases}, \tag{8.3}$$

$$N_{i,p}(u) = \frac{u - u_i}{u_{i+p} - u_i} \cdot N_{i,p-1}(u) + \frac{u_{i+p+1} - u}{u_{i+p+1} - u_{i+1}} \cdot N_{i+p-1}(u), \tag{8.4}$$

$$U = \{u_o, \ldots, u_m\}, u_i \le u_{i+1}, \\ V \ analogous \tag{8.5}$$

As the parameter values u and v can be chosen continuously, the resulting objects are mathematically defined in any point that show no irregularities or breaks.

There are several parameters that have to be adjusted for the approximation of given points, changing the view of the described object and, if necessary, an interpolation of all points can be achieved.

First, the polynomial order describes the curvature of the resulting curve or surface, which gives the mathematical function a higher level of flexibility. Secondly, the defining points can be weighted in accordance with their dominance with respect to the other control points. A higher weighted point influences the direction of the curve or surface more than a lower weighted one. Furthermore, knot vectors U and V define the local or global influence of control points, so that every calculated point is defined by a smaller or a greater array of points, which results in local or global deformations, respectively.

NURBS are easy to use for modeling as well as modifying the respective achievement by means of moving the control points that the user is able to adjust the objects simply by pulling or pushing the control points, as shown in Fig. 8.6.

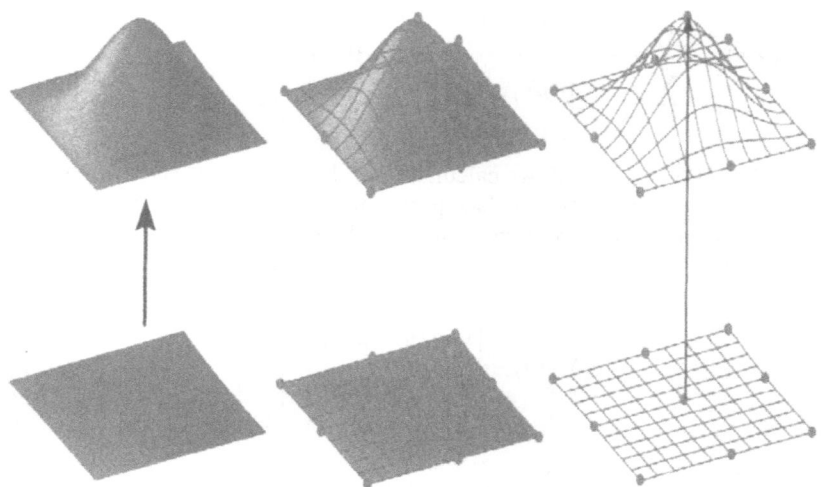

Fig. 8.6. Modeling and modification of a NURBS surface

Based on these concepts a methodology to interpolate a given set of points, i.e. the results of scanned data of a human face surface measurement, has been developed. As shown in Fig. 8.7, huge sets of scattered data points are used to generate the resulting object using 3D simulation.

Using multiple levels of surface morphing, the multi-level B-spline approximation algorithm (MBA) adjusts a predefined surface, i.e. a flat square or a cylinder. Constraints like the curvature or direction at specific points are given or can be evaluated with the algorithm, as shown in Fig. 8.8.

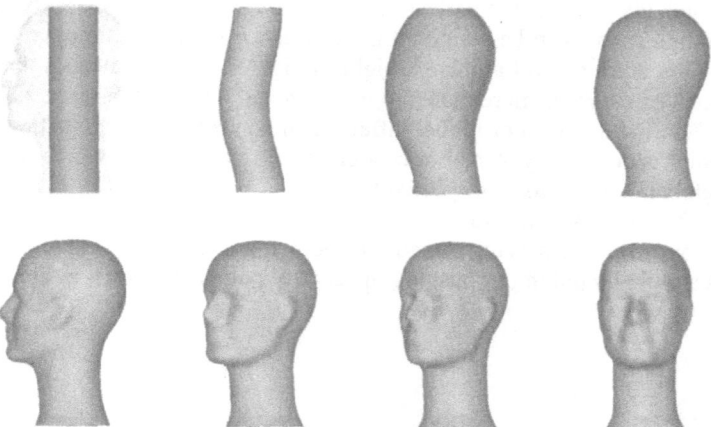

Fig. 8.7. Morphing based on a multi level B-Spline approximation

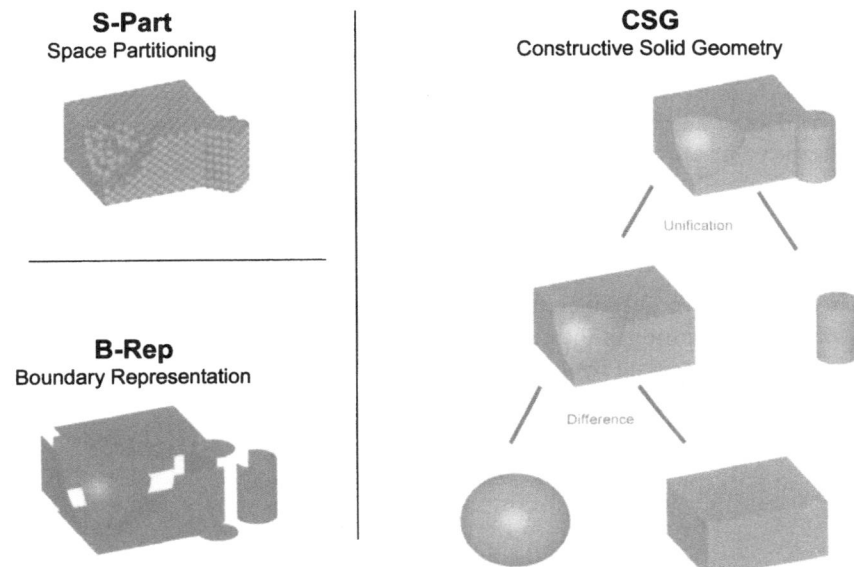

Fig. 8.8. Volumetric models for virtual reality representation, top left: space partitioning, bottom left: boundary representation, bottom right: constructive solid geometry, top to middle: unification, middle to bottom: difference

8.2.3 Deformable Models*

Mathematically, geometric subjects can be interpreted as embedded contours in an image plane, which can be written as a virtual reality framework concept

$$(x, y) \in \Re^2 . \tag{8.6}$$

The contour itself can be assumed to be

$$\exists(s) = (x(s), y(s))^T , \tag{8.7}$$

where x and y are the coordinate functions and $s \in [0,1]$, the parametric domain.

The shape of a contour subject to an image $I = (x, y)$ can be described by the functional

$$\Im(\exists) = E(\exists) + \Gamma(\exists). \tag{8.8}$$

The functional given above can be interpreted as representation of the energy of the contour, which means that the final shape of this contour corresponds to a minimum of energy. Due to that assumption the first term of the functional given above can be introduced as internal deformation energy

$$\Xi(\exists) = \int_0^1 \Lambda_1(s)\left|\frac{\partial \exists}{\partial s}\right|^2 + \Lambda(s)\left|\frac{\partial \exists^2}{\partial s^2}\right|^2 \cdot \partial s. \tag{8.9}$$

This equation describes the deformation of a stretchy, and flexible contour, with $\Lambda_1(s)$ as tension of the contour and $\Lambda_2(s)$ as rigidity.

In accordance with the calculus of variations, the contour $\exists(s)$, which minimizes the energy $\Im(\exists)$ must satisfy the Euler Lagrange equation

$$-\frac{\partial}{\partial s}(w_1 \cdot \frac{\partial \exists}{\partial s}) + \frac{\partial^2}{\partial s^2}(w_2 \cdot \frac{\partial^2 \exists}{\partial s^2}) + \nabla P(\exists(s,t)) = 0. \tag{8.10}$$

The vector partial differential equation, introduced above, describes the balance of internal and external forces when the contour rests at equilibrium. Therefore the first two terms represent the internal stretching and bending forces, respectively, while the third term represents the external forces that couple the contour to the image data.

8.2.4 Deformable Models for Surface Reconstruction in Medicine*

The treatment of patients with myocardial pumping insufficiency results in cardiac surgery cases that finally ends with a heart transplantation. Heart transplantation carries a high risk for the patient, a long-life postoperative specific lifestyle and a long-life drug dose regimen, as well as huge costs for the surgical as well as for the postsurgical treatment. Due to this situation, the possibilities of cardiac-assist systems and/or mechanisms are the focus of several medical research projects. One of which deals with surface reconstruction of the heart, which can then be converted into a model for a fiber elastic network, which could be pulled tight over the heart, inserted during cardiac surgery, to assist the insufficient heart muscle. The principle behind this is based on the mechanisms well known from the use of assist pantyhose for varicose veins. Together with the technique of minimal invasive surgery, the fiber elastic network could be used as an intra-corporal contractility assistance for the insufficient ventricles.

Measurements with the nuclear medical devices NMR and CT result in very good representations of the inner sections of the human body. Nevertheless, the segmentation of separate organs and the representation of real 3D reconstructions, based on soft shapes from the measurements, is a task that had not been solved sufficiently till now. The different segmentation methods in use are based on the assumption that the grey values of the measured dot space can be interpreted as the border between organs. Conventional methods are based on dot pursuit algorithms that have important disadvantages as follows:

- The resulting models are primarily voxel-based and limited to carrying on the signal processing procedures
- If grey values are within gaps they will be considered within the 3D model, hence an actual fit is necessary
- Vague data of in vivo as well as in vitro measurements of organs result in a significant deterioration, meaning the extracted 3D model will be inaccurate

Treating the disadvantages of the above-mentioned methods for shape reconstruction of the human heart, a specific morphing algorithm, which was previously used for applications in the virtual universe, the multi-level B-spline approximation for 3D modeling, was redesigned for the medical domain and implemented.

The data obtained from NMR or CT measurements are weighted in accordance to their grey scale values that, thereafter, being treated as a projection in a free-space allocation area, the non uniform rational B-spline (NURBS). The mathematical representation with the projection reference point in the direction of the vector of the projection, results in a space-domain description that will be deformed in a successive way, one after each other. The influence of the dot projection will be more severely weighted and more effective due to the respective free-space allocation. In a sequence of steps, the influence of weighting factors will be more and more reduced, hence the deformation will only affect the local areas of representations. Changing the order of the polynomials of the functions, and the number of iteration steps used, allows an application-domain-specific approximation of information of the dot space.

Example 8.2
Using appropriate weighting factors, and simultaneous distortion of the models through the dots, single and remote dots will be smoothed automatically, whereby, also from NMR or CT pictures, 3D deformable models can be extracted. Moreover, a rough approximation of the human heart can be extracted, which, together with the clinical expertise results in a much more adopted surface reconstruction. Due to the intrinsic power of the multi-level B-spline approximation for 3D modeling, an exact view of the hearts surface is possible after a few iteration cycles, as shown in Fig. 8.9, which can then be converted into a simulation mode of the fiber elastic network, which could be pulled tight over the heart, within the virtual-heart environment.

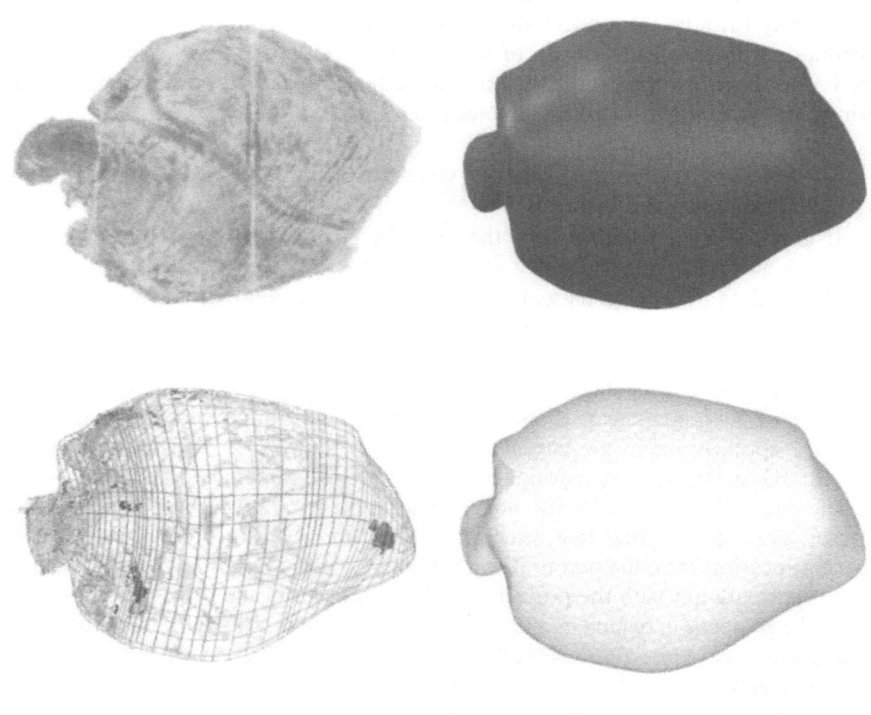

Fig. 8.9. Surface reconstruction with NURBS used in the virtual-pig-heart environment for simulation of the tailored fiber elastic net

Based on constraints of the most important dots, single subdomains of the outcome space domain can be manipulated. Based on the knowledge of the position of single organ compartments, e.g. the ventricle of the right heart, the target oriented deformation of the basis space domain is possible.

Example 8.3

Magnetic resonance imaging (MRI) data typically contain a number of slice planes taken through a volume, such as the human body. MATLAB includes an MRI data set that contains 27 image slices of a human head, which can be used as a virtual reality simulation environment, as shown in Example 8.1. Some useful techniques for visualizing this data include displaying the data as:

- Series of 2D images representing slices through the head
- 2D and 3D contour slices taken at arbitrary locations within the data
- Isosurface with isocaps showing a cross section of the interior

The MRI data are stored as arrays. The first step in this virtual reality simulation environment is to load the data and transform the data array from 4D to 3D based on the MATLAB specific commands

```
load mri
D = squeeze(D);
```

Displaying one of the MRI images, one uses the image command, indexing into the data array to obtain the respective image. Thereafter one has to adjust axis scaling, and install the MRI color map, which was loaded along with the data.

```
image_num = 8;
image(D(:,:,image_num))
axis image
colormap(map)
```

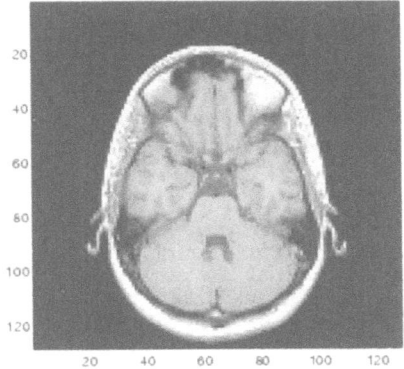

Now save the x- and y-axis limits for use in the next part of the example:

```
x = xlim
y = ylim
```

This MRI data, shown above, can be treated as a volume because it is a collection of slices taken progressively through the 3D object. The contour slice is used by displaying a contour plot of a slice of the volume. To create a contour plot with the same orientation and size as the image created in the first part of Example 8.3, one has to adjust the y-axis direction, set the x and y limits, and set the data aspect ratio:

```
contourslice(D,[ ],[ ],image_num)
axis ij
xlim(x)
ylim(y)
daspect([1,1,1])
colormap('default')
```

This contour plot uses the figure color map to map color to contour value.

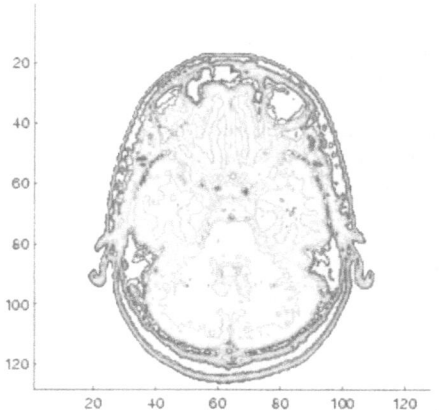

Unlike images, which are 2D objects, contour slices are 3D objects that can be displayed in any orientation. For example, one can display four contour slices in a 3D view. To improve the visibility of the contour line, one has to increase the line width to 2 points:

```
phandles = contourslice(D,[ ],[ ],[1,12,19,27],8);
view(3); axis tight
set(phandles,'LineWidth',2)
```

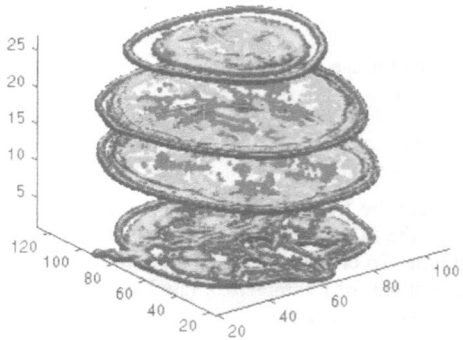

Isosurfaces can be used in MATLAB to display the overall structure of a volume. When combined with isocaps, this technique can reveal information about data on the interior of the isosurface.

First, the data have to be smoothed. Thereafter, the isosurface are used to calculate the isodata in MATLAB. Patch is used to display this data as a graphics object.

```
Ds = smooth3(D);
hiso = patch(isosurface(Ds,5),...
'FaceColor',[1,.75,.65],...
'EdgeColor','none')
```

Moreover, the isocaps are used to calculate the data for another patch that is displayed at the same isovalue (5) as the surface. The unsmoothed data (D) are used to show details of the interior. One can see this technique as the sliced-away top of the head.

```
hcap = patch(isocaps(D,5),...
'FaceColor','interp',...
'EdgeColor','none')
colormap(map)
```

Defining the view and set the aspect ratio:

```
view(45,30)
axis tight
daspect([1,1,.4])
```

Add lighting and recalculate the surface normals based on the gradient of the volume data, which produces smoother lighting. Increase the ambient strength property of the iso cap to brighten the coloring without affecting the isosurface. Set the specular color reflectance of the isosurface to make the color of the specular reflected light closer to the color of the iso surface; then set the specular exponent to reduce the size of the specular spot.

```
lightangle(45,30);
set(gcf,'Renderer','zbuffer'); lighting phong
isonormals(Ds,hiso)
set(hcap,'AmbientStrength',.6)
set(hiso,'SpecularColorReflectance',0,'SpecularExponent',50)
```

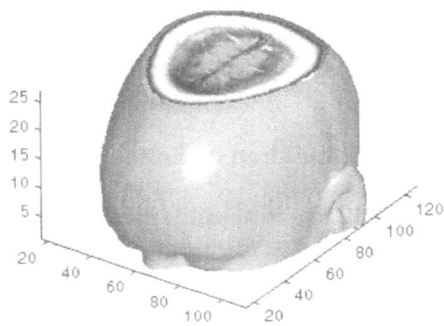

The isocap uses interpolated face coloring, which means the figure color map determines the coloring of the patch. This example uses the color map supplied with the data.

To display isocaps at other data values, try changing the iso surface value or use the sub-volume command. See the isocaps and subvolume reference pages for examples.

8.3 Virtual Reality in Geo Science*

8.3.1 Introduction*

The focus of geo science research nowadays concentrates on computer simulation and information systems for underground studies, i.e. in soil modeling, meaning that expensive insitu testing can be replaced by accurate computer simulations. This not only reduces the costs, because simulation can also calculate results within minutes to show first hints why certain effects occur. These hints then can be used to achieve even better simulation studies and can help to better understanding the complex spatiotemporal nature of geological processes. The information of virtual environments used in the geo scientific research can be categorized as follows:

- Spatial temporal information; the main purpose deals with an efficient storage, analysis and display of subjects that characterize the geometrical data. Geologists use spatial as well as temporal information to create rendered 3D models of technical entities and the surrounding underground structure to explore the environment. Real-time user interactions are not provided.
- Process information; obtained from fluid flows or chemical-process analysis, are mainly concerned with solving some kind of differential equations, which are normally specialized for the respective application domain, hence the input data can be extracted from real-world measurements.

Combining these two approaches with the virtual reality techniques results in a unified simulation system embedded in a virtual environment.

8.3.2 Modeling and Simulation of Space and Time*

The process of modeling in geo scientific problems differs from computer-aided design (CAD) where the user develops a new object based on ideas and given restrictions. In contrast to this construction, creating a 3D model of a geological underground deals with the reconstruction of complex objects of the real-world system, which are only known in a very small surrounding, i.e. through borehole drilling, etc. The geoscientist tries to remodel parts of the real-world domain as accurately and realistically as possible.

The next step is to integrate temporal aspects representing the dynamic processes and the development of the geological underground that finally leads to real-time virtual reality models. To accomplish this task temporal-data base concepts for time and scenario management have to be developed due to the geo scientific background. Temporal aspects can be seen as shown in Fig. 8.10

- Continuous-time by means of a single vectored parameter
- Time-dependent versioning allowing the simulation and adaptation of 3D spatial models based on various parameters

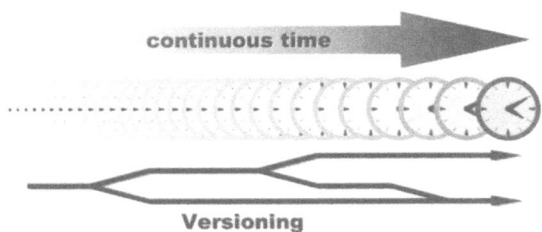

Fig. 8.10. Continuous time and versioning

The temporal concepts are embedded within the underlying virtual reality data base model allowing the user to build up 4D models as a combination of the 3D structures and the processes in these structures. Based on the integration of temporal information the basic concepts for simulating the processes and analyzing as well as predicting future developments under differing conditions are designed. For 4D models in space and time, consistency is of major concern. While the user tries to describe real-world objects, the virtual reality system has to prove whether user-defined constraints fit, also when all solids are placed nonoverlapping, not only in the 3D models, but furthermore in its changing through temporal developments.

In addition, the best results of simulation or analysis are worth nothing, if their calculated values can not be interpreted and remapped to the original real-world objects. For the evaluation of the simulations it is necessary to recheck the results with real-world measurements to ensure their correctness. Therefore, examples must be found that allow the determination of error values, which in fact will be a difficult task.

But any simulation result is inapplicable if the data presentation can not provide the new information to the user. Hence, first approaches for user-friendly data visualization use charts and diagrams. The problem arising here is the loss of the 3D background. The best approach presenting the results would be the integration in the abstract spatial model of the real-world domain, which can then be converted into an image by specifying information such as the objects color, position and orientation in space, and from what location it is to be viewed.

Example 8.4

A practical example is found in geology where detailed geometric descriptions are input to the CAD system of the virtual environment, which can then render perspective views of the scene. The benefits of such techniques are manifold, because the user then can identify different parameters and their distribution directly at their actual position in space and time, which is shown as a practical example in Fig. 8.11.

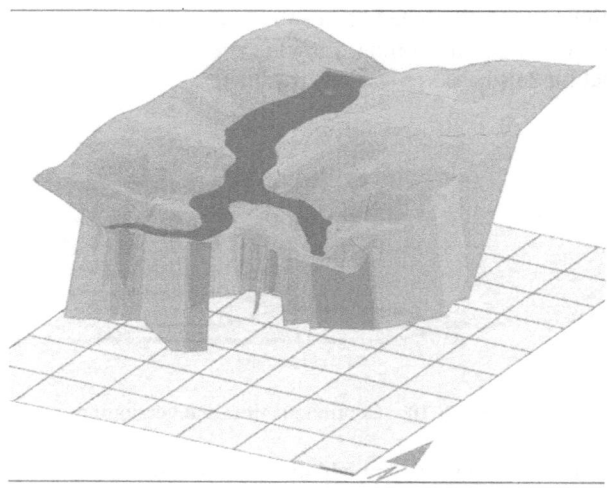

Fig. 8.11. Integration of the 3D virtual reality terrain model with subsurface fractures

8.3.3 Combined Virtual Reality System CoRe*

A system analysis for spatial information management has shown that movements in an artificial 3D model are quite difficult, especially for inexperienced users. Through the integration of virtual reality techniques for spatial information management the gap between a 4D model and the usually 2D user interface can be closed.

As mentioned before, virtual reality can be introduced as an embedded system of hardware and software components that allow users to view and interact within a virtual universe of space and time analyzing different scenarios, while changing the underlying simulation parameters.

The hardware components for a virtual environment consists of computer(s), 3D-input and output devices such as head mounted displays (HMDs), cyber gloves, head tracking devices, and for some applications some sort of measuring equipment for real-world data.

The software for a virtual environment consists of tools for 3D modeling, realistic rendering, shadowing, imaging, photogrammetry, simulation, etc., as well as an embedded data base system for storage and retrieval of spatial, temporal, and thematic data, object characterization, etc., which results in a multilayer data base, as shown in Fig. 8.12.

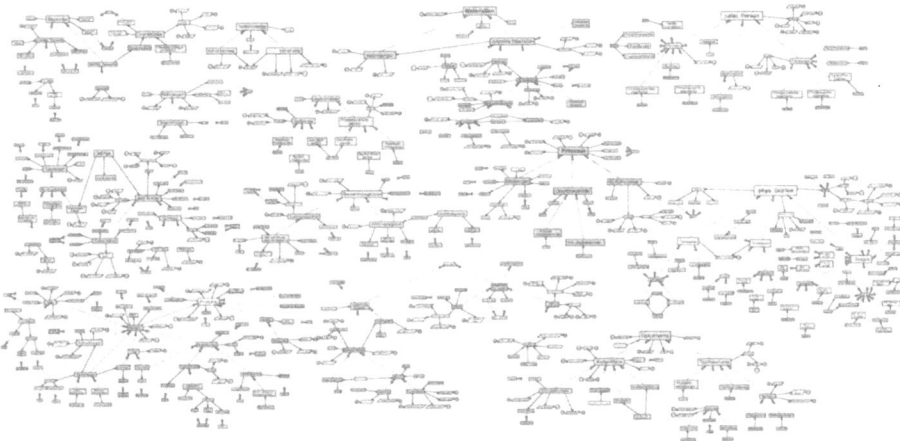

Fig. 8.12. Multilayer data base structure of a real-world geological process including the
metrical layer, the topological layer, and the thematic layer

The architectural concept of the virtual-environment system CoRe contains a
partition in three modular layers, shown in Fig. 8.13, which are the:

- Internal level; which represents the data base system, as shown in Fig. 8.12,
  including the essential concepts of data-management systems, such as transaction management, concurrency control, recovery mechanisms, and data retrieval.
- External level; which represents the integrated user interface that combines aspects of well-known user interaction through dialogs in conjunction with advanced virtual environments. Using virtual reality devices such as head mounted displays (HMD), head mounted tracking devices (HMTD), data gloves and
  3D mice (space ball, space mouse), and eventually some sort of measuring
  equipment for real-world data. The user can navigate through space and time
  of the virtual universe model of the geological real-world domain, retrieve information at any point in the model and directly interact, i.e. change parameters or the model itself.
- Conceptual level, which embeds the modeling and simulation tasks. Within
  this level temporal, spatial, and thematic information is represented with the
  aid of the proposed concepts. The object-relational approach allows any representation of the real-world domain to be changed through space and time. As
  these operations are achieved with methods, the objects themselves behave on
  the basis of their internal status, the overall model and the user interaction.

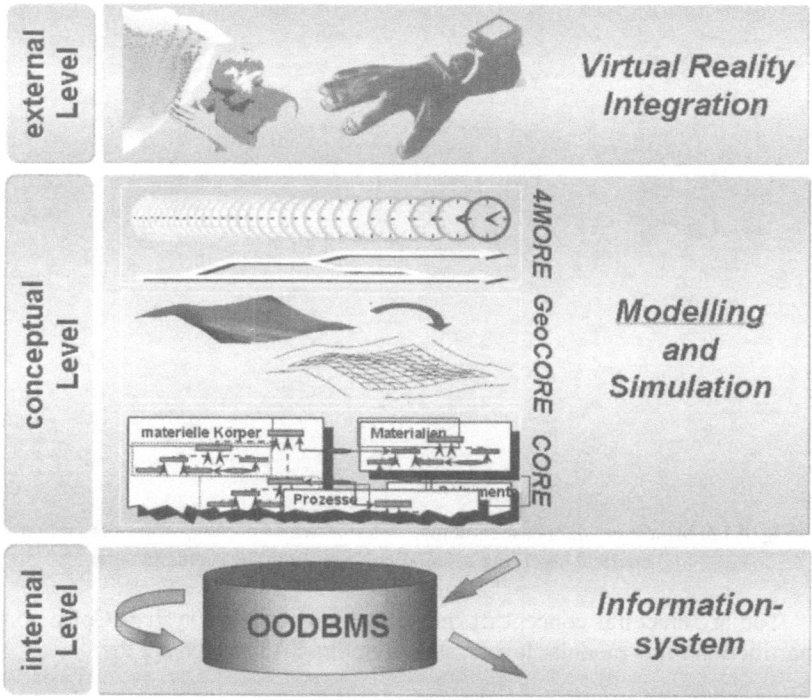

Fig. 8.13. The virtual environment CoRe

8.4 DDSim Prototyping Tool for Autonomous Robots

DDSim is a Java™-based system prototyping tool allowing users of mobile robots to develop and validate behavior in a simulated environment before running the strategies on a real robot[2]. DDSim can easily be configured and adapted to the manifold of specific needs, such as real-world and multiagent interaction, which can be simulated very appropriative. DDSim uses modern interface standards to configure all important parameters of the application. For example, the geometry of simulated objects and robots are specified using XML.

[2] I would like to thank Dipl Inf. Peter Schöll, Fraunhofer Institute of Autonomous Intelligent Systems, Schloß Birlinghoven, St. Augustin, Germany, for his support.

XML is a mark-up language for documents containing structured information. Extensible markup language (XML) is a simple, yet very flexible text format derived from SGML (ISO 8879). Originally designed to meet the challenges of large-scale electronic publishing, XML is also playing an increasingly important role in the exchange of a wide variety of data on the WWW.

Using XML is appropriate as it is hierarchically structured like a mobile robot. Each robot consists of several subsystems like sensors and actuators. These sensors have attributes, which can be described using XML-structured documents.

At the Fraunhofer Institute of Autonomous Systems at Schloß Birlinghofen, Germany, two different document type definitions (DTD) had been developed, which can be used for defining one´s own configurations, which are the definition file for the robot and the environment of the simulation.

The robot definition contains the description for arbitrary robots. Each robot can be equipped with various sensors, which are appropriate to evaluate the optimal configuration of the robot, representing the solution of a given problem. Using simulation it is possible to evaluate the behavior of the robot with respect to the given specified sensor configuration. Therefore, a suitable configuration for the desired behavior and accordingly the robot configuration can easily be found before building an actual physical prototype of the robot. An example of the document type definition (DTD) for a robot is

```
<!ELEMENT Sensor (Label?, DrawColor?,(TouchSensor | Infrared |
                        Camera | LaserScanner |
                        RelativeEncoder | AbsoluteEncoder |
                        Compass))>

<!ELEMENT Infrared (ScanRange, DistanceSensorDefaultReturnValue,
                        Position, Rotation)>
<!ELEMENT ScanRange EMPTY>
<!ATTLIST ScanRange
                        MinRange CDATA #REQUIRED
                        MaxRange CDATA #REQUIRED>
<!ELEMENT DistanceSensorDefaultReturnValue EMPTY>
<!ATTLIST DistanceSensorDefaultReturnValue
                        SmallerMinRange CDATA #REQUIRED
                        BiggerMaxRangeCDATA #REQUIRED>
```

An XML document with the corresponding content looks like:

```
<Robot RobotType="KURT2">
<CenterOfRotation          XCenterOfRotation="0"
                           YCenterOfRotation="0"/>
<Drive                     DistanceLeftRightWheel="30"
                           WheelDiameter="10"
                           MaxSpeed="150"/>

<Infrared>
<ScanRange
                           MinRange="20"
```

```
                              MaxRange="55"
          />
<DistanceSensorDefaultReturnValue
                              SmallerMinRange="-1"
                              BiggerMaxRange="0"
          />
<Position
                              XPosition="17"
                              YPosition="-15"
                              ZPosition="9"
          />
<Rotation
                              XRotation="-45"
                              YRotation="0"
                              ZRotation="0"
          />
</Infrared>
```

Fig. 8.14 shows the robot for the above-mentioned XML description in an office environment. In the upper left corner, the output of the laser scanner is displayed in a special trace window. Fig. 8.15 shows the same program with a different configuration. The environment depicts a soccer field with three robots equipped with color cameras. This scenario is called RoboCup.

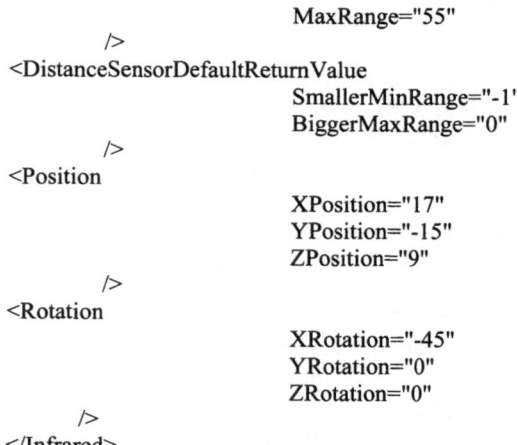
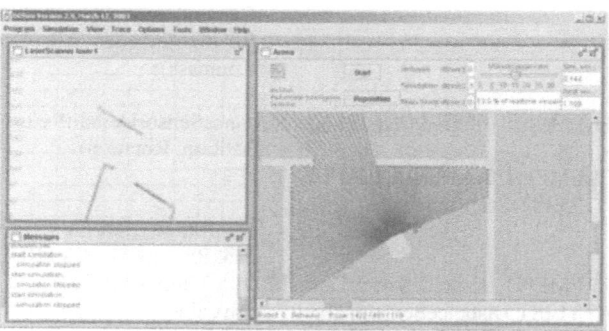

Fig. 8.14. Simulation of a robot in an office environment

RoboCup is an international joint project to promote artificial intelligence (AI), robotics, and related fields. It is an attempt to foster AI and intelligent robotics research by providing a standard problem where a wide range of technologies can be integrated and examined. RoboCup chose to use soccer as a central topic of research, aiming at innovations to be applied for socially significant problems and industries.

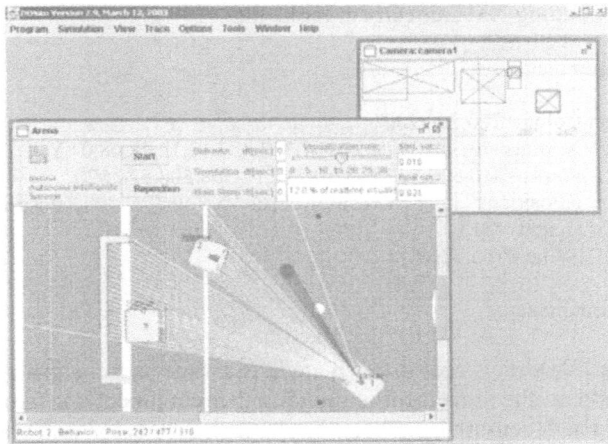

Fig. 8.15. RoboCup robots equipped with color cameras. In the upper right corner a special trace window is shown, which views the detected color-bounding boxes of the robot on the right. The red ball for example is depicted on the right side of the trace window as a red box

In an environment definition the user describes their own environment, e.g. an office environment in which they want to simulate the robots. This is the real-world for the robot in which they can interact.

The definition for the objects in the environment is based on 2D polygons. To make the scene more realistic it is possible to add textures to the objects. Also a height can be specified, thus actually a 2½ D scene description results, shown as follows. For more specialized implementation a 3D rendering algorithm can be used allowing the user a 3D view.

```
<!DOCTYPE Environment

<!ELEMENT Object EMPTY>
<!ATTLIST Object
          Color CDATA #REQUIRED
          XPoints CDATA #REQUIRED
          YPoints CDATA #REQUIRED
          XPosition CDATA #REQUIRED
          YPosition CDATA #REQUIRED
          ZPosition CDATA #REQUIRED
          XRotation CDATA #REQUIRED
          Height CDATA #REQUIRED
          Name CDATA #REQUIRED
      >
  >
```

It is possible to define one or more objects in the XML document. The attributes, which are declared in the DTD, have to be filled with content for each new object in XML document.

A corresponding XML document looks like:

```
<Environment>

<Object Color="200,200,200"
        XPoints="0,349,349,154,154,337,337,12,12,48,48,0" Y
        Points="0,0,520,520,508,508,12,12,508,508,520,520"
        XPosition="1242" YPosition="100" XRotation="0" ZPosition="0"
        Height="315"
        Name="room208"/>

</Environment>
```

Moreover, DDSim is able to calculate in real-time sensor signals for each robot, which depend on the movement in the virtual environment. Different sensors, like distance or touch sensors have been implemented. A color camera based on blob detection is often used in robotics environments and can be configured as a 360° omnidirectional camera. The color blob information is used in robotics for navigation or obstacle avoidance. Furthermore, a simulated laser scanner can be used to explore the self-localization or navigation algorithms. It is fairly simple to implement new sensors and integrate them into the system.

The dynamic model of the robot is simulated with a trained neuronal network. This model is generated from data recorded in actual training experiments, which can be done for each robot, via collecting and storing data from the odometry. The matrix of the neuronal network will be saved into files that are read by the simulator. The source files of the matrix are specified in the XML documents.

The robot's behavior is generated by the DD-designer, which is a graphical design program developed by the Fraunhofer Institute AIS. DD-Designer generates the specific code for each behavior system, which can be tested directly in the simulation.

The behavior system is embedded in a so-called behavior client. This client communicates via CORBA with the simulation engine. Therefore, it is possible to integrate any robot control program – independent of the general structure of the respective program. But other programming languages that supporting CORBA can be used for implementing the behavior client. This approach makes it simple to integrate existing algorithms.

For integration of existing behavior, it is necessary to create an architecture, allowing the community of mobile robot users an easy adaptation of their own constructs, which can be done using a CORBA communication layer. The CORBA communication is specified with the interface-definition language. This IDL file holds the specific set- and get-functions. One of the main advantages of introducing the communication layer is that the client can also interact with a real robot. At the Fraunhofer Institute AIS a CORBA server with two robots has been implemented. One is the new "VolksBot" , a modular platform for education. The other is a special outdoor robot called "Pegasus" . These robots run with specific behavior, developed and proved by simulation.

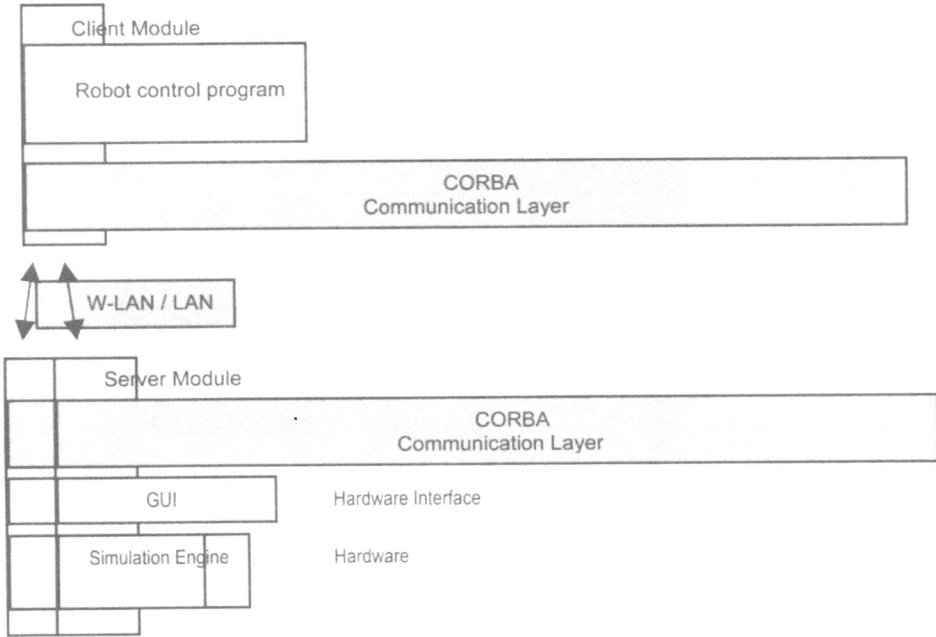

Fig. 8.16. Communication diagram that shows at the top the client module with the embedded-robot control program. At the bottom left is shown the architecture of the simulation DDSim, at the bottom right the architecture of a real robot

A 3D rendering program can supplement the 2D visualization of DDSim. Rendering was implemented to improve vision algorithms to produce more realistic scenes. With this add-on, complex scene representations can be shown from any arbitrary viewpoint. The program is developed in Java™ and based on OpenGL®. It is possible to use file formats like 3DS (3D Max) or VRML. The simulation sends the position of the moving objects via CORBA to the rendering program. Therefore, it is possible to use another computer for visualization of the simulation. A result of which is shown in Fig. 8.17.

Adaptation to different environments like office, RoboCup or garden or different robot types like KURT2, RoboCup or VolksBot successfully prove the flexible concepts behind DDSim.

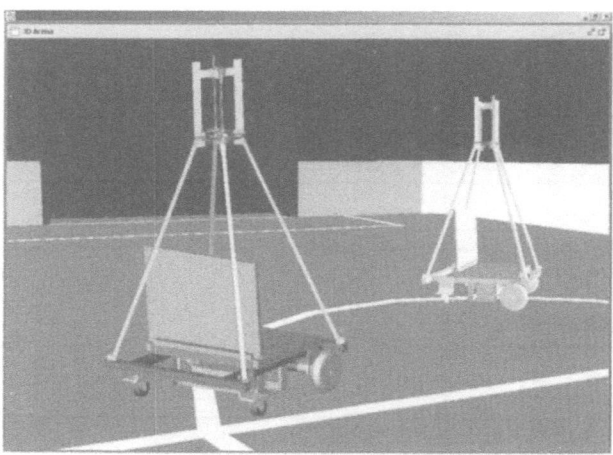

Fig. 8.17. 3D rendering

8.5 References and Further Reading

Abramowski S, Müller H, (1991), Geometric Modeling (in German), BI-Wissenschaftsverlag, Series Computer Science, Vol. 75

Aumann G, Spitzmüller K, (1993), Computer based Geometry (in German) Computerorientierte Geometrie, BI–Wissenschaftsverlag, Series Computer Science, Vol. 89

Bozinowski S, Schoell P, Engineering goalkeeper behavior using an emotion learning method, KI99: German Annual Meeting of AI Workshop on RoboCup, Bonn, September 13th and 15th, 1999

Bredenfeld A, Christaller T, Göhring W, Günther H, Jaeger H, Kobialka HU, Plöger PG, Schöll P, Siegberg A, Streit A, Verbeek C, Wilberg J, (2000), Behavior engineering with dual dynamics models and design tools, In: RoboCup-99: Robot Soccer World Cup III, Veloso M, Ed., Lecture Notes in Computer Science, pp. 231 – 242

Bredenfeld A, Christaller T, Jaeger H, Kobialka HU, Schöll P, (2000), Robot behavior design using dual dynamics, GMD report, 117, GMD Research Center Information Technology, St. Augustin, 23 pages

Bredenfeld A, Christaller T, Göhring W, Günther H, Jaeger H, Kobialka HU, Plöger PG, Schöll P, Siegberg A, Streit A, Verbeek C, Wilberg J, (1999), Behavior engineering with dual dynamics models and design tools, In: Sixteenth International Joint Conference on Artificial Intelligence IJCAI-99 Workshop ABS-4 Third International Workshop on RoboCup, Veloso M, Ed., pp. 57 – 62

Crilly AJ, Earnshaw RA, Jones H, (1993), Applications of Fractals and Chaos. Springer, Berlin

Earnshaw RA, Gigante MA, Jones HH, (1993), Virtual Reality Systems, Academic Press

Encarnacao J, Peitgen HO, Saka G, Englert G, (1991), Fractal Geometry and Computer Graphics, Springer, Berlin

Encarnacao J, Strasser W, Klein R, (1996), Graphic Dataprocessing I (in German), Oldenburg Verlag

Encarnacao J, Strasser W, Klein R, (1997), Graphic Dataprocessing II (in German), Oldenburg Verlag

Gilfillan L, Harbison K, (1998), Using distributed virtual environments (DVE) for collaborative program planning and management: Problems and potential. In: Proc. VWSIM'98 (Eds.: Landauer C, Bellman KL), SCS Publishers, San Diego, pp.39–46

Hoffmann C M, (1989), Geometric and Solid Modeling, Morgan Kaufmann Pub.

Kalawsky R, (1993), The Science of Virtual Reality and Virtual Environments, Addison-Wesley

Kesper B, (200), Cocept of a Geo-Data Model for the use of Free Form Volume Bodies based on Volume Non Uniform Rational B-Splines (in German) PhD Thesis, Hamburg

Kobialka HU, Schöll P, (2000), Quality Management for Mobile Robot Development, In: Intelligent autonomous systems, Pagello E, Ed., p. 698 –703

Kobialka HU, Schöll P, (2000), Fast Assessment of Robot Programs, In: Robotik 2000, VDI-Berichte, p. 293–298

Möller DPF, (1998), Virtual Reality: Simulation Synergy in Laboratories and Outer Space Domains. In: Simulation: Past, Present and Future (Eds.: Zobel R, Möller DPF), Vol. II, SCS Publishers, Delft, pp. 64–66

Schneider M, (1997), Spatial Data Types for Database Systems. Springer, Berlin

Singh A, Goldgof D, Terzopoulos, (1998), Deformable Models in Medical Image Analysis, IEEE Press, Los Alamitos, USA

Straßer W, Seidel HP, (1989), Theory and Practice of Geometric Modeling, Springer, Berlin

Vince J A, (1992), 3D Computer Animation, Addison-Wesley

Vince JA, (1995), Virtual Reality Systems, Addison-Wesley

Watt A, (1993), 3D Computer Graphics, Addison-Wesley

Watt A, Watt A, (1992), Advanced Animation and Rendering Techniques, Addison-Wesley

Yachik TR, (1998), Synthetic Scene Quality Assessment Metrics Development Considerations. In: Proc. VWSIM'98 (Eds.: Landauer, C. and Bellman, K.L.), SCS Publishers, San Diego, pp. 47–57

Links:
www.ddsim.de

8.6 Exercises

8.1   What is meant by the term virtual reality?
8.2   What is meant by the term openGL?
8.3   What is meant by the term spatiotemporal data?
8.4   What is meant by the term head mounted display?
8.5   What is meant by the term data glove?
8.6   What is meant by the term touch-and-feel sensation?
8.7   What is meant by the term CAVE?
8.8   What is meant by the term DMU?
8.9   What is meant by the term morphing?
8.10 What is meant by the term NURBS?
8.11 What is meant by the term MBA?
8.12 What is meant by the term constructive solid geometry?
8.13 What is meant by the term space partitioning?
8.14 What is meant by the term deformable model?
8.15 What is meant by the term surface reconstruction?
8.16 What is meant by the term isosurface?
8.17 What is meant be RoboCup?
8.18 What is meant by 3d rendering?
8.19 What is meant by the term CORBA?
8.20 What is meant by the term XML?

Appendix A

Numeric Integration

The digital simulation software systems determine values for the continuous signals by producing a series of discrete values, meaning the continuous function $x(t)$ becomes a sequence of discrete values $x(t_0)$, $x(t_1)$, $x(t_2)$, ..., $x(t_n)$, $x(t_{n+1})$ as noticed in Sect. 1.5. Usually, the time interval between adjacent values is constant and represented by $h = t_{n+1} - t_n$. Ideally $x(t_n)$, the produced discrete values of the function at a particular point in time, should be identical to its continuous equivalent at $t = t_n$, which depends on the accuracy of the computer, meaning the number of bits the computer uses to represent a value as well as by the accuracy of the simulation model. Furthermore, the discretization error of the methods used to calculate the derivatives, commonly referred to as a numeric integration, is often critical. It represents the primary source of error in a simulation variable $x(t)$ at $t = t_n$. A very crude but easily obtainable measure of the accuracy of an integration formula is the order of error. The order of error can be derived by comparing the error of a formula with Taylors series expansion x_{n+1} in terms of x_n, which can be written in the form

$$x(t_n + h) = x_n + h \cdot \dot{x}_n + \frac{h^2}{2!} \cdot \ddot{x}_n + \frac{h^3}{3!} \cdot \dddot{x}_n + ..., \qquad (A.1)$$

where h is the time interval and $\dot{x}$ is the derivative of x at $t = t_n$. Basically the series gives the values of x at $t = t_{n+1}$ in terms of x_n and its derivatives. The order of the first term, in which both differ, is said to be the order of error. Taylors series can be used to derive several numeric integration formulas, but more importantly, it is the criterion used for evaluating almost all numeric integration techniques. As an example of a numeric integration method, consider the approximation using only the first two terms of Taylors series

$$x(t_n + h) = x_n + h \cdot \dot{x}_n \, . \tag{A.2}$$

Solving for the derivatives gives

$$\frac{dx_n}{dt} = \frac{x_{n+1} - x_n}{h} = \frac{x_{n+1} - x_n}{t_{n+1} - t_n} \, , \tag{A.3}$$

which is commonly referred to as Eulers method, or the rectangular rule. This method can be simply demonstrated for first-order differential equations

$$\dot{x} + a \cdot x(t) = b(t) \, , \tag{A.4}$$

with $x(t=0) = 0$ and $b(t) = 1$, a unit step input. For the discrete equivalent, (A-3) is used in (A-4), which results in

$$\frac{x_{n+1} - x_n}{h} + a \cdot x_n = b_n \, , \tag{A.5}$$

which can be solved for x_{n+1} in terms of x_n, a and the input b, which yields

$$x_{n+1} = x_n - a \cdot h \cdot x_n + h \cdot b_n \, . \tag{A.6}$$

Again, it should be noticed that h is the time interval between adjacent discrete values and a is a parameter in the original equation, and (A-6) is a difference equation. Almost any programming language can be used to write a program to calculate successive values of x_{n+1} from the proceeding value x_n, the time interval h, the parameter a, and the input b_n.

There is a variety of numeric integration methods existing that require a mathematical translation of the original set of differential equations. This requires methods that can be inserted as numeric integration into a continuous system simulation. Integration schemes of that type can be classified into:

- Single-step integration methods
- Multi-step integration methods

In either case we can distinguish between

- Methods with constant step size
- Methods with variable step size

Any such numeric integration method is an approximation of true integration by a discrete difference scheme, the problem of which is to define an operator F such that

$$x_{n+1} = F(x_n,...,x_{n-k+1}; \dot{x}_{n+1},...,\dot{x}_{n-k+1}) , \qquad (A.7)$$

approximated over a given interval $(t_n, t_n+h,)$ the integral

$$x_{n+1} = \int_{t_n}^{t_n+h} \varphi(X,S,t)dt = \int_{t_n}^{t_n+h} \dot{x}dt , \qquad (A.8)$$

with h as the interval size, and F being a k-step integration formula.

Single-Step Formulae

The simplest single-step formulae, amongst others, are:

- Eulers numeric integration: $x_{n+1} = x_n + h \cdot \dot{x}_n + 0(h^2)$

- Trapezoidal numeric integration: $x_{n+1} = x_n + \dfrac{h}{2} \cdot (\dot{x}_n + x_{n+1}) + 0(h^3)$

- Heuns numeric integration: $\begin{aligned} x_{n+1} &= x_n + \dfrac{h}{2} \cdot (\dot{x}_n + \widehat{x}_{n+1}) + 0(h^3) \\ \widehat{x}_{n+1} &= x_n + h \cdot \dot{x}_n \end{aligned}$

These methods formulate an approximation x_{n+1} at time step t_{n+1} of the true integration $x(t)$ with a discrete difference scheme, compared to the solution obtained from the previous integration step x_n, t_n and the increment h, which is the time interval. For a Euler integration, the increment $x_{n-1} - x_n$ of the integral is simply determined by taking the derivative $\dot{x}_n$ at the beginning and multiplying it by the interval size h. The poor performance of this simple numerical integration scheme can be slightly improved using the average of the derivatives at the beginning and at the end of the interval and multiplying by h, which is a description of the trapezoidal scheme. With the value $\widehat{x}_{n+1}$ we have a predictive step first while using the Euler numerical integration scheme, which finally results in the Heun numeric integration, which is called the second-order Runge Kutta formula.

As mentioned before, the most important topic when using integration schemes, is the accuracy. Hence the approximation of the integration scheme has to be as accurate as possible. Again, this can be realized if a value of a state variable taken at $t = t_{n+1}$ is calculated as an expression of preceding values of the same state variable and its derivative. The integration scheme itself determines the approximation by taking the derivative at the initial step $x_n = F(x_n, t_n)$ and thereafter, moving one half-step ahead, calculating at time step $t_{n+\frac{1}{2}}$ the gradient

$$x^1_{n+\frac{1}{2}} = F\left(x^1_{n+\frac{1}{2}}, t_{n+\frac{1}{2}}\right).$$

(A.9)

From

$$x^1_{n+\frac{1}{2}} = x_n + \frac{h}{2} \cdot x_{n+\frac{1}{2}},$$

(A.10)

based on the stored values of x_n, and t_n. Moreover, the integration scheme evaluates the gradient, described by (A-9), for half a step ahead the derivative, yields

$$x^2_{n+\frac{1}{2}} = F\left(x^2_{n+\frac{1}{2}}, t_{n+\frac{1}{2}}\right),$$

(A.11)

from

$$x^2_{n+\frac{1}{2}} = x_n + \frac{h}{2} \cdot x^1_{n+\frac{1}{2}}.$$

(A.12)

Consider the gradient in (A-12) the next integration sequence starts one step ahead to approximate the integral, which results in

$$x^3_{n+\frac{1}{2}} = x_n + \frac{h}{2} \cdot x^2_{n+\frac{1}{2}}.$$

(A.13)

The approximations described through (A-10), (A-12), and (A-13), can be rewritten in a form that more clearly reveals the fact that a weighting average of the previous predictions of the derivatives are taken into account for the evaluation of the final approximation one step ahead to obtain a more accurate approximation of the integral as follows:

$$x_{n+1} = x_n + h \cdot \left(a_1 \cdot x_n + a_2 \cdot x^1_{n+\frac{1}{2}} + a_3 \cdot x^2_{n+\frac{1}{2}} + a_4 x^3_{n+\frac{1}{2}} \right). \tag{A.14}$$

A very crude but easily obtainable measure of the accuracy of the numeric integration scheme is to choose the weighting factors a_i, $i = 1, ..., 4$ such that the error of x_{n+1} is comparable to the error of a formula with Taylors series expansion of the genuine solution, given in (A-1), fits at the term of fourth order, with R as the respective rest term.

$$x_{n+1} = x_n + h \cdot \dot{x}_n + \frac{h^2}{2!} \cdot \ddot{x}_n + \frac{h^3}{3!} \cdot \dddot{x}_n +, ..., \frac{h^m}{m!} \cdot x_n^{(m)} + R. \tag{A.15}$$

The solution x_{n+1} is bounded from x_n, t_n and h as follows

$$x_{n+1} = x_n + h \cdot \Xi(x_n, t_n, h), \tag{A.16}$$

with Ξ as incremental function, which depends on the type of differential equation. Ξ can be chosen as follows

$$\Xi(x_n, t_n, h) = \sum_{n-1}^{m} a_n \cdot K_n, \tag{A.17}$$

with

$$K_n = F(x_n + h) \sum_{n=1}^{n-1} \beta_n. \tag{A.18}$$

The coefficients a_n, β_n, K_n with $n = 1, ... , n^{-1}$, in (A-17) and (A-18) are unknown. A solution can be obtained deriving Taylors series expansion compared due to the respective series terms of (A-17) and (A-18) with the coefficients a_n, β_n, K_n. The index m in (A-17) characterizes the order of the numeric integration scheme used. Hence, m represents a measure of the accuracy of the integration formula, meaning that the derivative order of the incremental function is unimportant compared with Taylors series expansion terms. The accuracy is better the higher the degree of m, because accuracy depends on the order of the discretization criterion, which is $0(h^{m+1})$. For practical reasons, proven values for m are 1, 2, 3, and 4 which finally results in the fourth-order Runge Kutta formula. This formula proves the increment h for each discrete integration step to evaluate the difference between the discretization error and the error criterion, which will be $\varepsilon <$ 0. For a given increment h the program calculates one integration step and the respective error function e. In the case that $e > \varepsilon$, the integration step will be repeated with one half-step length. For $e < \varepsilon$ for all $e = 1, \dots , n$, an integration step with twice the step length will be used.

Multi-Step Formulae

Single-step formulae, especially when they are of the complexity of the fourth-order Runge Kutta method and higher, are difficulty to handle. In the case of the fourth-order Runge Kutta method one has to calculate four discrete function values to obtain one value of the integral. Schemes, that use the results of preceding steps seem to be less difficult to handle. These types of schemes are the multi-step integration formulae, which can be written in the general notation as follows:

$$
\begin{aligned}
x_{n+1} = a_0 \cdot x_n + a_1 \cdot x_{n-1} +,\dots + a_m \cdot x_{n-m} + \\
h\left[b_{-1} \cdot \dot{x}_{n+1} + b_0 \cdot \dot{x}_n +,\dots,+ b_m \cdot \dot{x}_{n-m} \right]
\end{aligned} \tag{A.19}
$$

The iterative nature of an integration scheme given in (A-19), results in a problem of the multi step integration scheme, the inherent potential of instability. For this reason, Dahlquist[1] introduced a necessary condition for stability: In an m-step formula, not more than m-1 coefficients if m is even and not more than m coefficients if m is odd can be chosen freely for the purpose of minimizing the approximation error.

---

[1] Dahlquist G, (1975), Convergence and Stability in the Numerical Integration of ODEs, Math. Scand. 4, pp. 33–35

Appendix B

Laplace Transform

Chapters 2 and 3 are concerned with continuous-time systems and their mathematical models in terms of differential equations, transfer functions and state-space models, as well as discrete-time systems and their mathematical models in terms of statistical notations, queuing systems, Petri-nets, and parallelisms. In this section the Laplace transform (and its inverse) are introduced, a very convenient tool for the analysis of the response characteristics of models in terms of differential equations representing linear continuous-time systems. From a more general point of view a transform is a mathematical method that allows a much simpler solution with it.

A function $f(t)$ is called transformable if

$$\int_0^\infty |f(t)| e^{-\sigma \cdot t} dt \le \infty, \tag{B.1}$$

for some real, positive σ, known as the abscissa of absolute convergence. The Laplace transform converts or transforms a real function of a real variable into a complex function of a complex variable. In particular, let $f(t)$, be a real function defined for $t > 0$, then the Laplace transform of this function is defined by

$$F(s) = \int_0^\infty f(t) \cdot e^{-s \cdot t} dt = L[f(t)]. \tag{B.2}$$

Likewise, given $F(s)$, one expresses the inverse Laplace transform $f(t)$ as

$$f(t) = L^{-1}\{F(s)\}, \tag{B.3}$$

which is defined as

$$f(t) = \frac{1}{2\pi j} \int_{c-j\omega}^{c+j\omega} F(s)e^{s \cdot t} ds ,$$

(B.4)

where c is a real constant that is greater than the real part of all the singularities of $F(s)$, the Laplace transform, and $f(t)$ as a time-dependent function.

Example B1
Obtain the Laplace transform for $f(t) = e^{-a \cdot t}$ for $t \geq 0$, where a is a constant.

$$F(s) = \int_{0}^{\infty} e^{-a \cdot t} e^{-s \cdot t} dt = \frac{e^{-(s \cdot a) \cdot t}}{-(s+a)} \Big|_{0}^{\infty} = \frac{1}{s+a} .$$

(B.5)

The integrals in (B-2) and (B-4) are defined for complex values of s hence $F(s)$ represents a complex function for which the methods of complex analysis are applicable. The Laplace transform is defined for a wide class of functions, as long as the complex values s are restricted to a region in the complex plane for which the indefinite integral converges.

Under normal circumstances systems analysts are interested in the dynamic behavior of the system under test, starting, at time $t = 0$, which with the lower boundary of the integral is $t = 0$ instead of $-j\omega$, which results in the so-called one boundary Laplace transform

$$F(s) = \partial f(t) \cdot e^{-s \cdot t} dt = L[f(t)].$$

(B.6)

Let a be an arbitrary real number and let $f(t)$, $t > 0$, have a Laplace transform, then

$$L[a \cdot f(t)] = a \cdot L[f(t)] = a \cdot F(s).$$

(B.7)

Let $f(t)$ and $g(t)$ be defined for $t > 0$, each having a Laplace transform

$$L[f(t) + g(t)] = L[f(t)] + L[g(t)].$$

(B.8)

The properties given in (B-7) and (B-8) guarantee that the Laplace operator is linear; linearity is a critical property of the Laplace transform.

For $\frac{d}{dt}$ we obtain the respective Laplace operator $L\left[\frac{d}{dt}\right] = s$. Hence the Laplace transform of the first derivative of $f(t)$ is found to be

$$L\frac{df(t)}{dt} = s \cdot F(s) - f(0),$$ (B.9)

and for higher-order derivatives

$$L\frac{d^n f(t)}{dt^2} = s^n F(s) - s^{n-1} f(0) - s^{n-2} f(0) - ... - f^{n-1}(0).$$ (B.10)

If all the initial conditions of $f(t)$ and its derivatives $f'(t), f''(t), f^{n-1}(t)$ are zero,

$$L\frac{d^n f(t)}{dt^n} = s^n \cdot F(s).$$ (B.11)

Let a be an arbitrary real number and let $f(t)$, $t > 0$, have a Laplace transform $F(s)$, then

$$L[e^{-a \cdot t} f(t)] = F(s+a).$$ (B.12)

Suppose that $f(t)$, $t > 0$, has a Laplace transform $F(s)$ and that the limit

$$\lim_{t \to \infty} f(t)$$ (B.13)

exists. Then the final-value theorem holds

$$\lim_{t \to \infty} f(t) = \lim_{s \to \infty} s \cdot F(s).$$ (B.14)

Suppose that $f(t)$, $t > 0$, has a Laplace transform $F(s)$ and that the limit

$$\lim_{t \to 0} f(t),$$ (B.15)

exists. Then the initial-value theorem is

$$\lim_{t \to 0} f(t) = \lim_{s \to 0} s \cdot F(s).$$ (B.16)

Suppose

$$\frac{dF(s)}{ds} = \frac{d}{ds}\partial f(t)\cdot e^{-s\cdot t}dt = -\partial t\cdot f(t)\cdot e^{-s\cdot t}dt \qquad \text{(B.17)}$$

$$L[t\cdot f(t)] = -\frac{d}{ds}\cdot F(s). \qquad \text{(B.18)}$$

In a similar way we obtain

$$L[t^2\cdot f(t)] = \frac{d^2}{ds^2}\cdot F(s) \qquad \text{(B.19)}$$

$$L[t^2\cdot f(t)] = (-1)^2\cdot\frac{d^2}{ds^2}\cdot F(s). \qquad \text{(B.20)}$$

For $e^{-a\cdot t}$ we receive the L-operator $L[e^{-a\cdot t}] = \dfrac{1}{s+a}$, shown in Example B1.

The previous Laplace transform theorems, presented without proof, are often useful in developing the Laplace transform of a given function. However, the inversion problem is also of substantial interest and importance. Given the complex function $F(s)$ it is desired to find a function $f(t)$ defined for $t > 0$ such that

$$L[f(t)] = F(s). \qquad \text{(B.21)}$$

Sometimes this inversion property is written as

$$f(t) = L^{-1}[f(s)]. \qquad \text{(B.22)}$$

For a linear time-invariant system the ratio of the Laplace transform of the output to the Laplace transform of the input is called the transfer function. The transfer function is useful; however, it does not provide any information concerning the physical structure of the system.

Example B2
The transfer function of a simple RC network, shown in Fig. B-1, is to be derived.

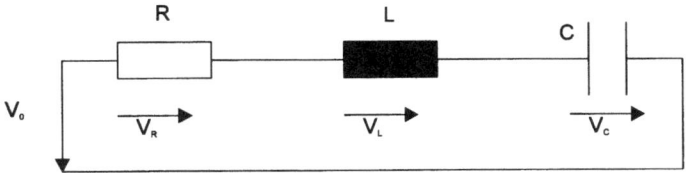

Fig. B-1. Simple RC network

Assuming

$$I(s) \Leftrightarrow L[i(t)] \tag{B.23}$$

$$V_0 = (s) \Leftrightarrow L[V_0(t)]$$

$$V_C(s) \Leftrightarrow L[V_C(t)]$$

and $\tau = R \cdot C$, as well as $\dfrac{dQ}{dt} = I$ the voltage-current relationships can be written as

$$V_C(s) = \frac{I(s)}{C \cdot s}, \tag{B.24}$$

$$V_0(s) = R \cdot I(s) + \frac{I(s)}{C \cdot s} \tag{B.25}$$

which results in

$$F(s) = \frac{V_0(s)}{V_C(s)} = \frac{\dfrac{1}{\tau}}{s + \dfrac{1}{\tau}}. \tag{B.26}$$

Example B3
For the RC network, introduced in Example B2, the ordinary differential equation can be written as follows

$$\frac{Q}{C} + I \cdot R = V_C + V_R = V_0(t) = V_C + \tau \cdot \frac{dV_C}{dt}, \tag{B.27}$$

which yields in the Laplace domain

$$(s \cdot \tau + 1) \cdot V_C(s) = V_0 \cdot \frac{1}{s} + \tau \cdot V_C(0). \tag{B.28}$$

With $V_C(0) = V_0$, we obtain

$$F(s) = V_C(s) = V_0 \cdot \left[\frac{1}{s(1+\tau \cdot s)} \right] + V_0 \cdot \left[\frac{\tau}{1+\tau \cdot s} \right]. \tag{B.29}$$

For the time domain we obtain

$$V_C(t) = V_0 \cdot \left(1 - e^{-\frac{t}{\tau}} \right) + V_0 \cdot e^{-\frac{t}{\tau}} = V_0 - (V_0 - V(0)) \cdot e^{-\frac{t}{\tau}}, \tag{B.30}$$

which results for the current $I(t)$ in

$$I(t) = C \cdot \frac{dV_C}{dt} = \frac{1}{R} \cdot (V_0 - V(0)) \cdot e^{-\frac{t}{\tau}}, \tag{B.31}$$

while $I = \dfrac{dQ}{dt} = \dfrac{V}{R}$. The graph of the voltage and the current behavior over time is shown in Fig. B-2.

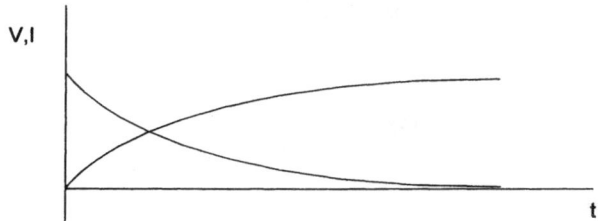

Fig. B-2. $V-I$ graph of the simple RC network from Example B2

Example B4

For the RC network in Example B2 the input quantity is the voltage V_i applied at the input ports. Assuming that the applied voltage source V_i has zero impedance. The output of the network is the voltage V_o across the resistor. Let Q denote the charge on the capacitor and let I denote the current through the capacitor; then the total voltage drop in the loop is

$$\frac{Q}{C} + R \cdot I - V_i = 0. \tag{B.32}$$

The charge-current relationship is

$$\frac{dQ}{dt} = I \, . \tag{B.33}$$

The charge Q is represented as state variable x; the input variable is $V_o = V_i$ and the output variable is $y = V_o = R \cdot I$. Thus the state model for the RC network results in

$$\frac{dx}{dt} = -\frac{1}{\tau}x + \frac{1}{R}V_0 \tag{B.34}$$

$$y = \frac{1}{C}x + V_0 \, . \tag{B.35}$$

The objective of Example B4 is to determine the response properties of the output voltage to the initial state of the RC network and to the input voltage, meaning examining the response properties. Determining the response of the RC network uses the Laplace transform of the state equation, assuming Q_0 is the initial state and V_0 (t), $t > 0$ is the input, hence we get

$$sX - Q_0 = -\frac{1}{\tau}X + \frac{1}{R}V_0 \, , \tag{B.36}$$

where $V_0(s)$ and X (s) are Laplace transforms of V_0 (t) and x (t), yielding

$$X = \frac{Q_0}{s + \dfrac{1}{\tau}} + \frac{V_0}{R(s + \dfrac{1}{\tau})} \, , \tag{B.37}$$

and

$$x(t) = e^{-\frac{t}{\tau}} \cdot Q_0 + \int_0^t e^{-\frac{t-\tau}{\tau}} \cdot \frac{V_0(\tau)}{R} \cdot d\tau \, . \tag{B.38}$$

Thus the output voltage is

$$y(t) = -\frac{Q_0}{C} \cdot e^{-\frac{t}{\tau}} - \frac{1}{\tau}\int_0^t e^{-\frac{t-\tau}{\tau}} \cdot V_0(\tau) \cdot d\tau + V_0(t) \, . \tag{B.39}$$

The transfer function for the RC network is

$$G(s) = \frac{s}{s + \dfrac{1}{\tau}} \, . \tag{B.40}$$

Since $\tau > 0$ the system is stableThe response to a sinusoidal input voltage $V_0(t)$ $=V_i{\cdot}\cos\omega{\cdot}t$ for $t > 0$ results in an output as follows

$$y(t) = -\frac{Q_0}{C}\cdot e^{-\frac{t}{\tau}} - \frac{V_i}{\tau}\cdot\int_0^1 e^{-\frac{t-\tau}{\tau}}\cdot\cos\omega\cdot\tau\cdot d\tau + V_i\cdot\cos\omega\cdot\tau . \qquad \text{(B.41)}$$

This integral can be evaluated with some difficulty. An alternative solution is based on the frequency-response function

$$y(t) \rightarrow M(\omega)\cdot V_i\cdot\cos[\omega\cdot t + \phi(\omega)], \quad as\ t \rightarrow \infty , \qquad \text{(B.42)}$$

with the transfer function,

$$M(\omega)e^{j\phi(\omega)} = \frac{j\omega}{j\omega + \frac{1}{\tau}} , \qquad \text{(B.43)}$$

$$M(\omega) = \frac{\omega}{\sqrt{\omega^2 + (\frac{1}{\tau})^2}} , \qquad \text{(B.44)}$$

yielding

$$\tan\phi(\omega) = \frac{1}{\omega\cdot\tau} . \qquad \text{(B.45)}$$

Thus the frequency-response function has the following properties: If $\omega = 0$, $M(\omega) = 0$ and $\phi(\omega) = \frac{\pi}{2}$; for $\omega > 0$, $0 < M(\omega) < 1$, $0 < \phi(\omega) < \frac{\pi}{2}$; as $\omega \rightarrow \infty$, $M(\omega) \rightarrow 1$, and $\phi(\omega) > 0$.

Appendix C

Online Resources

Intentionally this textbook has been designed to be independent of any particular simulation software package. This decision was made largely because of the growing popularity of modeling and simulation itself, and the continuous introduction of new-generation simulation software packages has been accompanied by tremendous diversity. The days when most courses on modeling and simulation used one or two simulation software packages are quickly giving way to the situation of tremendous diversity in other possible efficient solutions using modeling and simulation as a generalized methodology. Some courses integrate the use of MATLAB SIMULINK, others using ACSL, and so on. This diversity, coupled with the evolution of modeling and simulation into a discipline, make the need to decouple lecture material from more advanced case study material quite evident.

However, we have not simply left the instructor and students entirely on their own with respect to the use of simulation software packages. Instead we have used the World Wide Web to supplement this book with extensive Case Study materials. In fact, using the Web, we can provide even more than a typical modeling and simulation textbook might be able to provide.

The Web site accompanying "Mathematical and Computational Modeling and Simulation: Fundamentals and Case Studies" can be found at:

http://www.springer.de/cgi/svcat/search_book.pl?isbn=3-540-40389-2

and

http://www.informatik.uni-hamburg.de/TIS/M&S

It currently includes items like:

- Case studies in the various domains of chemical engineering, electrical engineering, environmental systems, mechanical engineering, medicine, physics, telecommunication, etc.

- Links to related Web sites, and simulation software industry documents
- Power-point tutorials,
- Power-point lecture slides (to be added in future).

Of course, the Web site will continually evolve, so more items may be added in future.

Index

Druck und Bindung: Strauss Offsetdruck GmbH